MATERIALS AND PROCESSING – MOVE INTO THE 90's

SOCIETY FOR THE ADVANCEMENT OF
MATERIAL AND PROCESS ENGINEERING

MATERIALS SCIENCE MONOGRAPHS, 55

MATERIALS AND PROCESSING – MOVE INTO THE 90's

Proceedings of the 10th International European Chapter Conference of the Society for the Advancement of Material and Process Engineering, Birmingham, United Kingdom, July 11–13, 1989

Edited by

S. BENSON
British Aerospace (Military Aircraft) Ltd, Preston, U.K.

T. COOK
British Aerospace (Dynamics) Ltd, Stevenage, U.K.

E. TREWIN
Courtaulds Advanced Materials, Coventry, U.K.

R. M. TURNER
ICI, Wilton Materials Research Centre, Cleveland, U.K.

ELSEVIER
Amsterdam – Oxford – New York – Tokyo 1989

ELSEVIER SCIENCE PUBLISHERS B.V.
Sara Burgerhartstraat 25
P.O. Box 211, 1000 AE Amsterdam, The Netherlands

Distributors for the United States and Canada:

ELSEVIER SCIENCE PUBLISHING COMPANY Inc.
655 Avenue of the Americas
New York, NY 10010, U.S.A.

Disclaimer
Responsibility for the contents and security clearance of papers published herein rests solely upon the authors and not upon SAMPE or any of its members.

ISBN 0-444-88005-4 (Vol. 55)
ISBN 0-444-41685-4 (Series)

This book is printed on acid-free paper.

Printed in the Netherlands

WELCOME

As we assemble for our tenth SAMPE European Chapter Conference with a developing sense of active maturity, it is appropriate that we shall meet in Birmingham; 'The Big Heart of England' and second city of Britain.

This year Birmingham is celebrating its centenary as a city. Having in the recent past, suffered from declining industries and major change, it is now poised to move into the 90's as a leading International Business, Cultural, Sport and Conference Centre.

Here at the National Exhibition Centre, we are conveniently located between Birmingham City Centre with its industrial might and Coventry with its international reputation for Carbon Fibre and Aero Engines. Cultural and social activities are many and varied. The Bull Ring Market Area has the biggest concentration of shops in England and has everything from large department stores to beautiful arcades. Birmingham boasts its own symphony orchestra and the world-renowned Sadlers Wells Royal Ballet Company is about to make its home here. To the South of Birmingham is Stratford-Upon-Avon, the birthplace of William Shakespeare, a piece of Elizabethan England. Coventry has a magnificent 20th Century cathedral and in contrast there is Medieval Warwick Castle.

Our Conference Theme 'MOVE INTO THE 90'S' is well represented by the transition from heavy metal engineering mass production to the application of high technology in the Aston Science Park and the University of Warwick. Even the roads are taking on a European flavour with Britain's only street Motor Racing Event, called the Superprix, which takes place annually. While we shall be concentrating the conference on the technical papers and exhibits, we hope you will also find time to enjoy some of the Warwickshire countryside, possibly a trip to the Severn Valley Railway or Ironbridge Gorge, both within easy reach.

The Organising Committee has made a selection of papers with high technical content and covering a wide range of subjects. As in previous years all sessions will be open to all delegates and we hope you will find your attendance at this conference a stimulating and rewarding experience.

I look forward to seeing you all in Birmingham.

JACQUES DE BOSSU

S.A.M.P.E. ORGANISING COMMITTEE

T. Cook (Chairman)	British Aerospace
S. Benson	British Aerospace
C. Rochester	AIM Aviation
E. Trewin	Courtaulds
M. Turner	Ciba-Geigy
R.M. Turner	ICI

EUROPEAN CHAPTER OFFICERS

J. De Bossu	Brochier SA, President
H.G. Benninghoff	Du Pont de Nemours, Vice President
H. Meyer	Ciba-Geigy, Secretary
E. Schiantarelli	Swissair, Treasurer and National Delegate for Switzerland
L. Van Arkel	Hexcel, Delegate Benelux
T. Cook	British Aerospace, Delegate United Kingdom
G. Lidor	Israel Aircraft Ind., Delegate Israel
H.J. Semrau	Dornier GmbH, Delegate Germany
D.P. Singh	Senter F.Industrif, Delegate Scandinavia
R.J. Schliekelmann	Student Programme Co-Ordinator

CONTENTS

Materials and Processing – Move into the 90's
edited by S. Benson, T. Cook, E. Trewin and R.M. Turner
Elsevier Science Publishers B.V., Amsterdam, 1989

THERMOPLASTIC STRUCTURAL COMPOSITES : EVOLUTION AND REVOLUTION

Neil COGSWELL

ICI Wilton Materials Research Centre, PO Box 90, Middlesbrough, TS6 8JE

From a 'Cinderella' subject seven years ago thermoplastic structural composites have become prominent in the field of Advance Materials and Process Engineering. This paper charts their evolution and in particular the scientific understanding of their structure, performance and processing and indicates the revolution to which they may lead.

1. INTRODUCTION

Fibre reinforced plastics are a representative of the family of designed materials wherein the properties can be specified by the user. In this class of materials the fibre determines the basic property profile in partnership with the resin, which shields the structure from attack by hostile environment and determines the route by which structures are fabricated. This partnership is an equal one: improvements in fibre properties demand complementary advances in resin technology if the full potential of the reinforcement is to be realised.

The development of carbon fibres brought with it new opportunities for the development of light weight structural composites. Because of the very large surface area of the fine fibres, and the absolute need for total wetting of those fibres by the resin if their full potential was to be realised, it was natural to impregnate the fibres with monomers which could be subsequently polymerised once the structure had been assembled into shape: thermosetting epoxy resins became widely adopted. This was not to the exclusion of interest in linear chain thermoplastic polymers, but the early experience with those resins was that the only way to preimpregnate the fibres was from a low viscosity solution[1]. The need for solution prepregging for thermoplastics brought with it a problem, removing the residual solvent, and identified a weakness of such systems, if they could be prepregged from solution they could be attacked by solvents in service. The urge to use thermoplastic resins, because of potential advantages in processing cycle time, remained and a range of ingenious solutions to the problem were proposed. Film stacking, wherein layers of thermoplastic polymer films were interleaved with layers of fibre and them compression moulded in a protracted process to carry out both impregnation and shaping simultaneously, a technique pioneered by Rolls Royce[2] and RAE Farnborough[3], was commercialised by Specmat and has been used for the production of radomes. Film stacking thus kept the promise of thermoplastics open but was unable to offer the full advantage of rapid processing. Thermosetting epoxy resins became the industry standard.

This perspective changed about seven years ago. At that time the major aerospace companies were becoming increasingly concerned by two problems with epoxy resins: their resistance to water and their tolerance of damage. While either problem could be alleviated, an improvement in one appeared, inevitably, to lead to a deterioration of the other. At this time of great need there were two significant advances in thermoplastic technology. First was the evolution of the poly(ether ketone) family of resins[4] and the commercial development of "Victrex" PEEK (polyether etherketone), a linear chain semicrystalline polymer with a glass transition temperature of 140°C. The ability for the entanglements of the linear chains to slip past one another provided a mechanism for energy absorption without catastrophic failure, and the capability of the linear chains to pack closely together and form crystalline regions, which can be likened to physical crosslinks, provided a resin which was unaffected by such aggressive agents as paint stripper, hydraulic fluid and water: as a matrix, PEEK is tough and environmentally tolerant. The second significant advance was the ability to preimpregnate continuous fibres with thermoplastic polymer melts effectively totally wetting each individual fibre to produce the first continuous fibre reinforced thermoplastic[5]. The combination of these two developments provided a solution to the two problems then facing the aerospace users of composites and, in addition, offered new opportunities for the development of mass production technology. In 1982 ICI introduced the first of a new generation of thermoplastic Aromatic Polymer Composites, APC-1.

The fact that an apparently insoluble problem, how to combine damage tolerance with environmental resistance, could be demonstrated to have a solution, provided a stimulus for further research. Thus one of the first successes of thermoplastic composite materials, on behalf of the reinforced plastics industry at large, has been the development of improved thermosetting resin systems. Not surprisingly, the more successful of those developments have been based on modification of epoxy resins by thermoplastic polymers[6]. Thermoplastic and thermosetting resins are now offering enhanced performance and complementary product and process advantages.

2. EVOLUTION

Continuous carbon fibre tapes preimpregnated with PEEK resin created extensive interest in the aerospace industry. The first of the range, APC-1, was widely evaluated and, as a result of feedback from that evaluation, an optimised product was evolved. Introduced in 1983, APC-2 became the industry standard, not only for new thermoplastic composites, but also that against which new thermosetting system are compared.

High strength carbon fibres were the reinforcement originally selected for this new class of materials. Prepregs based on the new family of intermediate modulus fibres are also available and these are preferred for stiffness critical applications. Ultra high modulus pitch based fibres have also been prepregged as an experimental product for space applications. For some applications, especially those requiring radar transparency, glass fibres are preferred. Similar electrical properties to glass fibre, while retaining the stiffness of carbon fibre, is achieved by prepregging alumina fibres with PEEK.

<u>Comparative Performance of Structural Composites Based on PEEK Resin</u>

Fibre	Designation	Axial Flexural Modulus GN/m^2	Axial Flexural Strength MN/m^2	Transverse Flexural Strength MN/m^2	Interlaminar Shear Strength MN/m^2
High Strength Carbon	APC-2/AS4	121	1880	137	105
Intermediate Modulus Carbon	APC-2/D/IM8	176	1969	166	112
Pitch based Carbon	P75/PEEK	278	728	52+	53*
S-glass	APC-2/S-2 glass	54	1551	157	90
Alumina	APC-2/D/Alumina	120	1516	186	90

+ Low result due to fibre failure
* Sample does not fail in shear

In this table the axial flexural modulus and strength reflect very high utilisation of the fibre properties. The transverse flexural and interlaminar shear strength values are a reflection of the excellent adhesion between the resin and the reinforcement.

The range of thermoplastic structural composites has also been expanded with different resins. There are two key service criteria: the glass transition temperature, Tg, which limits the serviceability of the system as a structural material, and the crystalline characteristic of the resin, crystalline resins generally being considered to have superior environmental resistance.

Resin Tg °C	Semi-Crystalline Matrix	Amorphous	Composite Supplier
260		"Victrex" HTA	ICI Fiberite
260		"Avimid" K III*	DuPont
260		"Torlon" C*	Amoco
250		"Radel" C	Amoco
220		"Victrex" PES	Specmat, BASF, ICI Fiberite
220		"Ultem" PEI	American Cyanamid, Ten Cate
220		"Ryton" PAS-2	Phillips
220		"Radel" X	Amoco
205	"Victrex" HTX		ICI Fiberite
145		J II Polymer	DuPont
140	"Victrex" PEEK		ICI Fiberite, BASF
95	"Ryton" PPS		Phillips

*"Avimid" K III and "Torlon" C may be more properly described as linear chain thermosetting polymers in that they are preimpregnated as prepolymers and then polymerised during processing into the final component. All the other systems noted are fully polymerised true thermoplastic systems capable of repeated processing.

Several other polymers, particularly polyketone variants such as PEK, PEKK and PEKEKK, having intermediate properties between PEEK and HTX, have been proposed. In addition to these continuous fibre reinforced structural composite materials, preimpregnated fibre reinforced products have been prepared based on less stiff resins, including polycarbonate, nylon and polypropylene. Typical of there materials, designed for general industrial use, is the "Plytron" development product range of materials from ICI. Where, only seven years ago preimpregnated continuous fibre reinforced thermoplastics was an empty field, today we can select from within the entire range of this versatile family of resins.

The understanding of thermoplastic composites and their role in the field of advanced composites can be followed by an analysis of the published literature for fibre reinforced polyether etherketone[7] - Table 1. This table presents the publications by year under a series of headings: General Description, Product Form, Morphology, Processing, Service Performance, and Design. These headings are further subdivided to allow a more detailed analysis.

With respect to the category of general papers, there is an early peak in 1984 followed by a steady series of review papers. About half of these reviews are publications from materials suppliers, but a broad ranging review prepared by a group of academic and industrial authors for the National Science Foundation is available[8].

Significant variations in product form did not begin to appear until 1985, but this is an area of very great activity at the present time. Even more than the case with general papers, this is an area where publications have been dominated by the raw material supply industry and only a few objective comparisons between the different product forms have been published[9]. The development of an abundance of product forms offers enhanced choice to the fabricator but also creates some confusion. The philosophy of the original Aromatic Polymer Composites was that wetting the fibres with the resin is the duty of prepreggers and creating the interface between these two elements is the responsibility of a chemical company, while the proper business of fabricators is making shapes. Many of the product form variants seek to carry forward the first two stages into the final shaping process. At the present time, the original fully impregnated uniaxial tape product remains the standard against which other systems are judged.

In respect of morphology less than one third of publications derive from raw material suppliers, and most of those publications appeared relatively early on. The major interest has been the field of crystallinity: this reflected an anxiety which the structural composite user felt when encountering semi-crystalline polymers for the first time and a concern that incorrect processing might lead to inferior performance. The initial peak occurred about 1986 as people learnt to evaluate this unfamiliar property and the continuing high level of interest[10] reflects attempts to significantly modify the performance of these composites by using processing histories which are outside that which is recommended by the manufacturers. Today it is generally recognised that it is easy, and convenient to stay within the recommended processing window and that excursions beyond that range, for the fully impregnated product form at least, produce only minor variations in property[11]. Other aspects of morphology, such as interface and fibre distribution, have received relatively low levels of attention. This absence of detailed study may be taken as indicating that these aspects of the material are considered satisfactory.

Publications in the field of service performance, with only one sixth of the papers in this group emanating from raw materials suppliers, are dominated by contributions about the toughness of this class of materials: confirming the most outstanding single property. This has proved a popular field for academic research in fracture mechanics but the evidence suggests that a peak may have been reached in 1987 and that interest is now waning: there are only so many ways to measure toughness, and PEEK has resisted the most vigorous assaults against it. The interest in toughness is being replaced by the proving of creep and fatigue performance and performance at extremes of temperature both high and low as evidence of a growing commitment to long term performance. The level of interest in environmental resistance, the other outstanding property of semi-crystalline thermoplastic polymers, by no means matches the attention to mechanical performance, possibly because the properties cannot be represented in a dramatic fashion. In this area, as well as outstanding resistance to aviation fluids, PEEK composites have been praised for their resistance to fire and radiation. Of significance is the collection of papers which have recently appeared describing tribological properties and demonstrating that structural composite materials can give valuable service in other fields.

By far the most dramatic development in publications about thermoplastic composites is the surge in papers dealing with processing. This began in 1986 and appears to be still accelerating. Most significantly only one in eight of the papers originate from the materials suppliers. Where, up to 1985, less than one in ten papers dealt with this subject, last year nearly half the total publications were concerned with manufacturing science, and that interest still continues[12]. A minor key to this processing theme is the recent appearance of papers detailing optimised methods for assembling structures from fabricated components. One final area remains to be charted: we have seen very little of composite design for thermoplastics. As a starting point this dearth may be because standard laminate analysis has been found adequate. However, with a new, designed, material, we may confidently expect that optimised design strategies will be created[13].

This analysis of the open literature represents only the tip of the knowledge iceberg. The existence of such knowledge is a firm basis for confidence in application.

3. REVOLUTION

Structural composites have, so far, been largely the preserve of the aerospace industrial sphere. Nevertheless, since the earliest days of high performance carbon fibres, there has been the expectation and intention that that technology will be translated for the benefit of a much wider audience. That translation would bring with it the economies of scale for material production and a general reduction of costs. This reduction of material cost and the greater efficiency of manufacturing, especially of stock items, which would be developed for mass production in the general industrial sphere should, in turn, significantly reduce the costs in the aerospace industry. As long as advanced structural composites depended upon thermosetting resins, the possibility of achieving the critical translation to high rate production remained remote, but thermoplastics, where components can be moulded in a matter of a few minutes or even seconds alter the perspective entirely. In the aerospace environment this capability for mass production cannot be exploited to the full: the number of aeroplanes which one wishes to make in a day is small. Even with that constraint, air frame manufacturers[14] are already reporting cost savings with thermoplastic matrix composites in comparison to thermosets and aluminium. The growing interest in thermoplastic composite materials for industrial products is the herald of a revolution which promises to change the world of Advanced Materials and Process Engineering.

4. DESIGNING THE FUTURE

Continuous fibre reinforced thermoplastic preimpregnated tapes are now an established part of the materials scene. They extend the field of thermoplastic composites which includes short fibre injection moulding materials, long fibre injection moulding materials -"Verton"- with enhanced toughness and dimensional stability, and glass mat stampable sheet products. Not least of the advantages of continuous fibre reinforced thermoplastics is that any offcuts from mouldings can be reclaimed by the addition of extra resin as a high value moulding material. Continuous fibre reinforced thermoplastics also provide a partnership with their longer established thermosetting cousins: inevitably there will be some structures and applications for which thermosetting systems will be preferred and the need will be to select the appropriate material to each application. In particular, thermoplastic matrix systems offer the key to mass production technology with high performance composite materials. Mass production will bring with it economies of scale which will lead to reduced raw material costs. As well as the markets of aerospace and sporting goods, where the properties of composites have always been valued, thermoplastic structural composites have excited interest in the field of industrial products and particularly in areas involving reciprocating mass. Such components are frequently found in the field of machine tools for manufacturing other products, and in such applications the benefits of high stiffness and low weight can be directly translated into increased output and lower power consumption. It is with such tools that mankind shapes his world.

5. REFERENCES

1. J McAinsh "The Reinforcement of Polysulphones and other Thermoplastics with Continuous Carbon Fibre" BPF 8th International Reinforced Plastics Conference 1972.

2. D J Lund and V J Coffey "A Method of Manufacturing Composite Material" British Patent 1 485 586 1977.

3. L N Phillips "The Fabrication of Reinforced Thermoplastics by Means of Film Stacking Technique" HMSO London 1980.

4. P A Staniland "Poly (ether ketone)s" in "Comprehensive Polymer Science" edited by G Allen, and J Bevington 5, 483-497 Pergamon Press 1989.

5. G R Belbin, I Brewster, F N Cogswell, D J Hezzell and M S Swerdlow "Carbon Fibre Reinforced PEEK: A Thermoplastic Composite for Aerospace Applications "Stresa Meeting of SAMPE" (1982)

6. G R Almen, R K Maskell, V Malhotra, M S Sefton, P T McGrail and S P Wilkinson "Semi-IPN Matrix Systems for Composite Aircraft Primary Structures" 33rd International SAMPE Symposium 979-989 (1988).

7. L A Carlsson (editor) "Thermoplastic Composite Materials" Elsevier Applied Science Publishers in press 1989.

8. National Materials Advisory Board "The Place for Thermoplastic Composites in Structural Components" National Research Council NMAB-434 (1987).

9. E M Silverman and R J Jones "Property and Processing Performance of Graphite/PEEK Prepreg Tapes and Fabrics" SAMPE J 24, 4, 33-42 (1988).

10. A M Maffezzoli, J M Kenny and L Nicolais "Modelling of Thermal and Crystallisation Behaviour of the Processing of Thermoplastic Matrix Composites" this conference.

11. E P Corrigan and D C Leach "The Influence of Processing Conditions on the Properties of PEEK Matrix Composites" this conference.

12. J H C Rowan and R N Askander "Filament Winding of High Performance Thermoplastic Prepregs" this conference.

13. D R Moore, I M Robinson, N Zahlan and F J Guild "Mechanical Properties for Design and Analysis of Carbon Fibre Reinforced PEEK Composite Structures" this conference.

14. G R Griffiths "Thermoplastic Composites - Past, Present and Future" this conference.

TABLE 1

PAPERS DESCRIBING CARBON FIBRE REINFORCED POLYETHER ETHER KETONE 1982-1988

SUBJECT	1982	1983	1984	1985	1986	1987	1988	TOTAL
General	1	1	5	-	4	7	5	23
Product Form	-	-	-	1	4	3	13	21
Morphology								
- crystallinity	-	1	2	5	8	5	8	29
- molecular weight	-	-	-	1	-	-	-	1
- interface	-	-	-	-	2	2	-	4
- internal stress	-	-	-	-	-	1	2	3
- fibre distribution	-	1	-	-	-	-	1	2
Service Performance								
- stiffness & strength	1	-	4	1	3	6	2	17
- toughness	-	5	2	6	11	19	14	57
- creep & fatigue	-	-	1	-	2	4	3	10
- temperature	-	-	-	-	1	5	-	6
- environment	-	1	2	1	4	2	2	12
- friction & wear	-	-	-	-	1	3	1	5
- thermal properties	-	-	-	-	-	1	-	1
- specialist uses	-	-	-	-	2	-	7	9
Processing								
- shaping	-	1	2	1	7	19	42	72
- bonding	-	-	-	-	-	3	3	6
Design	-	-	-	-	1	-	1	2
	2	10	18	16	50	80	104	280

Materials and Processing – Move into the 90's
edited by S. Benson, T. Cook, E. Trewin and R.M. Turner
Elsevier Science Publishers B.V., Amsterdam, 1989

FILAMENT WINDING OF HIGH PERFORMANCE THERMOPLASTIC COMPOSITES

J H C ROWAN and R N ASKANDER

Courtaulds Research, P O Box 111, 72 Lockhurst Lane, Coventry, CV6 5RS

ABSTRACT

Thermoplastic matrix composites offer performance and potential cost advantages to aerospace and other industries. An account is given of some of the requirements and methods for filament winding with these materials. An experimental programme at Courtaulds Research is described in which APC-2 (carbon fibre reinforced PEEK) was wound under controlled conditions to produce NOL ring test specimens. Evaluated by ASTM methods, the mechanical and physical properties of these components compared well with published data on compression moulded flat laminates of the same material.

1. INTRODUCTION

Thermoplastic matrix composites are of increasing interest to the aerospace and other industries because of cost and material property factors such as toughness and environmental resistance. Filament winding is one area of application of these materials where they are expected to compete with thermosetting composites on grounds of performance and economics.

Among the engineering thermoplastics, PEEK (polyetheretherketone) has been a popular choice for filament wound carbon fibre composites because of its excellent properties and its availability in the form of suitable continuous length feedstock.

This paper reports work done at Courtaulds Research to develop a process for filament winding thermoplastic composites of PEEK and carbon fibre. The aims of the work were to establish processing techniques and control equipment for the production of good quality composites.

2. PROCESSING CONSIDERATIONS

The essence of filament winding with thermoplastics is the consolidation of successive layers of feedstock to form the required composite structure. In this consolidation process, uneven prepreg surfaces must flow until complete interlaminar contact is achieved, and matrix fusion occurs. This process is dependent on three factors: pressure, temperature and time.

Consolidation can be achieved at or near the point where the feedstock arrives at the component surface, or alternatively in a separate operation after filament winding is complete. The former method, known as in situ consolidation, is one of the keys to the potential economic advantage of using thermoplastics in filament winding, since no further process is required. In situ consolidation is applicable to many, if not most, filament wound structures, see Figure 1.

The pressure required can be generated by feedstock tension or by direct mechanical force. Tension is relatively simple to use, but is adequate only within a certain range of mandrel diameters and winding angles. Pressure applied via rollers or sliding shoes is more universally applicable, but requires accurate guidance, and uniformity of pressure across the feedstock width on all surface contours.

Many techniques have been described for raising the matrix temperature above melting point to achieve in situ consolidation. Methods using conduction, convection, radiation and ultrasonics have been reviewed elsewhere[1,2,3]. The heat source should provide adequate and controllable energy density, be readily moveable, and geometrically compatible with both the winding equipment and the component.

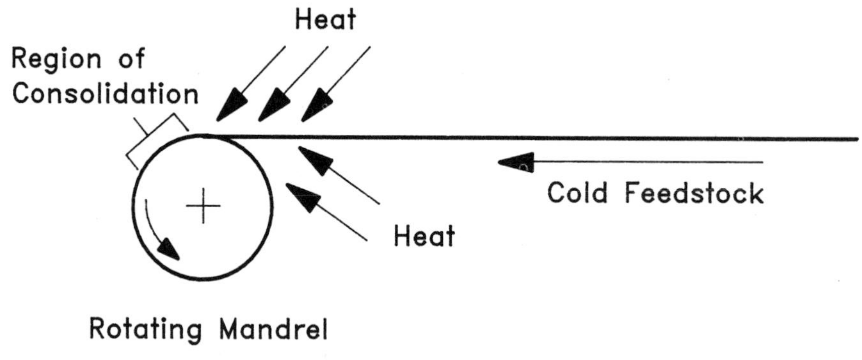

Figure 1.

ESSENTIALS OF IN SITU CONSOLIDATION, USING FEEDSTOCK TENSION OR APPLIED PRESSURE

Consolidation time is dependent on the dynamic interplay of temperatures (of mandrel, component and consolidation region), heat flow rates and material throughput or winding speed.

Based on these considerations, the process developed by Courtaulds Research uses in situ consolidation by tension, and a focused heat source selected on grounds of controllability and ease of adaptation to the winding process.

3. EXPERIMENTAL

3.1 Programme Outline

Courtaulds' experimental work was structured to provide an understanding of process parameters and their control, and hence to generate test structures for quantitative evaluation of microstructure and mechanical properties. Hoop-wound NOL ring specimens and ASTM methods were chosen for this purpose.

3.2 Materials

The materials used in this programme of work were 3 and 6 mm wide APC-2 filament winding tow acquired from ICI's Fiberite with approximately 60% fibre volume fraction. The tow contained unidirectional carbon fibre (AS4 from Hercules) embedded in a thermoplastic PEEK matrix.

The 3 or 6 mm wide tow had an ideal length of several hundred metres suitable for filament winding.

3.3 Equipment

The equipment mainly consisted of a stationary creel with a facility to apply moderate tension (0.5-4.0 kgf), and a two-axis McClean Anderson computer controlled filament winding machine loaded with a steel mandrel 0.146 m in diameter.

In addition a specially designed tensioning device was used to apply tension onto the tow in excess of 4 kgf when required. The equipment described above is shown in Figure 2 together with the positions of the heat source and the temperature measuring device.

3.4 Winding Preparation

The computer on board the filament winding machine was fed with the necessary information for the construction of a variable thickness hoop wound composite tube. The spool of APC-2 was placed on the creel and the tow was fed through the tensioning device and pay-out-eye, and finally was attached to the mandrel; see Figure 2. The tension in the tow was measured and recorded before the start of the wind.

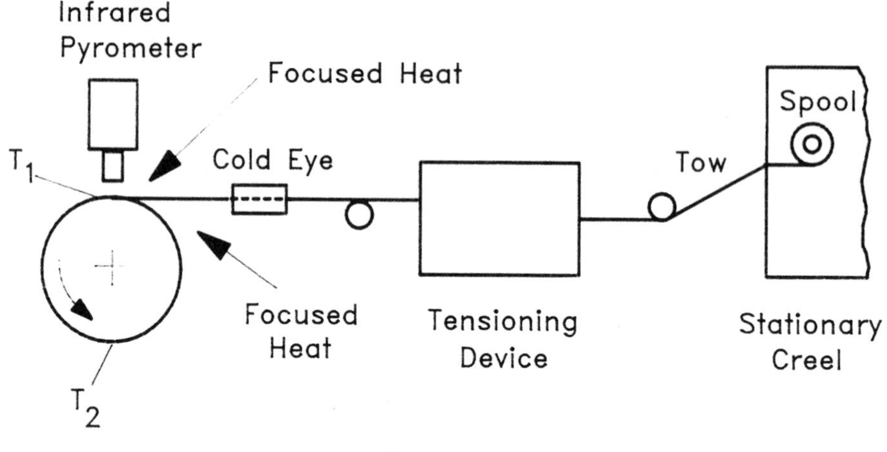

Figure 2.

LAYOUT OF COURTAULDS PROCESS FOR FILAMENT WINDING WITH THERMOPLASTICS

3.5 Temperature Measurement

The temperature of the tow (T1) in the consolidation region (see Figure 2) was measured using a water cooled, infrared pyrometer capable of focussing on a spot 1 mm in diameter. The temperature was monitored continuously during the wind. The processing temperature window of the tow was chosen to be between 380-400°C. The mandrel was always heated to a temperature well below the melting point of PEEK before the start of the wind. The temperature of the tube (T2) was measured by focussing a hand held infrared pyrometer at the surface of the material.

3.6 Consolidation

The consolidation of the filament wound tube was achieved on the mandrel during the winding using positive tow tension applied partially by the creel and mainly by the tensioning device. The tension value ranged between 4 and 20 kgf depending on the width of the tow. During the wind there was sufficient evidence that reasonable matrix flow was taking place under the effect of temperature and the consolidation pressure.

3.7 Process Control and Monitoring

The main process parameters under control or continuous monitoring were:

1-tow temperature (T1)) see Figure 2
2-tube temperature (T2))
3-winding speed

The tow temperature (T1) was measured using the infrared pyrometer described earlier and the output signal from the pyrometer (4-20mA) was then passed to a fast response PID temperature controller linked to the focused heat source. Depending on the actual temperature of the tow in the consolidation region the output from the heat source was adjusted accordingly by the action of the temperature controller in order to keep the tow temperature within the processing window. The system was able to achieve good and effective control of the tow temperature within +-5°C scatter.

The use of a feedback control system in accurately controlling T1 was essential in this work since it was clear from the beginning that the heat transfer boundary conditions in the consolidation region were varying with time. This was caused mainly by the increase in T2 during the wind as a result of which a lower heat flux was required to be focused on the tow in order to achieve T1. The winding speed was monitored by the computer on board the filament winding machine and was adjustable in 0.005 m/s increments.

3.8 Speed

It was possible to achieve linear winding speeds in excess of 0.3 m/s and maintain a good degree of consolidation. This was considered as an important achievement since it would help in reducing the operating costs for the production of filament wound thermoplastic composites.

At the beginning of the wind the speed employed was lower than the normal winding speed since the tow material needed a longer heating time in order to overcome the problem of heat losses mainly by conduction from the molten tow to the relatively colder mandrel. As the thickness of the tube increased so did T2 (up to 120°C difference) which mainly was caused by the difference between the thermal conductivity of steel and the transverse thermal conductivity of unidirectional carbon fibre composites. The latter is lower than the former and hence the composite on the mandrel acted as an insulator preventing further heat from escaping into the mandrel.

3.9 Cooling Rate

The cooling rate of the composite was governed by several factors among which were T2, the mandrel temperature, the tube thickness and the speed of winding.

As shown in Figure 2 the tow was heated by a source of focused heat to a temperature within the processing window at the point of contact between the incoming tow and the mandrel, and maintained at this temperature (T1) for a few centimetres after the point of contact on the mandrel for full consolidation to take effect. Once the consolidated tow had passed this region its temperature started to drop considerably (in the region of 150-300°C/s on average) until the consolidated tow reached thermal equilibrium with the rest of the composite on the mandrel. On finishing the winding process the composite was left to cool down freely on the mandrel from a temperature higher than the Tg of PEEK matrix to room temperature in a very slow manner (cooling rate of around 0.05°C/s).

3.10 Testing

NOL ring specimens were cut from tubes wound according to ASTM D 2291-83, and tested for tensile strength (ASTM D 2290-87), apparent inter-laminar shear strength (ILSS) (ASTM D 2344-84), density (ASTM D 792-86), fibre volume fraction (ASTM D 3171-76 (1982)), and void content (ASTM D 2734-70 (1985)). Cross sections were also examined microscopically.

4. RESULTS

The results obtained from the various tests on the NOL ring specimens are given in Tables 1 and 2, and Figure 3. In Table 1 the effect of winding speed on the tensile strength, ILSS and void content is illustrated, whereas in Table 2 results are given for 3 and 6 mm wide APC-2 tows. The tensile strength and ILSS are averaged over a group of rings with the standard deviation figure given in the brackets.

Figure 4 shows a cross-section of one of the NOL rings.

5. DISCUSSION

The results given in Table 1 show that it is possible to increase the winding speed up to 5.5 times with no real depreciation in the tensile strength and ILSS or an increase in void content. Specimen B has the advantage over specimen A of shorter production time due to faster winding speed (around 0.4 m/s), which is comparable with the speed used in filament winding of thermosetting resins.

Table 1.

EFFECT OF SPEED ON COMPOSITE QUALITY

Specimen Reference	Tow Width $m \times 10^{-3}$	Speed m/s	Tensile Strength MPa	ILSS MPa #	Void Content % v/v	Fibre Content % v/v
A	3	0.07	1643 (±89)	113 (±1)	1.5	58
B	3	* 0.07 – 0.39	1667 (±45)	108 (±2)	1.7	58
Fiberite Data (flat UD laminate)	–	–	2130	105	–	61

* Outer 2/3 of tube wound at the higher speed
\# All specimens failed by shear
Figures in brackets are standard deviations

Table 2.

COMPARISON OF DIFFERENT FEEDSTOCK WIDTHS

Specimen Reference	Tow Width $m \times 10^{-3}$	Speed m/s	ILSS MPa #	Void Content % v/v	Fibre Content % v/v
C	3	0.07	111 (±4)	1.9	60
D	6	0.07	108 (±1)	1.7	60

\# All specimens failed by shear
Figures in brackets are standard deviations

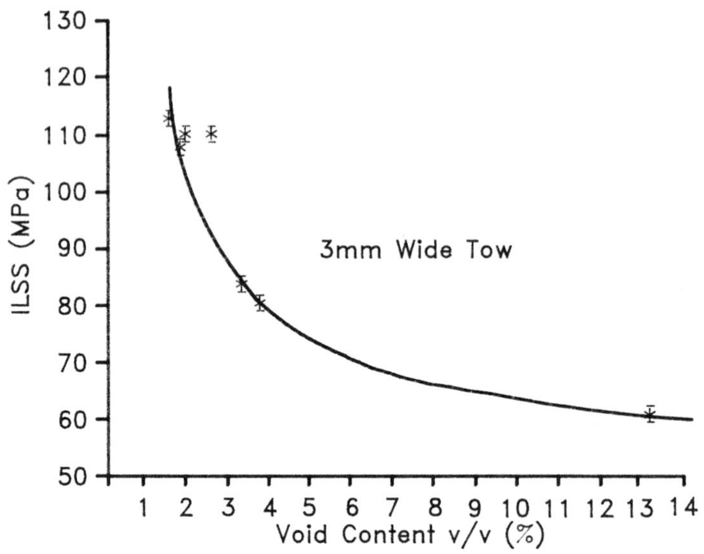

Figure 3.

EFFECT OF VOIDS ON ILSS IN FILAMENT WOUND APC-2 COMPOSITES

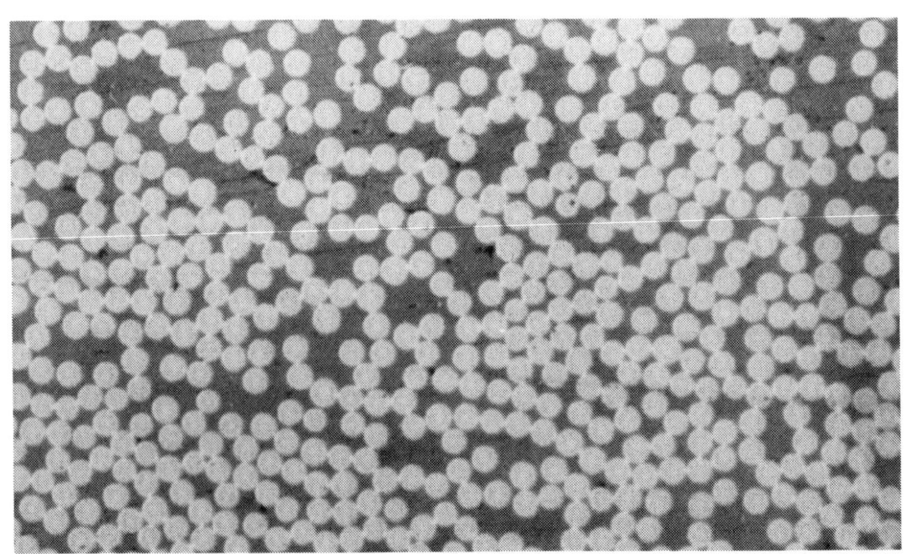

Figure 4.

CROSS SECTION OF APC-2 NOL RING

Good agreement is achieved between Courtaulds' results and Fiberite's flat laminate results as shown in Table 1. The reason why there is a noticeable difference in the tensile strength figures is due to lower fibre volume fraction in the wound composite compared to the flat laminate and perhaps also due to the fact that the split disc method (ASTM D 2290-87) gives an underestimated value.

For a given winding speed the production rate can be significantly increased if 6 mm wide tow is used rather than a 3 mm one without affecting the quality of the composite. This point is clearly seen in Table 2 where the properties of NOL ring specimens made from 3 and 6 mm tows show very good agreement.

Higher production rates achieved by faster winding speed or wider tows will undoubtedly require a correct combination of heat flux and consolidation pressure to obtain optimum results. However, if this balance cannot be achieved then it is inevitable that poor bonding and insufficient welding will take place, resulting in void formation. The presence of voids is known to affect the ILSS most dramatically as shown in Figure 3. For optimum quality, void content of less than 2% is needed in the composite. Figure 4 illustrates the low void content and excellent fibre wet-out present in Courtaulds' filament-wound APC-2 composite.

6. CONCLUSIONS

The mechanical and physical properties of NOL rings made by Courtaulds' process have demonstrated very good translation of APC-2 properties into the filament-wound structure.

The use of an automated feedback system to control the process temperature under widely varying conditions of speed and thermal environment aided the achievement of these composite properties, and paves the way for product quality assurance under future shop-floor conditions.

The winding speeds already achieved indicate the potential competitiveness of thermoplastic composites in filament winding.

The combination of high processing rates, automated production conditions, and excellent composite properties from materials such as APC-2, should ensure the future commercial application of filament winding with thermoplastics.

REFERENCES

1) M W Egerton, M B Gruber "Thermoplastic filament winding: demonstrating economics and properties via in-situ consolidation", 33rd Int. SAMPE Symp., Vol. 33, pp 35-46, 1988.

2) J A S Whiting, "Automated manufacture of advanced thermoplastic composites", 2nd Int. Conf., Automated Composites 88, Plastics and Rubber Institute, Noordwijkerhout, Netherlands, 1988.

3) D Hauber, "Filament winding with thermoplastics", Fabricating Composites '87 Conf., SME, Philadelphia, 1987.

Materials and Processing – Move into the 90's
edited by S. Benson, T. Cook, E. Trewin and R.M. Turner
Elsevier Science Publishers B.V., Amsterdam, 1989

FIBRE REINFORCED THERMOPLASTIC INTEGRAL CONSTRUCTIONS IN MODULAR BUILD-UP TECHNOLOGY THE 'THERMOPLASTIC IN-SITU-TECHNIQUE'

Gerhard MUSCH, Erich WINTERMANTEL

Institut fur Konstruktion und Bauweisen, Swiss Federal Institute of Technology, Ramistr. 101. CH-8092 Zurich, Switzerland.

Current applications of thermoset matrices in fibre reinforced structures do not allow a wider use of prefabricated elements with long shelf life that could easily be bonded together to form complex parts. This lead to a manufacturing engineering research programme at our Institute in order to apply knowledge gained from thermoset integral build-up technologies to thermoplastics. The objective is to get prefabricated standardized and storable elements that can easily be bonded to complex integral structures by melting of the thermoplastic matrix itself and to avoid as far as possible the application of a complete autoclave cycle.
As a model of complex integral structure a method of manufacturing stringer reinforced curved panels as a substructure of an airframe was developed. T-shaped and angled profiles had been prefabricated and integrated with all substructures to the complex definitive part in one heating cycle. Results shows that a scaled-down (1:6) section of a highly integrated airframe structure could be manufactured with the described method. Peel and adhesive properties concerning the joint-area between stringers and panels show satisfying results. Manufacturing methods of the standardized elements, the processing cycle and thermoplastic build-up technologies as applied here are presented.

1. INTRODUCTION

Processing-cycles for current high performance thermosets range in the order of hours whereas for thermoplastics they could be shortened considerably. In addition to this fact thermoplastics offer better impact properties, higher resistance against environmental influences and a practically unlimited storability at room temperatures.

These attractive and new properties require new methods of manufacturing or the further development, modification and adaptation of processes known from thermosets. Of particular importance is the need to develop techniques whereby thermoplastic components can be assembled into complex structures. This includes a requirement to supply a uniform distribution of pressure onto the complete part in order to achieve complete consolidation of laminates by means of elastomeric and fibre reinforced pressure pads, i.e. AIRPAD by Airtech

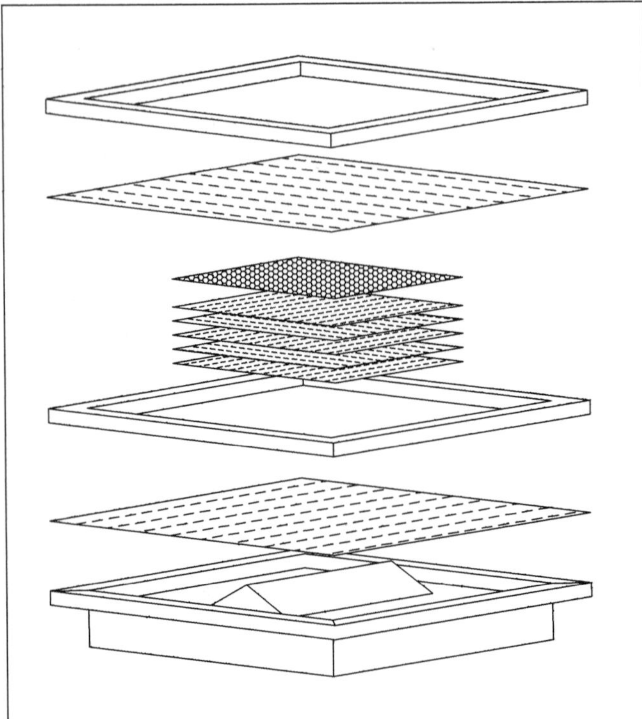

Fig. 1
shows the diaphragm method developed for autoclave technique. This technique has been modified and applied to vacuum-forming machines thus avoiding the autoclave cycle.

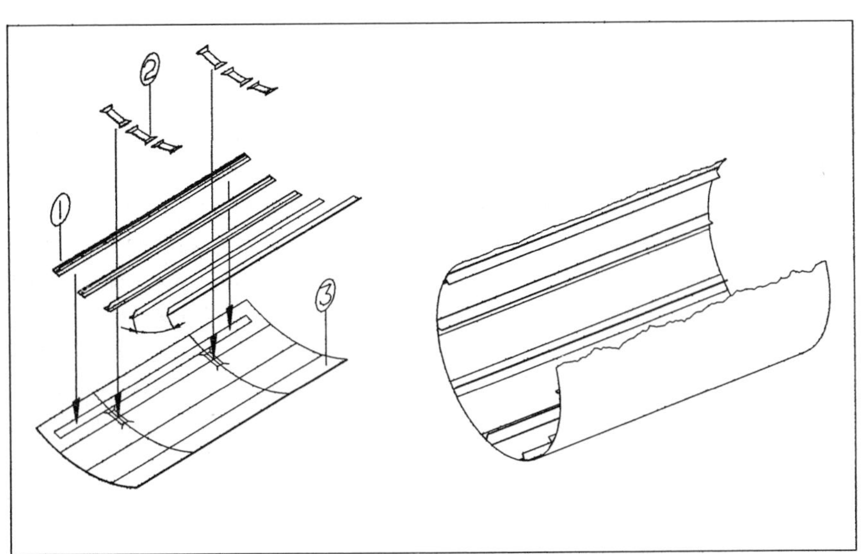

Fig. 2 Scaled down version (1:6) of an aircraft fuselage with 3 basic modules; longitudinal right angled T-Profiles (1), circumferential reinforcements (2) and filament wound outer shell (3).

Fluorocarbonic elastomerics permit higher temperature ranges i.e. 350 degrees C and thus may be applied to PEI and PPS for an integral structuring process. In-sheet-forming of thermoplastics by vacuum-forming results in processing cycles of less than ten minutes leaving open the additional application of pressure if the atmospheric pressure is not sufficient for consolidation of homogeneous laminates. This cycle time advantage can also be exploited when substructures, i.e. profiles, sheets, plates, frames etc. are bonded into the complex structure by melting of the thermoplastic matrix on one shot.

2. MATERIALS AND METHODS RESULTING IN A STRINGER REINFORCED AIRFRAME STRUCTURE

2.1. The 'Thermoplastic in-situ-technique'

In the diaphragm forming method the fibres are maintained under tension by sandwiching them between plastic diaphragms which provide a sealed package allowing the evacuation of air and volatiles from the moulding (2,3). Diaphragm and laminate are formed together into the mould (cf. Fig. 1). This technique originally developed for use in an autoclave, can be modified for use in the vacuum-forming technique. Infra-red heaters have been used to heat up the assembly to the forming temperature which is subsequently formed on a heated and coolable mould. Basic knowledge about the process was achieved by using carbon or glass fibres with a Polyamide PA 12 matrix. In a first step the tool was not heated resulting in an insufficient thickness of the laminate in a lateral area and in parts of the central area of the part. As such a profile has inadequate qualities for structural applications these findings lead to the following requirements for obtaining uniformly consolidated laminates and complex structures. According to our method this consolidation of the structure has to be performed together with the bonding of the substructures. This procedure is entitled 'Thermoplastic in-situ-technique'.

2.2. From individual parts to a complex structure (scaled down version of an aircraft fuselage, Fig. 2). Fig. 2 represents a modular build-up design of the fuselage. Longitudinal right-angles T-profiles, circumferential reinforcements and an outer filament-wound shell are the modules that had been moulded and consolidated against the outer tool forming the complex structure in a final autoclave cycle. The tooling system fulfilled the following demands:

1. The vacuum-preformed stringers (right-angled profiles) are not completely consolidated when leaving the vacuum-forming machine. They have to be consolidated in the final moulding step.

2. The tooling system distributes the pressure isostatically onto the connecting surfaces of the structure during the moulding process.

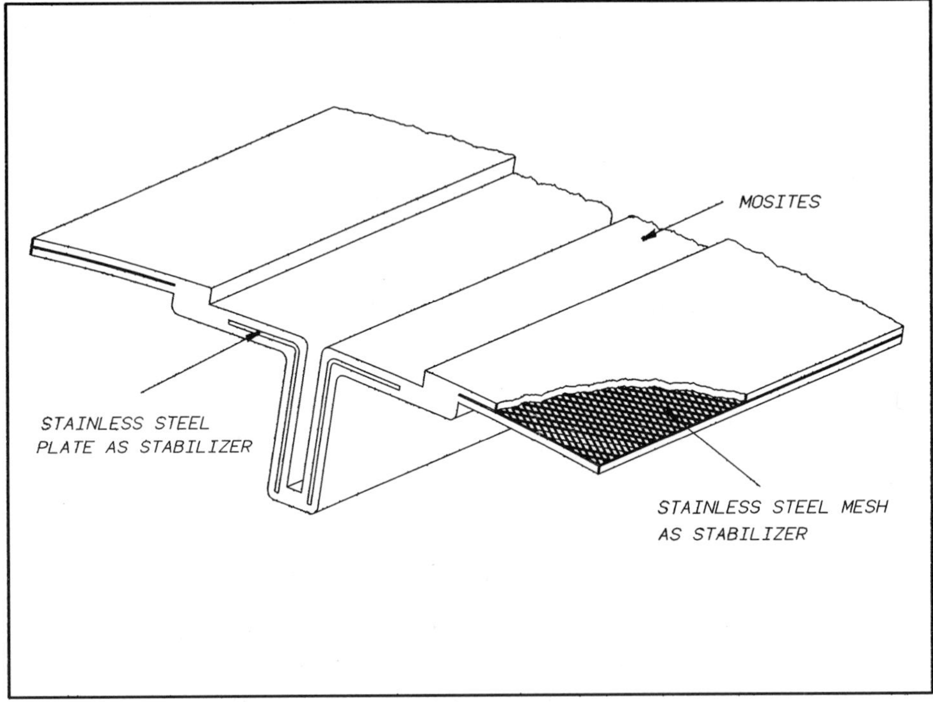

Fig. 3 Mosites build-up representing the inner tool with an elastomeric core reinforced by a stainless steel mesh. A stainless steel tape gives additional stability.

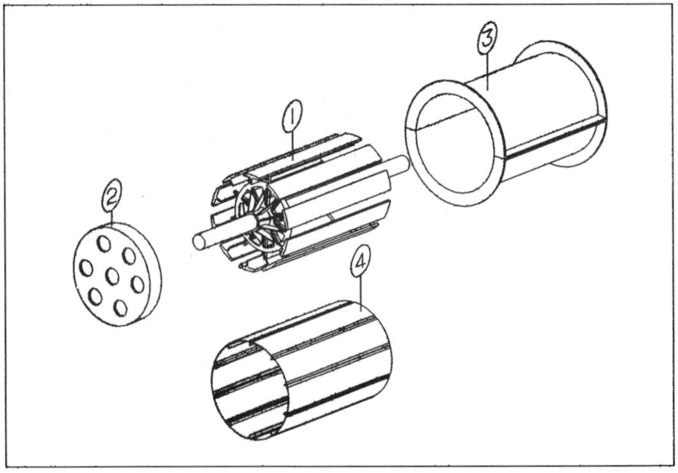

Fig. 4

Tooling concept in detail:

supporting structure (1), pole shapes (2), castable ceramic outer shell (3), elastic inner tool (4).

3. The elastic formfit part fixes the components relative to each other and assures the required precision of the contour. An example of an elastic formfit part fixing the basic modules (Mosites 2902, registered trade mark of Mosites Inc.) is shown in Fig. 3. It represents a fluorocarbonic elastomer ranging up to a processing temperature of 350 degress C. To ensure satisfactory results in respect of dimensional tolerance in repeated mouldings and good mould release the elastomeric structure is reinforced with steel mashes and tapes.

4. The partially consolidated fibre laminate fixed on the flexible inner tool, is pressed against the non-flexible outer tool. Note that the incomplete consolidation of the laminate enables it to conform fully to the precisely defined tool.

2.3. Tooling (Fig. 4)

1. An aluminium supporting structure (detail 1) secures the elastic inner tool (detail 4) and assures the required diameter.

2. Gaps within the elastic inner tool position the prefabricated stringers.

3. Pole shapes guide the filaments around the ends of the cylindrical structure (detail 2).

4. The outer shell can be divided into two symmetrical parts and is made from castable ceramics (detail 3).

5. Mosites, commercially available as calendered sheets in various thicknesses up to ¼ inch, forms the inner tool and have to be manufactured in segments. As a non-tack material its handling is difficult and requires special fixing methods as well as a separate autoclave cycle to cure a single segment. The bonding process of segments is described in detail in 1. Fig. 5 illustrates the bonding process of the preformed segments. Fig. 6 shows the application of the vacuum forming machine to a thermoplastic fibre reinforced part.

2.4. The manufacturing process of a fuselage section.

Four process steps can be distinguished:

1. Preparing the right-angled (T-section) profile on the vacuum forming machine according to the diaphragm method using atmospheric pressure.

2. Filament winding of the outer shell with the following layer distribution (in degrees): 90/+30/-30 // +30/-30/90.

3. Curing the in autoclave.

4. Mould-release.

Fig. 7 shows typical processing cycles in the vacuum forming machine and in the autoclave.

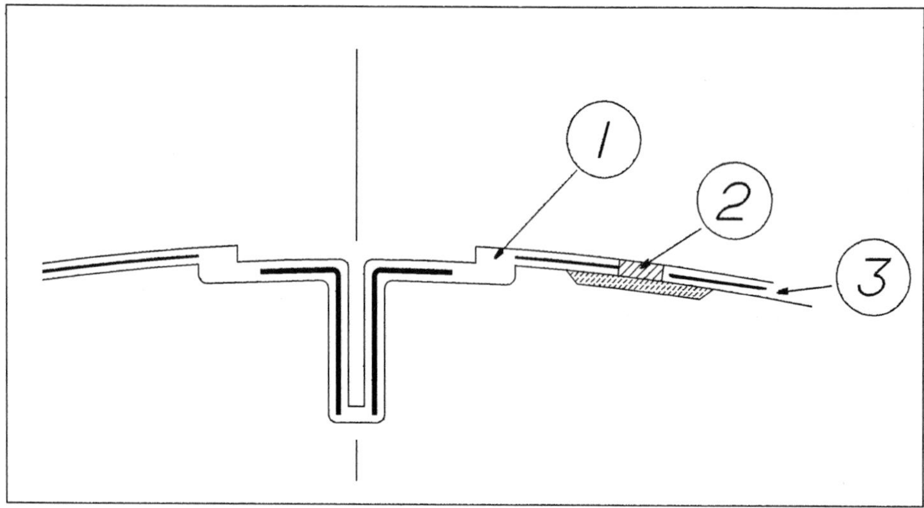

Fig. 5 Segementsof Mosites are bonded together:
Mosite elements (1) and (3), uncured Mosites Nr. 2902 sheets (2).

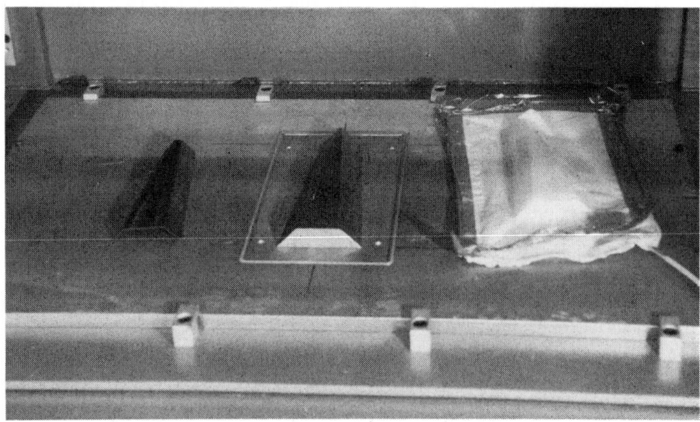

Fig. 6 Application of the vacuum-forming machine to a thermoplastic fibre reinforced part (example: wing-rib for a trailing edge).

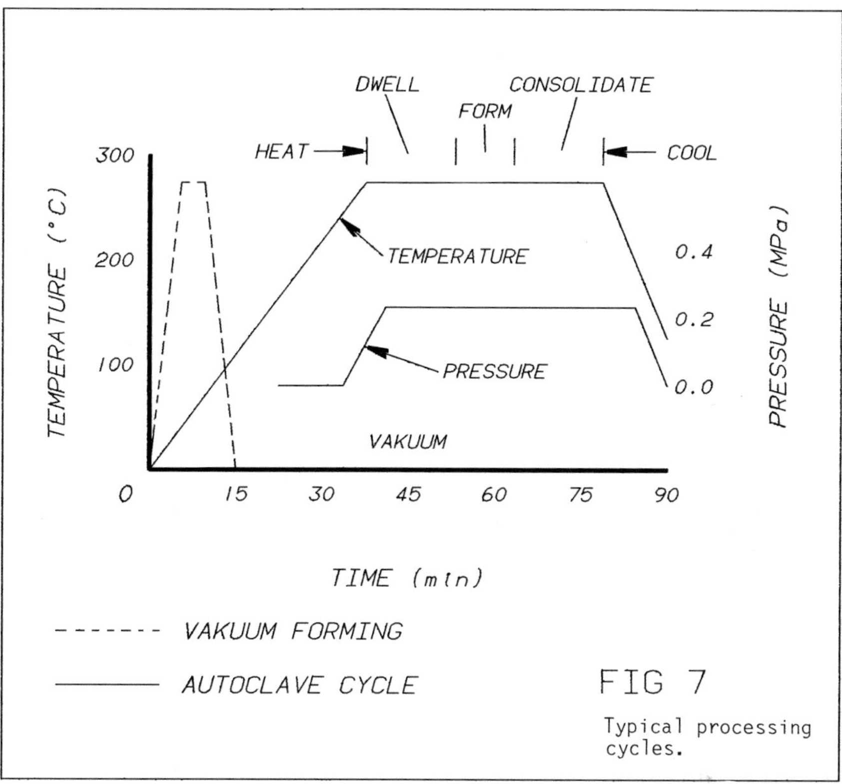

FIG 7

Typical processing cycles.

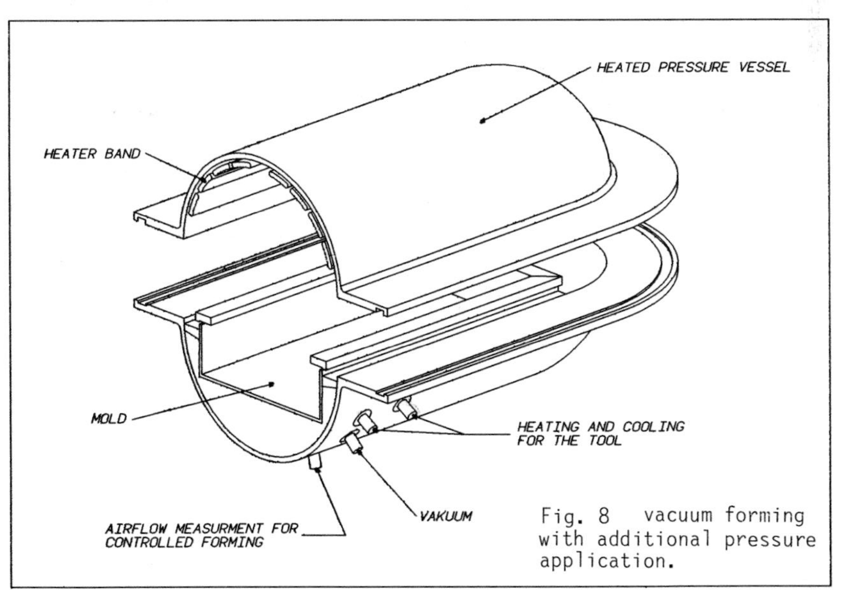

Fig. 8 vacuum forming with additional pressure application.

3. CONCLUSION

This method, describing geometric, processing and structural demands, shows that prefabricated thermoplastic components can be integrated to complex structures in a highly economic way. Processing times can considerably be reduced compared to the normal autoclave techniques. Additional application of pressure within a vacuum forming machine is a desirable additional process step and should result in a high quality part manufactured in a short time. The method can easily be applied to larger structures with more complex details. These limits have to be evaluated in future separate experimental steps. Another, existing limit are pressure pads available only up to temperature levels suitable for PEI and PPS.

Utilizing the advantages of thermoplastic matrices allows the manufacture standardized semifinished parts in a highly economic process. We have currently developed a vacuum forming machine combined with a separable pressure vessel in order to additionally apply high pressure to the structure to be formed (Fig. 8) this would result in a cycle time of less than 15minutes when applied to PEI and PPS.

4. BIBLIOGRAPHY

1) Data sheet and processing instructions for Mosite 2902 by Mosites Rubber Company, Forth Worth, Texas, U.S.A.

2) Mallon, P.J., O'Bradaigh, C.M., Pipes. R.B. Polymeric diaphragm forming of complex-curvature thermoplastic composite parts. Composites 20 (1988). 48 - 56.

3) Mallon, P.J., O'Bradaigh, C.M., Development of a pilot autoclave for polymeric diaphragm forming of continuous fibre reinforced thermoplastics. Composites 19 (1988). 37 - 47.

Materials and Processing – Move into the 90's
edited by S. Benson, T. Cook, E. Trewin and R.M. Turner
Elsevier Science Publishers B.V., Amsterdam, 1989

EFFECT OF TRIGGER GEOMETRY ON CRUSHING OF COMPOSITE TUBES

D.HULL and J. C. COPPOLA

Department of Materials Science and Metallurgy, University of Cambridge, Pembroke Street, Cambridge CB2 3QZ

Woven glasscloth-epoxy tubes have been used to evaluate the effect of trigger geometry on the characteristics of progressive crushing during axial compression. In particular, the changes in average crush load, specific energy absorption and crush mechanisms with the shape of the internal mandrel trigger and the chamfer angle have been determined. Maximum energy is absorbed when the mandrel trigger forces the material through a sharp radius of curvature. There is a close correlation between the load-displacement response and the fragmentation which occurs in the crush zone. It is concluded that in contrast to other types of trigger the average crush load is dependent on the shape of the mandrel trigger.

1. INTRODUCTION

The effectiveness of fibre composite tubes in crashworthy structures can be characterised by the force-displacement (P-d) response during axial crushing. Thus a tube with a P-d curve of the form illustrated in figure 1 will produce a constant deceleration and the total energy absorbed in a collision is the area under P-d curve, which is Pd.

It is usually necessary to design a 'trigger' into one end of the tube which will initiate the progressive crushing process at loads below the limiting buckling load or the brittle fracture load [1]. A variety of triggers have been evaluated including internal and external chamfers [2-5], tulip triggers [6], hole triggers [7], internal mandrels [8] and curved sections [9]. The trigger geometry can have a significant effect on the initial part of the P-d response and may also control the type of progressive crushing which develops. This, in turn, will affect the progressive crush stress (σ) and the specific energy absorption (E_S), which is the energy absorbed per unit mass of material crushed.

The internal mandrel trigger is of particular interest. It has been used by Ford US [8] in prototype designs for front end collision protection in a full-scale crash evaluation programme. The basic geometry is illustrated in figure 2 for a circular tube. The onset of crushing is controlled by the shear strength of the bond between the tube and the internal mandrel. In this paper the effect of two geometrical parameters, i.e. chamfer angle ø and mandrel radius R on the crush load are examined and compared to data on the same material using a more conventional chamfer trigger against a flat platen.

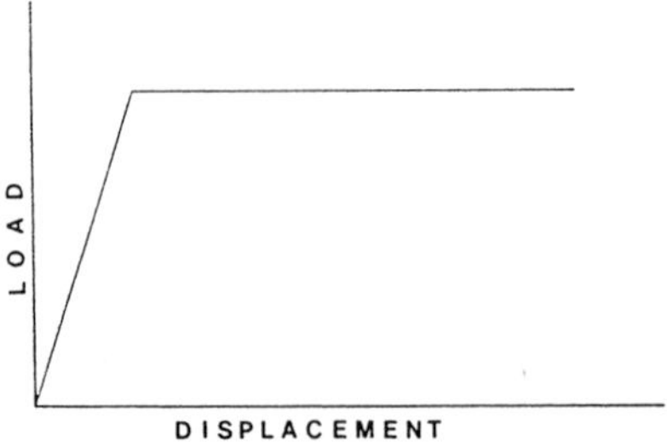

Fig. 1. Idealised load-displacement curve for axially compressed tube.

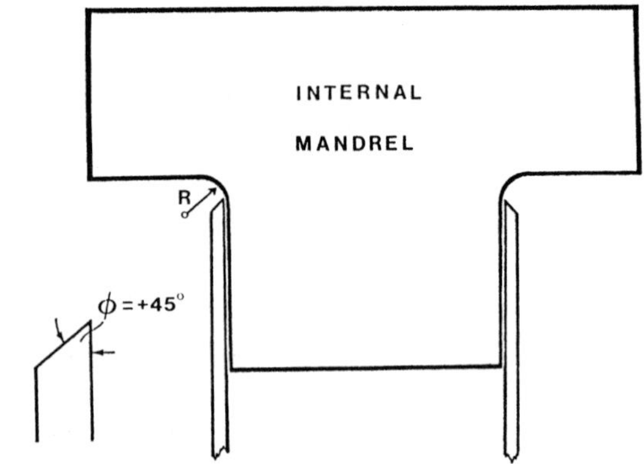

Fig. 2. Internal mandrel trigger illustrating radiussed section and chamfer on end of tube

2. EXPERIMENTAL DATA

All the tests were made on woven glasscloth/epoxy resin tubes in which the warp and weft directions in the cloth were aligned in the hoop and axial directions of the tube. The volume fraction of fibres was 0.48. The tubes had an internal diameter of 50mm and a wall thickness of 2.5mm. Tubes with a range of chamfer angles ø, see figure 2, were prepared by machining on a lathe.

Internal mandrels were made from stainless steel. The diameter of the mandrel gave a close tolerance fit with the inside of the tube. Four mandrels were made with shoulder radii R equal to 1.4, 2.4, 3.2 and 4.0mm.

The tubes were crushed in axial compression using either internal mandrels or a flat platen. The cross-head speed was 0.1mm/s.

The appearance of the crushed zone was examined by preparing polished sections normal to the wall of the tube through the crush zone. The end of the tube was embedded in a potting resin before sectioning to avoid break up of the crush zone and to facilitate grinding and polishing prior to microscopic examination.

3. RESULTS

A detailed description of the sequence of events leading to the formation of a stable crush zone when tubes with different values of ø are crushed against a flat platen has been given elsewhere [5]. The basic features of crushing for a tube with ø = 45° are illustrated in figures 3 and 4. The load-displacement curve shows that, after the initiation of crush, the crushing load is independent of crush distance, with $\overline{P}$ = 48kN. The equivalent value of E_s is 62kJ/kg. The stable crush zone, figure 4, is typified by crushed material which has splayed to the inside and outside of the wall of the tube. A wedge of debris has formed in the centre of the wall.

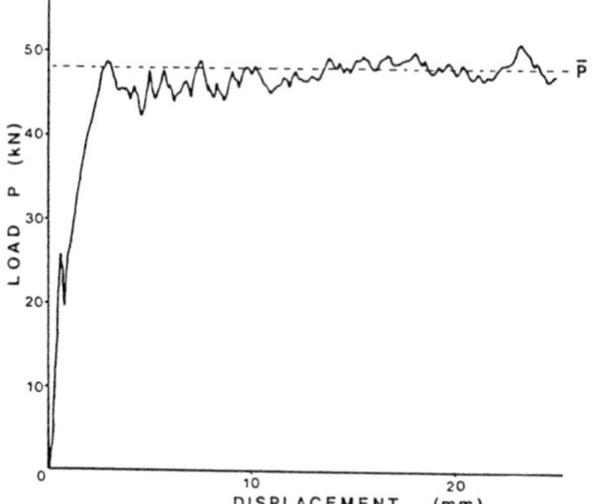

Fig. 3. Load-displacement curve for +45° chamfer tube crushed on flat platen.

Fig. 4. Crush zone of tube corresponding to Fig. 3.

A completely different form of crush zone developed with the internal mandrel trigger. A typical example for a tube with a +45° chamfer and a mandrel radius R = 2.4mm is shown in figure 6. The material is forced outwards at the base of the trigger. The load-displacement curve, figure 5, is similar to figure 3 with $\overline{P}$ = 39kN. The effect of the initial chamfer angle on the mean crush load $\overline{P}$ is shown in Table I. There is some scatter in the data but it may be concluded that the chamfer angle does not affect $\overline{P}$. However, the chamfer angle has a significant effect on the initial part of the load-displacement curve, cf. figures 5 and 7. Thus, for example, for ø = -45° a large peak occurred at P = 60kN before the load dropped to the progressive crush load. For ø = +45° (figure 5) no peak was observed.

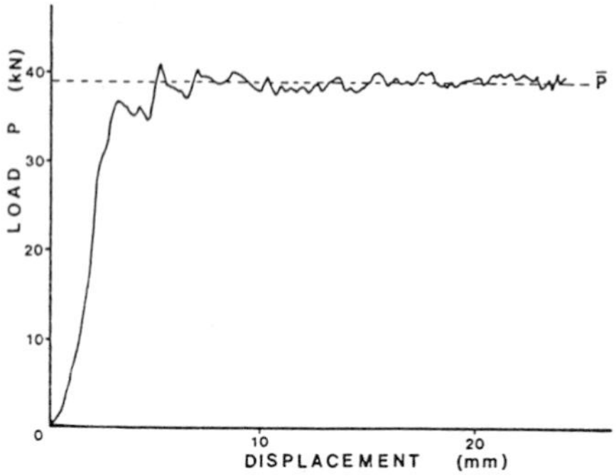

Fig. 5. Load-displacement curve for tube crushed with internal mandrel, (R = 2.4mm, ø =+45°).

Fig. 6. Crush zone of tube corresponding to Fig.5.

Table 1: Effect of chamfer angle on mean crushing load during progressive crushing for internal mandrel trigger with R = 2.4mm.

Chamfer Angle ø	Mean Crush Load $\overline{P}$ (kN)
90°	34, 35
-30°	42
-40°	33, 42, 36
-45°	35, 37
+45°	39

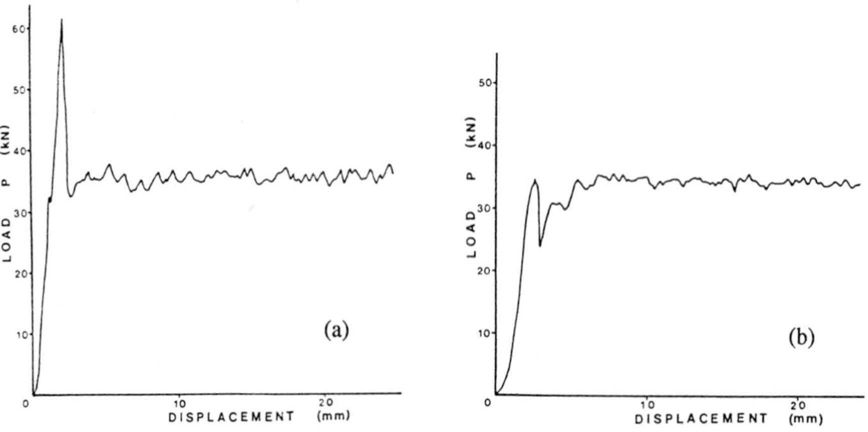

Fig. 7. Load-displacement curves for tubes crushed with internal mandrel (R = 2.4mm), (a) ø = -45° , (b) ø = 90° .

The effect of R on the load-displacement response for a tube with ø = +45° is illustrated in figure 8. There is a systematic change in the form of the curve and in the value of $\overline{P}$. There are corresponding changes in the appearance of the crush zones figure 9. $\overline{P}$ increased as R decreased, figure 10 and Table 2, to a maximum of 44kN with R = 1.4mm. For R = 4.0mm crushing occurred in a very erratic manner. As the tube expanded over the curved sections of the mandrel large pieces of the wall broke away (figure 9d). This resulted in a sharp drop in the load, particularly in the initial stages of crush. This sequence was repeated throughout the crush and there was a rough correlation between the size of the fragments and the separation of the

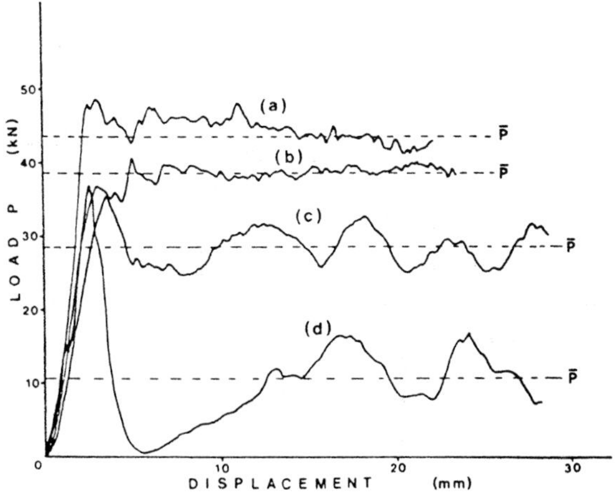

Fig. 8. Load-displacement curves for tubes crushed with internal mandrels with different values of R. (a) R = 1.4mm, (b) R = 2.4mm, (c) R = 3.2mm, (d) R = 4.0mm.

Table 2: Effect of Mandrel Radius on Crush Loads

Mandrel Radius (mm)	Initial Peak Load P_{max} (kN)	Mean Crush Load $\overline{P}$ (kN)	Mean Load Fluctuation ΔP (kN)	Specific Energy Absorption E_s (kJ/kg)
flat platen	49	48 ± 2	3	62
1.4	53	46 ± 2	3	60
2.4	35	36 ± 3	3	47
3.2	37	29 ± 1	5	38
4.0	34	11 ± 1	8	14

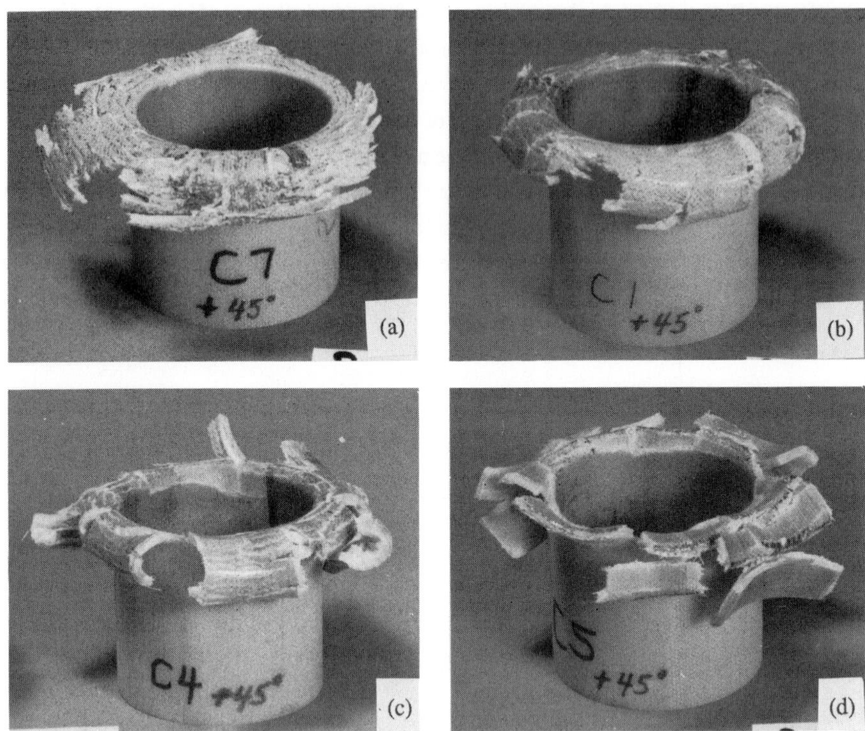

Fig. 9. Crush zones of tubes corresponding to Fig. 8, (a) R = 1.4mm, (b) R = 2.4mm, (c) R = 3.2mm, (d) R = 4.0mm.

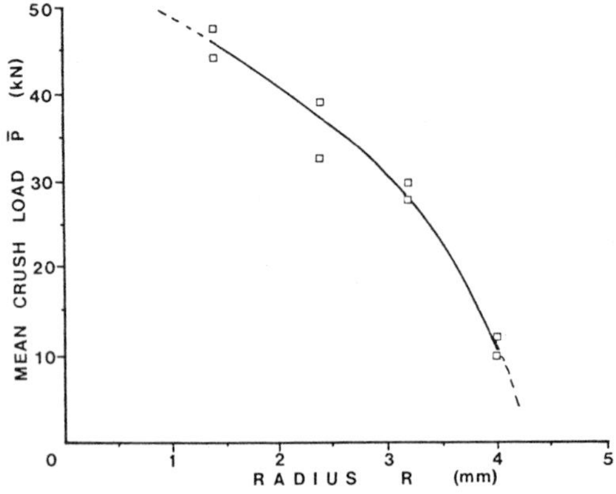

Fig. 10. Effect of R on mean crush load.

oscillations on the load-displacement trace. The size of the fragments for R = 3.2mm was less than for R = 4.0mm and the serrations on the load-displacement trace were less marked. This trend continued for R = 2.4 and 1.4mm. The correlation between the size of the fragments and the displacement between serrations was maintained. Thus for R = 1.4mm the average spacing between rings of fragmented material was 0.9mm, figure 9a, which compares well with an average distance between serrations of 0.7mm.

Photomicrographs of polished sections cut through the crush zone are shown in figure 11. They confirm the main features shown in figure 9 and provide a detailed picture of the microfracture processes involved. It is clear that R has a very significant effect on the damage micromechanisms. However, a detailed analysis of these effects is beyond the scope of this paper.

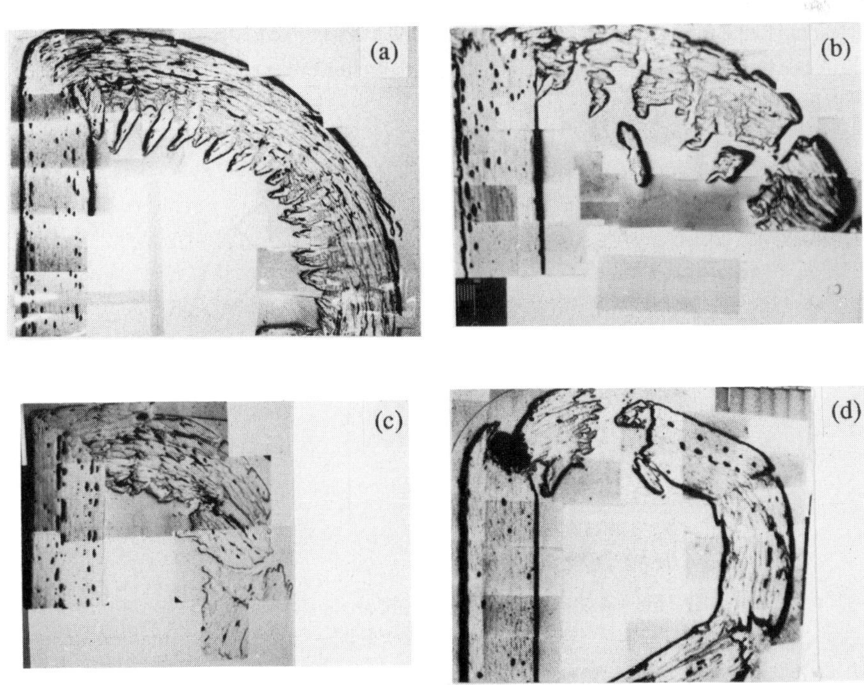

Fig. 11. Photomicrographs of sections of crush zones from tubes crushed with internal mandrels, (a) R = 1.4mm, (b) R = 2.4mm, (c) R = 3.2mm, (d) R = 4.0mm.

4. DISCUSSION AND CONCLUSIONS

Two aspects of trigger geometry have been demonstrated by this series of experiments. Firstly, the influence of the trigger on the initial stages of crushing before the establishment of a stable crush zone. Previous work [5] has shown that, in crush tests against flat platens, the chamfer angle has a marked effect on the initial form of the load-displacement curve, but does not affect the eventual progressive crush load. This is confirmed by the present work using a mandrel trigger with a fixed value of R. The difference in the load-displacement curves evident in figures 5 and 7 for ø = +45°, 90° and –45° can be accounted for in terms of the fracture processes which occur in the radiussed section of the trigger and the way that the load on the chamfered end of the tube is supported. This is illustrated in figure 12. Thus, with ø = +45° (figure 12a) the chamfer is readily deformed and fractured in the radiussed region and large peak loads do not develop before the stable zone is formed. In contrast for ø = –45° (figure 12b) the chamfer fits closely to the radiussed section and high loads can be supported before shear fracture occurs leading to the formation of the stable zone. Thus, it is possible to design-in the optimum load-displacement response at the onset of crushing by selection of the chamfer angle. For most applications a high peak load must be avoided.

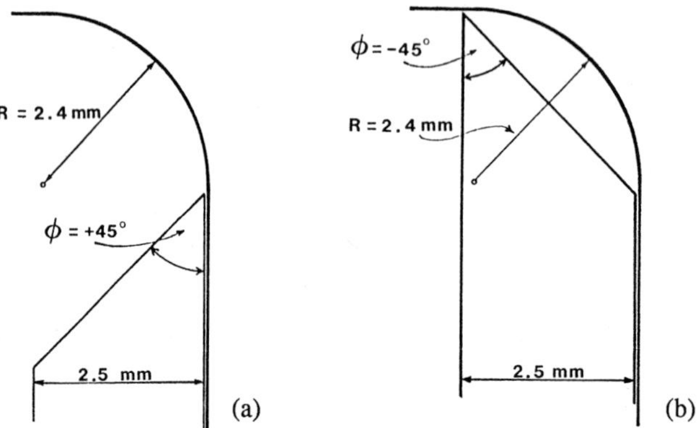

Fig. 12. Inter-relation between chamfer angle and radiussed section of mandrel wedge.

The second aspect is the importance of R on the progressive crush mode and consequently on the mean crush load and specific energy absorption. The increase in $\overline{P}$ shown in figure 10 suggests that further improvements in E_s could be achieved by further reductions in R. Systematic experiments have not been done to establish this but some preliminary tests indicate that other effects intervene. Thus a test was carried out with an internal mandrel with R less than 0.5mm. It was found that $\overline{P}$ = 36 kN compared with $\overline{P}$ = 46kN for R = 1.4mm. Examination of the crush zone and mandrel after testing provided a simple explanation, illustrated in figure 13, for this apparent anomaly. In the initial stages of crush the radiussed region was filled with densely compacted debris of crushed material which acted as a new radius for the crushing tube. In other words the machined radius became ineffective in the crushing process. The crushed debris is closely analogous to the debris wedge (figure 4) observed in tubes crushed against flat platens.

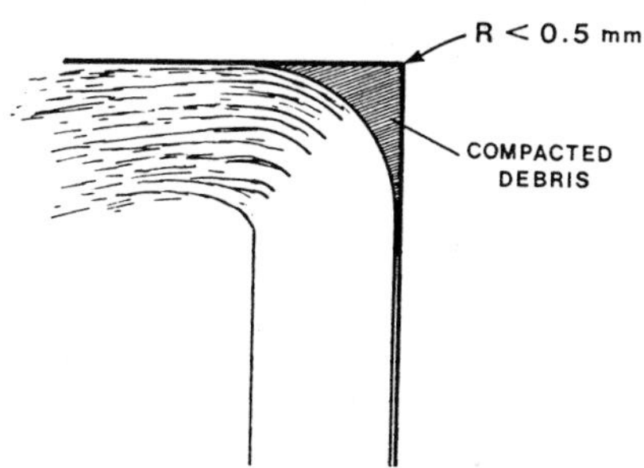

Fig. 13. Compacted debris formed in mandrel wedge with small values of R.

The type of progressive crushing which occurs at R = 3.2mm and 4.0mm is not likely to be acceptable in crashworthy applications because of the large fluctuations in load. It is also important to recognise that optimum values of R cannot be translated to tubes made from different composite materials with the same dimensions or tubes made from the same material with different dimensions. These scaling type rules have still to be identified. The close correlation between crush zone micromechanics and $\overline{P}$ is a clear indication that identification of scaling rules will not be achieved without a detailed understanding of the micromechanics of failure of composite materials and the way that these relate to structural form and dimension. An additional factor, which has a major effect on all the parameters involved in crushing is the role of friction[10]. The frictional contributions will depend on trigger geometry, material characteristics and crush zone morphology.

This work shows that choice of trigger geometry can provide a range of different crush responses which allows some degree of optimisation. The maximum values of $\overline{P}$ or E_s for the mandrel trigger ($\overline{P}$ = 46kN, E_s = 60kJ/kg) are close to the values obtained for a chamfer trigger against a flat platen, ($\overline{P}$ = 48kN, E_s = 62kJ/kg). This agreement is probably fortuitous although in both cases extensive micro-fracture and splaying occur at the crush front.

ACKNOWLEDGEMENT

This work is supported by SERC and Ford Motor Company.

REFERENCES

1) D.Hull, Axial crushing of fibre reinforced composite tubes, in: Structural Crashworthiness, eds. Jones and Wierzbicki (Butterworths, 1983) Chp.5, pp. 118-135.

2) D.Hull, Energy absorption of composite materials under crash conditions, in: Progress in Science and Engineering of Composites, Vol. 1, eds. Hayashi, Kawata and Umekawa (Japan Soc. for Composite Materials, 1982) pp. 861-870.

3) P.H.Thornton, J.J.Harwood and P.Beardmore, Comp. Sci. Technol., 24 (1985) 275-298.

4) G.L.Farley, J.Comp.Materials, 17 (1983) 267-279.

5) J.Sigalis, M.Kumosa and D.Hull, (to be published).

6) P.H.Thornton, Effect of trigger goemetry on energy absorption of composite tubes, in: ICCM-V, eds. Harrington, Strife and Dhingra (Metallurgical Society, Pennsylvania, 1985) pp. 1183-1199.

7) H.Vogt, P.Beardmore and D.Hull, Crash energy absorption of fibe-reinforced plastics in vehicle construction, in Kunststoffe als Problemloser in Automobilau (VDI Verlag, 1987) pp. 231-226.

8) M.J.Czaplicki, P.H.Thornton and R.E.Robertson, Collapse triggering of polymer composite energy absorbing structures, in: How to Apply Advanced Composite Technology (ASM International, 1988) pp. 39-46.

9) G.L.Farley, Energy absorption in composite materials for crashworthy structures, in: ICCM and ECCM, Vol. 3, eds. Matthews, Buskell, Hodgkinson and Morton (Elsevier Applied Science, London, 1987) pp. 3.57-3.66.

10) A.H.Fairfull and D.Hull, Energy absorption of polymer matrix structures: frictional effects, in: Structural Failure, eds. Jones and Wierzbicki (Wiley 1989) to be published.

Materials and Processing – Move into the 90's
edited by S. Benson, T. Cook, E. Trewin and R.M. Turner
Elsevier Science Publishers B.V., Amsterdam, 1989

DEVELOPMENT OF FRP REAR AXLE COMPONENTS

Ingo KUCH

Daimler-Benz AG, Basic and Product Research

ABSTRACT

The various stages of development for a composite rear axle will be illustrated. Starting with the verification of the industrial applicability of fibre-reinforced plastics (FRP) in the automobile and the substitution of individual components, a design concept for an integrated composite rear axle will be explained.

A description will be given of the design, manufacture and testing of CFRP spring links made by hand laminating process, filament winding and thermoforming of long fibre-reinforced thermoplastics.

The following development stage is the design of a leaf spring link which combines the load bearing function of the spring link with the spring-loaded property of a helical spring.

A first design concept for an FRP rear axle will be presented.

CONTENTS

1. INTRODUCTION

In the last few years fibre-reinforced plastics (FRP) have experienced a breakthrough by being used for standard parts in civil aerospace applications (e.g., vertical tail and aileron of the Airbus). In addition to weight savings, great advantages have been achieved by drastic reductions in the number of fastening agents. Experience gained in the area of aviation, however, is only partially transferable to automobile engineering, since other technical and economic requirements prevail. The large-surface aircraft components are mainly subjected to tensile and compressive stresses, whereas automotive components are also subjected to very complex loads such as flexure, shear stress and torsion.

The use of fibre composites in the automobile are of interest in those particular fields of application where, beyond weight savings, their spring-loaded and damping properties are utilised or where a lighter design of components can be obtained through weight reductions of extremely accelerated parts.

In order to obtain an economic utilisation of fibre in future automotive applications, trade-offs in fibre alignment are almost inevitable. The specific weight-related stiffness and strength properties of CFRP, GFRP and SFRP (carbon fibre, glass fibre and synthetic fibre-reinforced plastics) with pure unidirectional reinforcement are considerably superior to those of steel and aluminium. On the other hand, quasi-isotropic laminates are not suitable for high-performance applications (Fig. 1).

Series applicability of fibre-reinforced plastics mainly depends on the extent to which their positive properties can be transferred to the components through design and manufacturing processes. Moreover, appropriate quality assurance concepts have to be established.

2. DEVELOPMENT CONCEPTS

The FRP rear axle was selected as a functional prototype with the objective of obtaining advantages through reductions in unsprung masses, making use of the composites´ special spring-loaded properties and damping qualities, in order to improve handling even further.

This involves the challenge of developing mechanically highly stressed parts and proving their ability to withstand various loads such as water, oil, fuel and chippings.

One of the features of the Mercedes-Benz multi-link independent rear axle, fitted identically to the compact and mid-range models, is its precise wheel suspension and high standard of handling quality. The forces are transferred over five links to ensure minimum change in track and camber angle when mobile (Fig. 2).

A first step towards the realisation of a FRP rear axle was the design of a composite spring link, which represents the most extremely loaded axle component. The term "spring link" means that, in addition to the shock absorber and the stabilizer, the helical spring acts upon this axle component. The spring link was designed as a rigid part; it forms the joining element between the axle support and the wheel carrier.

We are aware of the fact that mere substitution of <u>one</u> metallic part does not allow to make optimum use of the composites´ positive characteristics, and that a great deal of effort is required to satisfy the geometrical boundary conditions. The most important task, however, which is to prove the suitability of fibre composites for automotive applications, is fully attainable.

Now that proof was furnished by experience gained in manufacturing and operational trials in the first phase, greater modifications to the vehicle are justified in further stages, requiring the substitution of <u>several</u> metallic parts for one composite component. The leaf spring represents such an

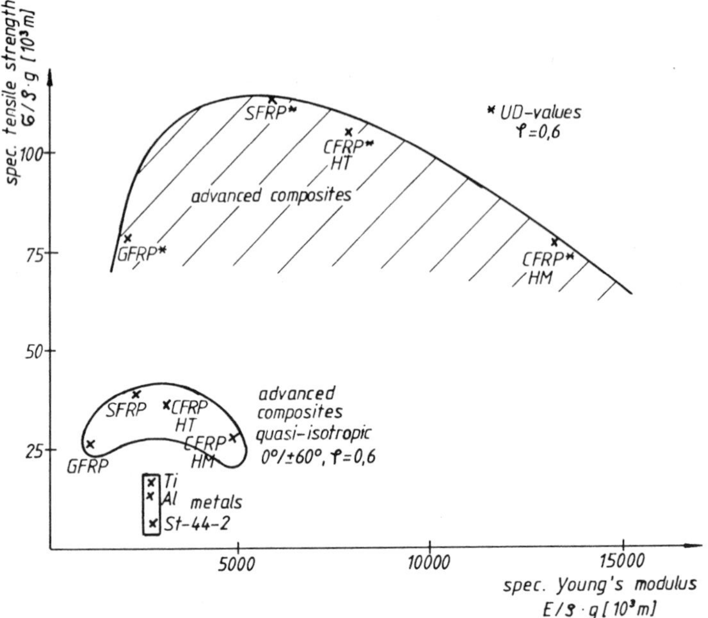

figure 1 comparison of materials

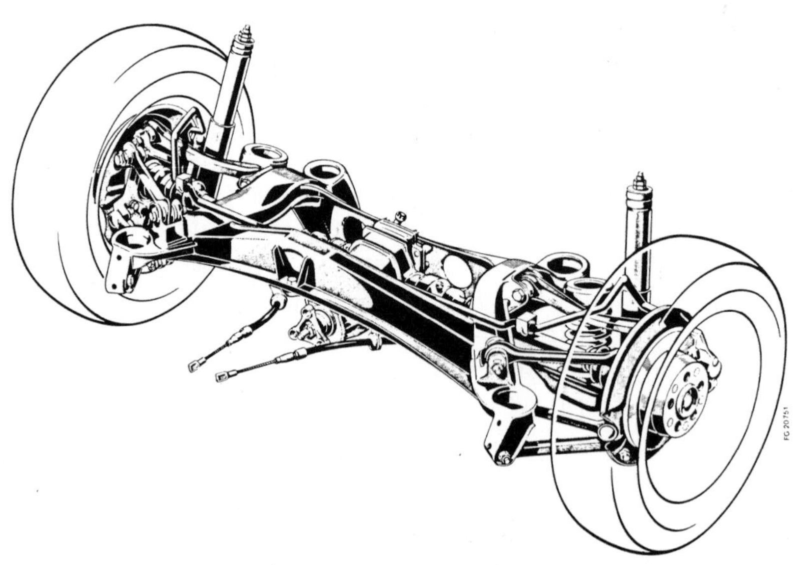

figure 2 Mercedes-Benz multi-link independent rear axle

intermediate stage, combining the load bearing function of the spring link with the spring-loaded properties of a helical spring. The hinge points of the axle are retained, and axle kinematics are only slightly modified.

However, the objective of developing an overall composite concept can only be achieved by a complete departure from the metallic design, and by including the axle and wheel carriers. It is only after the realisation of this concept that a reasonable comparison of the cost/benefit ratios of steel and composite materials can be made.

3. SPRING LINK MADE OF GFRP

3.1 Hand-laminated version

Fig. 3 depicts the structure of the hand-laminated version. To absorb shearing stress, layers of fabric are wrapped around a temperature-resistant rigid foam core so that they lie at an angle of + 45° to the side walls. Tensile and compressive loads caused by bending are absorbed by roving tapes laid endlessly in the longitudinal direction on the top and bottom of the component.

Sleeves for connecting the axle and wheel carriers are retained in the end faces. The shock absorbers and the stabilizer are screwed into disks embedded in the laminate.

The spring is supported by a sandwich wedge linked to the upper and lower flanges by a cap-shaped carrier basket insert.

The component is cured in two stages in the autoclave.

The two spring links later fitted to the car were statically and dynamically pre-tested. The components fitted to a complete axle were tested in a simulation test run over 2000 km with a total load equivalent to the real life of a passenger car. No damage occurred.

The components were then fitted to a Mercedes 190 E 2.3 and tested over 3500 km of poorly surfaced and gravel roads; afterwards the vehicle was released for internal use. Since July 1987 the parts have covered a total of 110,000 km (as per January 1989) (Fig. 4).

3.2 SPRING LINK MADE BY FILAMENT WINDING

The spring link was selected as a demonstration component to investigate the possible economic production methods using the filament winding process for complex geometries.

At present, a robot with a portal architecture and a synchronised external axis is being procured. It is capable of winding non-rotational symmetrical cores. The structure of the core geometry and collision considerations are generated using the CATIA CAD system. The data generated are transmitted to a personal computer which performs the laminate calculations and filament-wound simulations.

The design of the component provides for two different structures which absorb operating and crash loads. The single-part or multi-part crash structure is covered by a rigid foam and forms the core for the operational structure.

3.3 SPRING LINK MADE OF LONG FIBRE-REINFORCED THERMOPLASTICS

Long fibre-reinforced thermoplastic semi-finished products were used as an alternative for spring links by the German Aerospace Research Establishment, Institute for Structural Research and Design (DFVLR), Stuttgart.

To gain experience in deformability of the material, the original tool of the steel spring link was used.

Very good results were obtained with carbon fibre structures in satin weave pre-impregnated with polyetherimide (PEI) (Fig. 5). This type of weave has good deformability and allows wrinkle-free lay-up. To ensure a uniform contact pressure, a solid silicone rubber punch was used for the male mould.

A spring link manufactured from this material was statically - under the maximum loads - and dynamically tested by being subjected to staged operating stresses equivalent to 2000 km of poorly surfaced roads. No damage occurred. Because of this positive result, the slightly modified original tool was also utilised for further tests.

ICI, Wilton, U.K., has used UD (unidirectional) prepregs made of polyetheretherketone (PEEK) for the test production of a spring link. The laminate was placed between sheets with high extensibility, heated and pressed against a solid mould by applying vacuum and excessive pressure. Using this procedure, good surface qualities were obtained, but large spherical deformations can only be achieved under certain conditions.

4. LEAF SPRING LINK MADE OF GFRP

By substituting the metallic spring link and the helical spring by an FRP component, both operational loads requirements and spring-loaded properties must be fulfilled. The best properties for leaf spring element applications are provided by GFRP due to its high energy storage capacity. At the same time, the material price is much lower compared to CFRP.

Initial considerations, which were to replace the two spring links, the helical springs and the stabiliser by a continuous transverse leaf spring while retaining the hinge points, failed by reason of the position of the rear-axle differential.

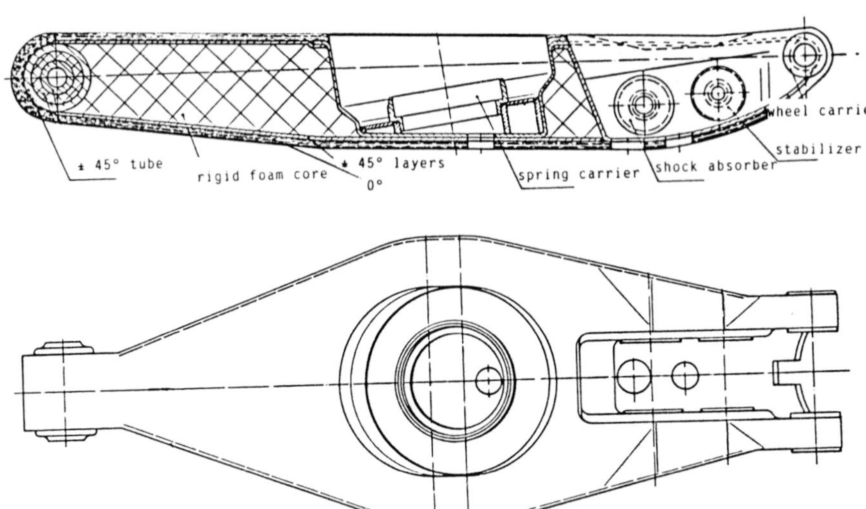

figure 3 Design drawing CFRP spring link

figure 4 CFRP spring link (thermoset), mounted in the vehicle

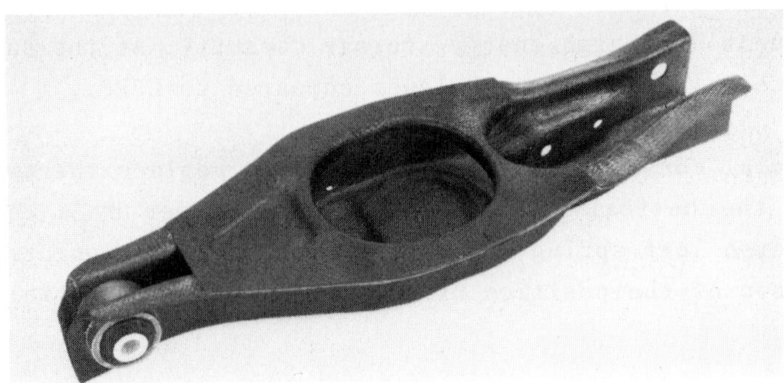

figure 5 Graphite-reinforced thermoplastic spring link

In order to achieve the required spring deflection with two separate leaf spring segments, they were extended over the axle carrier hinge points specified by the spring link to the vehicle centre-line and supported. Fig. 6 shows the rear view with the conventional structure on the left and the structure with the leaf spring link on the right.

The leaf spring link was designed as a beam under uniform bending load in order to achieve good material utilisation. The installation space available only permitted the use of a parabolic spring. However, due to its variable thickness and cross-section, it is relatively costly to produce even by automated means.

At the spring ends on the wheel carrier side, a fork-shaped division is required to connect the shock absorber and the stabilizer. The fork-shaped division was not completed, however, when the spring characteristic was tested experimentally.

Fig. 7 shows the leaf spring link during static testing.

The version intended for installation in a car will be available at the end of 1989.

5. OVERALL CONCEPT

The overall concept departed from the passenger car rear axle, and the rear-axle of the off-road vehicle was selected instead. It is better fitted for the utilisation of FRP since the installation space is larger and a greater optimisation potential is obtainable by breaking away from the previous rigid axle concept. Additionally, advanced composite constructions have greater advantages with low-volume production due to relatively lower tooling costs.

At present several axle constructions for off-road vehicles are being tested by their suitablility for use with fibre-reinforced plastics. The first concept sketches for an FRP axle have been made. A possible design is a double-wishbone axle.

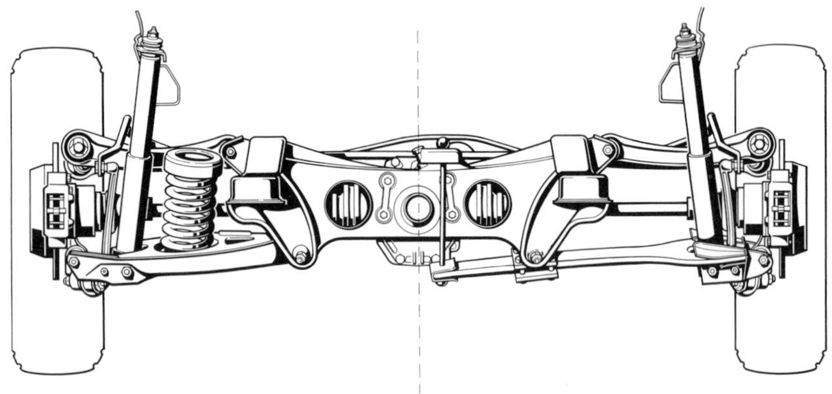

figure 6 mounting plan leaf spring link

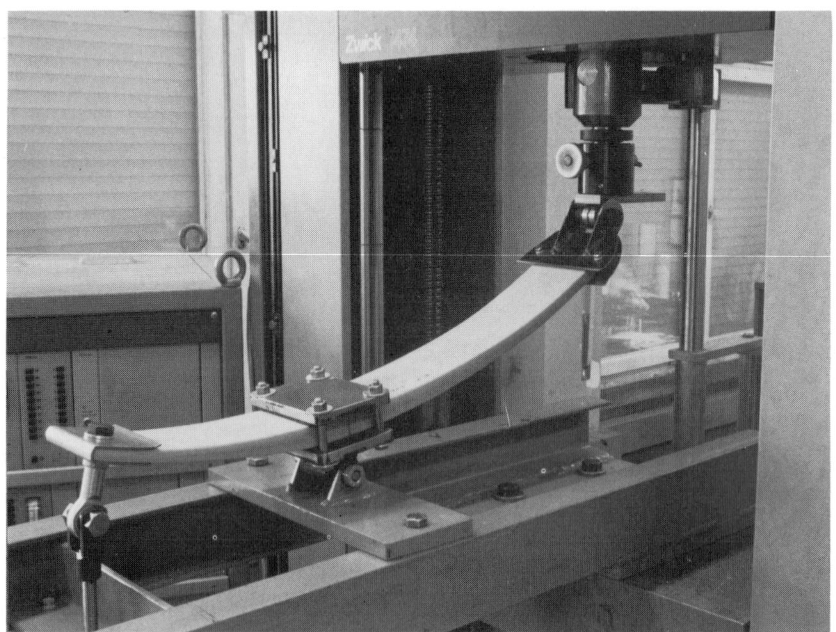

figure 7 leaf spring link, static test rig

The parallel links will be integrated to form a single assembly governing track and camber. In this concept, a longitudinally mounted leaf spring link will support brake and drive torques and will take change of the suspension function.

Several more years of research work will be needed on a broad basis to gather sufficient experience for manufacturing and operating FRP components.

Materials and Processing – Move into the 90's
edited by S. Benson, T. Cook, E. Trewin and R.M. Turner
Elsevier Science Publishers B.V., Amsterdam, 1989

STUDY & DEVELOPMENT OF A CFRP LIGHTWEIGHT LOWER LIMBS ORTHOTICS FOR MUSCULAR DISTROPHY

A. DE LOLLIS *   L. PIANCASTELLI **

ABSTRACT

In some situations of motor handicap and particularly in primitive or secondary myopathies with neurological damage the lower limbs are unable to sustain body weight. This problem is currently met by orthotics made of traditional materials that allow the patient to stand and to walk.
AS the patient grows and his weight increases there is a progressive degeneration of muscular tissue and the orthotics has consequently to bear increasing loads without significant deformation. Even the best designs using traditional materials cannot bear the necessary loads while at the same time remaining light enough for the patient's energy capacity, and the patient thus finds himself confined to a wheelchair with all its consequent limitations.
In order to avoid or delay this undesirable situation, an orthotics made of composite material has been designed and manufactured. The coice of a composite with a low temperature polymerization thermoplastic matrix was determined by the need to make slight modifications to the orthotics even after the curing process. After six months study a PMMA composite reinforced with standard carbon fibre was selected for this application. For the design of the structure a specially elaborated FEM program was used, able to calculate the interlaminar shear stress and to simulate inserts made of different materials.
The model for this FEM program was developed from an existing orthotics made of traditional materials so as to permit correct simulation of the new lightweight structure and obtain the best possible initial design. The orthotics was then constructed: the elements subject to the highest loads were constructed first and then built into the complete structure during a second curing process at room temperature.
Experimental tests were then carried out in order to evaluate the theoretical results and check the whole design process.

---

* Officine Ortopediche Rizzoli S.p.A. - Bologna Italy
** Dipartimento di Ingegneria delle Costruzioni Meccaniche, Nucleari, Aeronautiche e di Metallurgia - Università degli Studi di Bologna - Italy

INTRODUCTION

Muscular dystrophy progressively damages the tissue and causes a motor handicap condition that worsens as the patient grows, until it causes total immobility.
In order to limit this phenomenon and to prolong the muscular dystrophy patient's independence of movement, external orthotics are produced. These block the knee's articulation an facilitate walking by limiting muscle involvement.
These orthotics are created manually in traditional "light" materials such as polyethylene, aluminium and high strength steel at the points under the greatest stress.
In order to decrease the weight and increase the rigidity of the orthotics,which is closely related to the feeling of safety thereby granted to the patient, it was decided to use composite materials. Since composite materials are not suitable for sustaining concentrated loads and since it was desirable to avoid the use of inserts, the orthotics had to be entirely redesigned, in view also of the need to construct it using technologies suited to these materials.
Traditional orthotics are more the result of an empirical refining process than the fruit of an actual engineering analysis. Accordingly, the design process had to be entirely revised, starting with a study of the gait in order to determine the loads and aptimizing the sizing of the orthotics so as to obtain a degree of standardization per class of patient with shared characteristics.

CURRENT ORTHOTICS

Orthotics are created from a plaster cast which is an exact copy of the patient's limb with certain corrections made by the orthopaedic expert. This cast (positive) is obtained with a plaster casting from a negative mould taken directly from the patient's limb.
Polypropylene plates are heat-formed on this positive cast so as to obtain the half piece of the thigh and leg. The shape of the upper half is such as to permit support of the ischium through which most of the load applied to the orthotics is trasmitted.
The connection is made by special high strength steel screws and shaped aluminium rods.
A polypropylene patellar support is also fixed to these rods, their function being to keep the limb aligned. Leather fasteners and securing belts complete the structure.

The orthotics constructed in this way is tested on the patient and final adjustements are made to the shape of the halves so that the structure is as comfortable as possible.

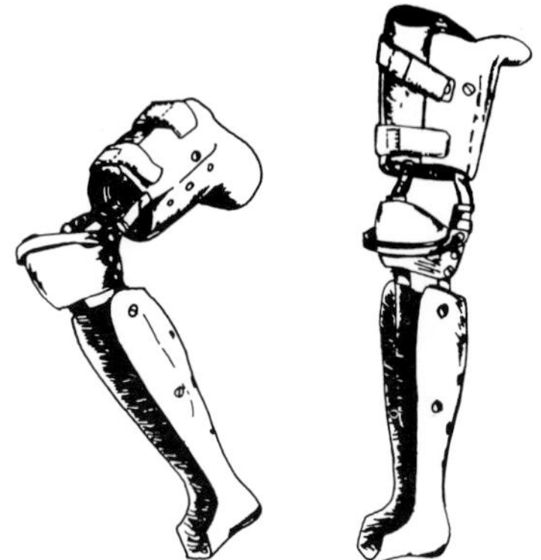

Fig. 1 Current orthotics constructed with traditional materials

LOADS ANALYSIS

The equilibrium of the orthotics is ensured by the forces bearing down on it and deriving from the different interactions: ground-orthotics and patient-orthotics.

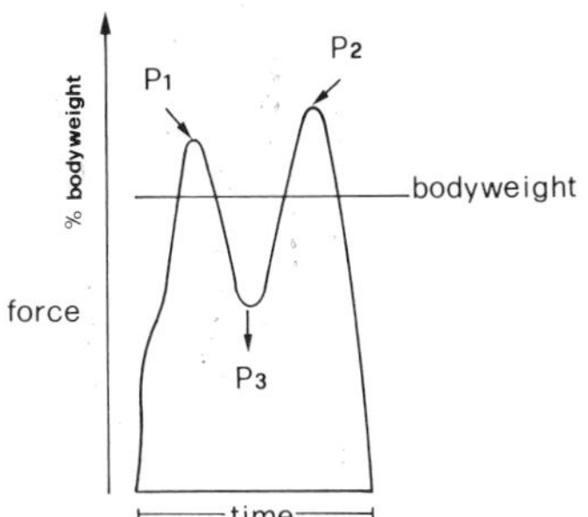

		P_1	P_2	P_3
Normal	Mean	19.8	18.9	19.9
boys (n=22)	SD	9.8	8.2	5.1
DMD	Mean	18.8	17.9	15.7
boys (n=9)	SD	8.6	9.9	8.8
Normal+DMD	Mean	19.3	18.6	18.8
boys (n=31)	SD	9.5	8.7	6.6

In the table: Mean and standard deviation (SD) for normal Duchenne muscular dystrophy (DMD) and the combination of normal and DMD boys

Fig. 2 Vertical force curve: identifies 1st peak (P_1), 2nd peak, trough (P_3)

With regard to the former, graphs have been taken from the literature on the subject (1) (2) (3) containing the behaviour over time of the components of the resultant of the forces that the ground exerts on the orthotics when the leg is on the ground, as obtained by measurements made on a force platform (see Figure 2).
The patient-orthotics interaction comes about through the ischiatic, patellar and plantar supports according to the diagram in Figure 3.

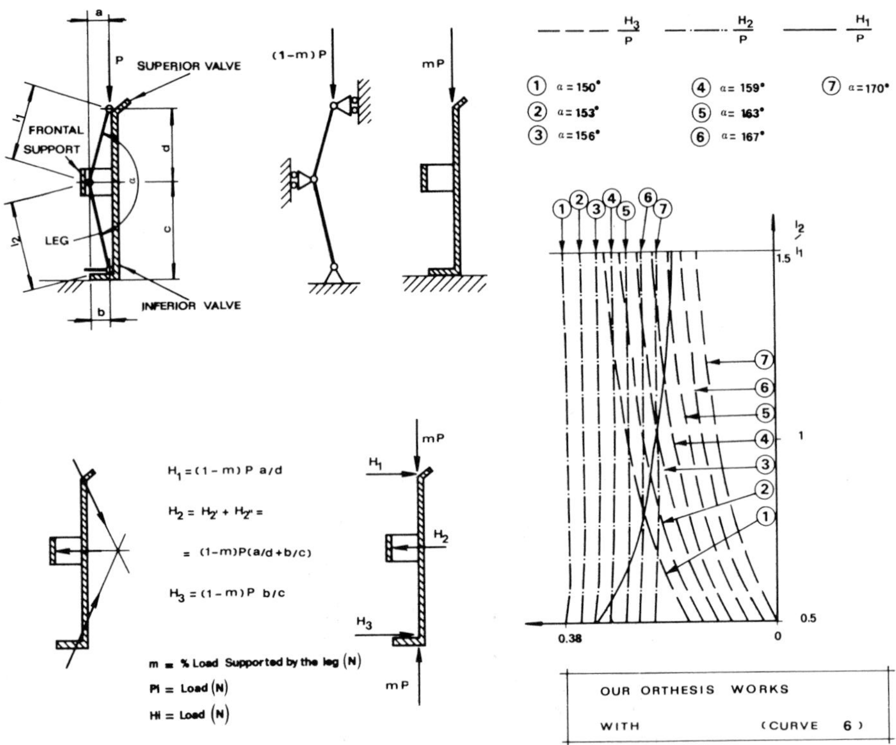

Fig. 3 Static analysis of the orthesis (H_3 does not depend on alpha)

The intensity of these forces is closely related to the condition of the patient; hence, the greater the resistance to bending of the limb exerted by the muscles, the lower the force bearing down on the patellar support.
Therefore statistical data were used to determine the best conditions for the patient.

	METAL ALUMINIUM 2024–T3	GRAPHITE 2423 PMMA* VACUUM BAG TECHNOLOGY Vf 50 %	GRAPHITE 2423 PMMA* HOT PRESS TECHNOLOGY Vf 50 %	GRAPHITE 2423 PES Vf 50 %	GRAPHITE 2423 POLICARBONATE Vf 50 %
FAILURE LOAD	2300	4200	5128	4650	—
POST FAILURE LOAD	—	2720	3215	3630	—
LONGITUDINAL TENSILE STRENGTH (MPa)	324	450	550	360	334
LONGITUDINAL TENSILE MODULE (GPa)	72.3	58	75	48.2	43.2
	* Polimer matrix composites fabricated by Officine Ortopediche Rizzoli				

Tab. 1 Comparative data: aluminium beams-thermoplastic composites beams

CHOICE OF MATERIALS

For technological reasons metal or ceramic matrix composites and short fibre composites were rejected and a composite material strengthened with thermoplastic matrix carbon fibres was chosen to ensure the shapeability required for the construction and final adjustment of the orthotics.
A comparative analysis of the mechanical characteristics (3) of the various resins available on the market and the need to keep the shaping temperatures relatively low (see table 1) led to the choice of an acrylic resin: a Poly Methyl Methacrylate with a monomer-polymer ratio of 80/20.
Two composite polymerization technologies were developed, the first producing the best mechanical characteristics by means of hot surface pressing, the second at ambient temperature and in a vacuum bag by means of a catalyzer (see Table 2).

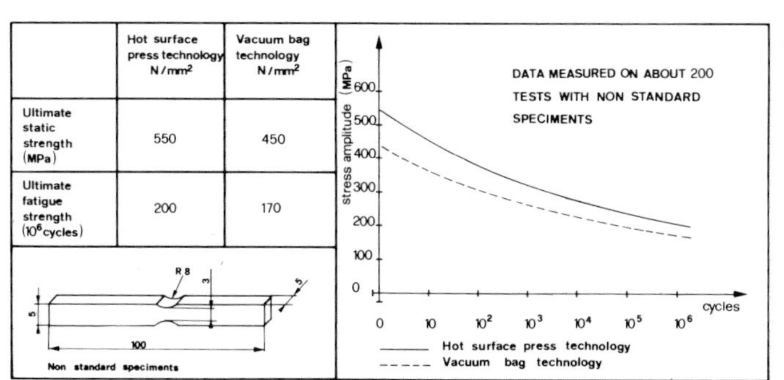

	Hot surface press technology N/mm²	Vacuum bag technology N/mm²
Ultimate static strength (MPa)	550	450
Ultimate fatigue strength (10^6 cycles)	200	170

Tab. 2 Fatigue behaviour of GRAPHITE-PMMA composite (V+50%)

Further, experimental tests were carried out to check resistance in a saline solution simulating human perspiration. Unidirectional samples were exposed to the above-mentioned environmental conditions for a minimum period of 3 months without any significant variations being found in the material's mechanical characteristics.
In addition, the tension-compression fatigue curve was determined for the material created using the two different technologies.

DESIGN OF THE COMPOSITE ORTHOTICS

In order to size the composite orthotics a finite element was developed simulating the deformative flexural and membranous condition of the structure, with which it was also possible to evaluate the shear stress between adjacent layers of tissue. The finite element is a three-dimensional 9 node element of the plate type in which the overall characteristics of the laminate are determined by a variant of the Classical Lamination Theory, known as the YNS theory (4) put forward by Young, Norris and Stawksy, supposing a movement range of the type:

$$u = u_0(x,y) + z\varphi_x(x,y)$$
$$v = v_0(x,y) + z\varphi_y(x,y)$$
$$w = w_0(x,y)$$

that permits introduction of the interlaminar shear. The internal congruence equations become the following:

$$\begin{Bmatrix} \varepsilon_x \\ \varepsilon_y \\ \gamma_{xy} \end{Bmatrix} = \begin{bmatrix} du/dx \\ dv/dy \\ du/dy + dv/dx \end{bmatrix} ; \begin{Bmatrix} \chi_x \\ \chi_y \\ \chi_{xy} \end{Bmatrix} = \begin{bmatrix} d\psi_x/dx \\ d\psi_y/dy \\ d\psi_x/dy + d\psi_y/dx \end{bmatrix} ; \begin{Bmatrix} \gamma_{xz} \\ \gamma_{yz} \end{Bmatrix} = \begin{bmatrix} dw/dx - \psi_x \\ dw/dx - \psi_y \end{bmatrix}$$

thus it is possible to calculate the two interlaminar shear stresses $\bar{T}_{ZX}$ and T_{ZY} determined by assuming that deformation due to shearing is constant in the individual lamina.
On the basis of these assumptions a 9 node isoparametric element was developed (4) (5) with forty-five degrees of freedom which was shown to be reasonably efficient in studyng the deformative membranous, flexural and shear behaviour of flat and moderately curved plates.
This element uses the Gauss-Legendre formulae for integration

and passes the patch test.
The strength check is conducted by calculating the ideal tension of each lamina according to the distortion energy rupture assumption in the Assi Tsai formulation:

$$\frac{\sigma_1^2}{R_1^2} + \frac{\sigma_2^2}{R_2^2} + \frac{\sigma_1 \sigma_2}{R_1 R_2} \leq 1$$

Checks are also carried out to ensure that the interlaminar shear stresses T_{ZX} and T_{ZY} do not exceed the maximum stresses permitted by the matrix.
The results obtained showed a satisfactory simulation of the deformative condition of the parts and an acceptable order of magnitude with regard to the stresses.
The element permits attempted simulation of inserts, sandwich panels and glued joints; however, experimental data are paching for these particular applications.
This element was used first of all to simulate the behaviour of the orthotics in traditional material so as to evaluate the correctness of the schematization of the problem, developing meshing, load organization and constraints.
Next the material was replaced with the composite chosen on which the optimization was carried out, altering the shape and the thickness of the various parts (see Figure 5).

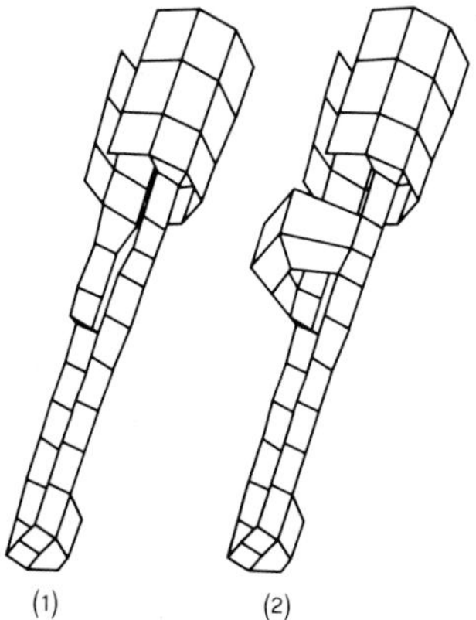

Fig. 5 Finite element mesh of the composite orthotic:
1) final design with flexible patellar support
2) initial design with rigid patellar support

The first stage entailed a schematization of the orthotics, giving it a composite patellar support: next this element working by traction was replaced with a belt in polyamide tissue. The final sizing was carried out by limiting the maximum sag under load and the interlaminar stress at the critical points; the latter parameter, given the relatively low mechanical characteristics of POly Methyl Methacrylate, was shown to be critical in sizing the orthotics. The prototype of the orthotics thus designed (see Figure 8) brought about a weigth saving of approximately 40%.

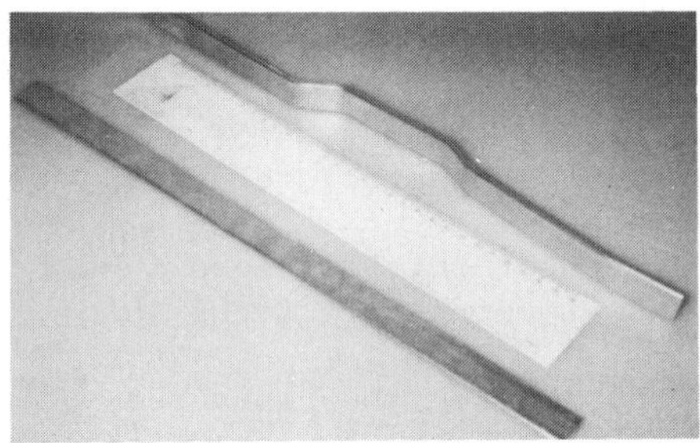

Fig. 6 Shaped aluminium rods

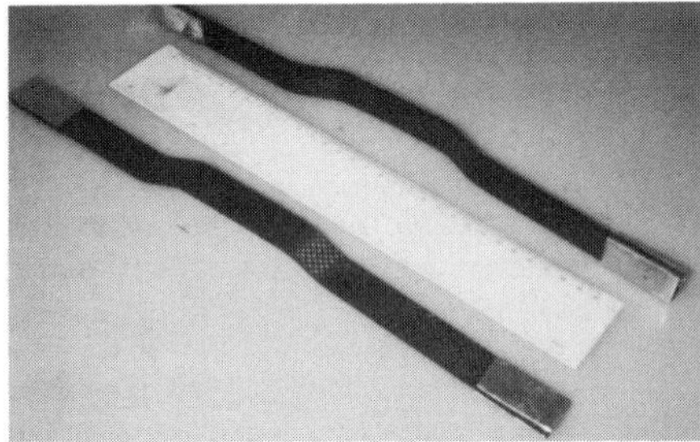

Fig. 7 Shaped composite rods

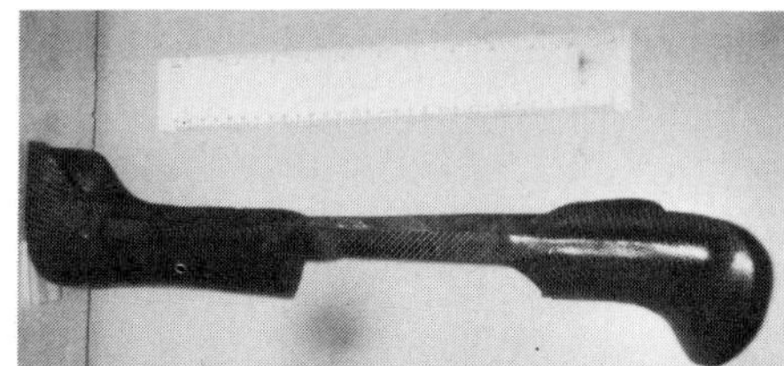

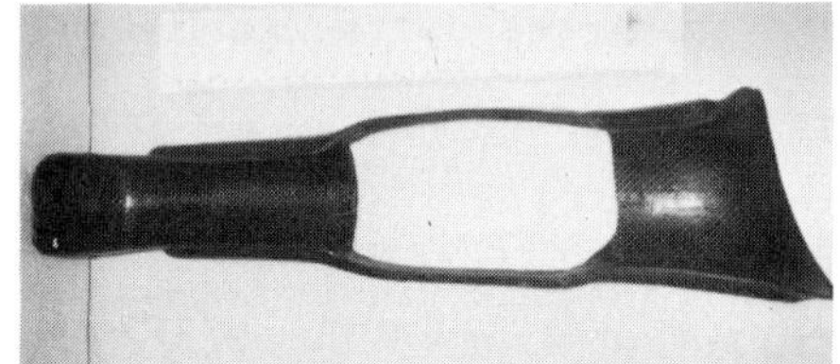

Fig. 8 Composite orthesis frontal and lateral view

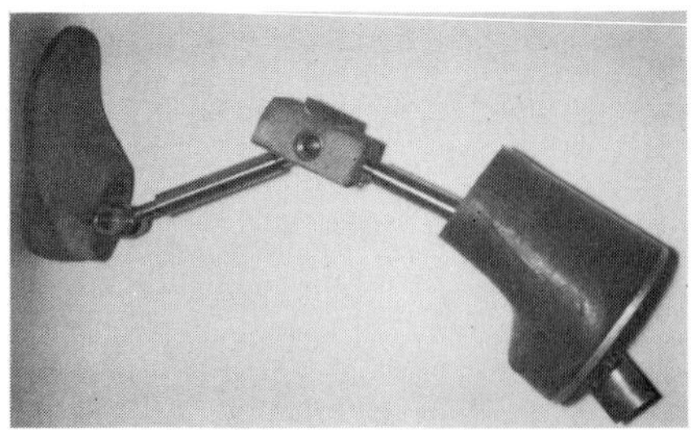

Fig. 9 Mock up of lower limb

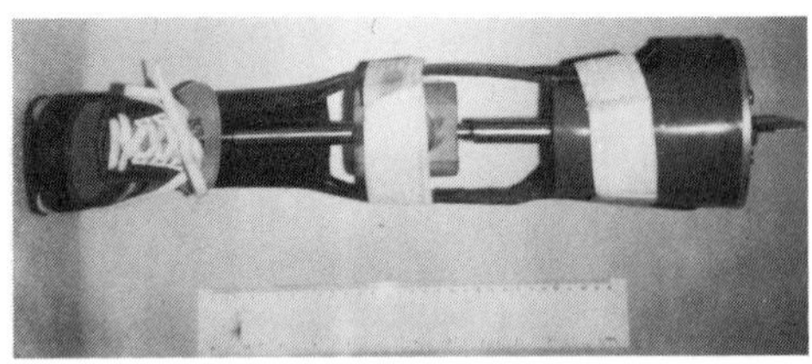

Fig. 10 Assembly

EXPERIMENTATION

Experimentation was carried out in order to compare the traditional orthotics with the new design type. First of all the static fatigue resistance of the aluminium rods (see Figure 6) was compared with the resistance of the composite rods (see Figure 7), considered to be fundamental structural elements in the orthotics.
For the fatigue test the mean number of work cycles of this type of orthotics was reckoned at approximately 350,000 cycles, and checks were carried out to see whether the designed orthotics could withstand this fatigue cycle with a pulsating load of an intensity equal to that given in the literature.
This test was carried out by creating a mock-up of the lower limb (see Figure 9) with which it was possible to stress for the same number of cycles both the composite prototype and, for comparison, the orthotics in traditional material (see Figure 10).
The experimental results, summarized in Table 3, are relatively close (+/- 10%) to those obtained with the finite element simulation.

CONCLUSIONS

The stresses applied to the experimental part did not lead to fracture of either the components or the orthotics prototype, indicating the basic correctness of the process adopted. The composite orthotics, though achieving a considerable weight saving (40%), nevertheless revealed substantial oversizing by standing up to load cycles in excess of those envisaged.
It is believed that this result is due to quality improvements in the composite material as a result of the refining of the manufacturing technique, compared to what was obtained initially. Therefore the orthotics will be progressively lightened as indicated by the finite element simulation and the halves will be made using materials with a lower elasticity modulus so as to ensure greater ergonomic efficiency of the orthotics, especially during gressing. A composite joint has already been designed which will be experimented with and introduced into the final composite orthotics.

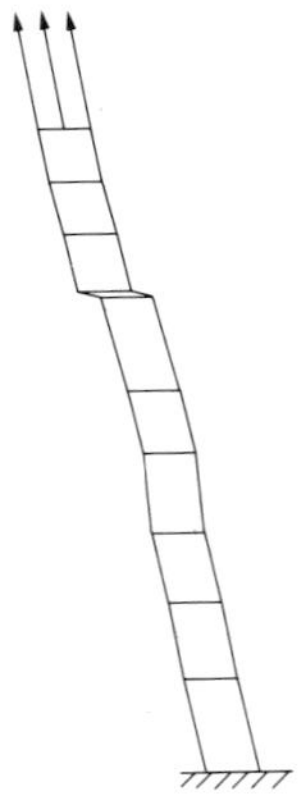

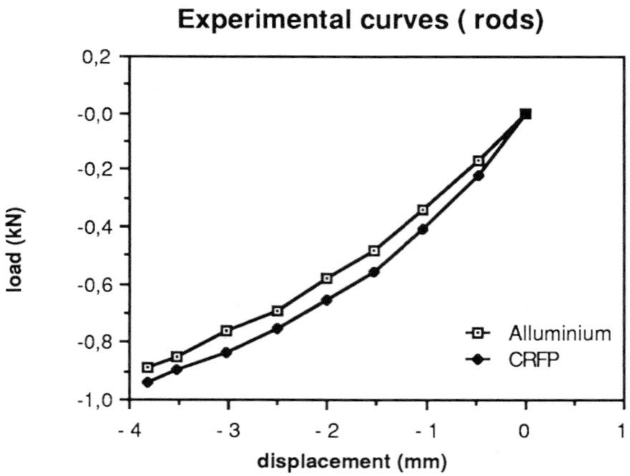

Experimental curves (rods)

experimental curves (orthesis)

EXPERIMENTAL DATA OF RODS
(Compression test {1 cycle})

Displacement (mm)	Alluminium (*)	CRFP(**)	FEM(****)
-0.798	-0.075	-0.110	-
-1.544	-0.130	-0.200	-
-2.294	-0.165	-0.310	-
-3.014	-0.210	-0.450	-
-3.794	-0.255	-0.590	-
-4.564	-0.315	-0.740	-
-5.314	-0.415	-0.910	-
-5.784	-0.475	-1.05(***)	-0.498/-0.968

(*) load (kN), Average of 10 tests, specimens not broken
(**) load (kN), Average of 20 tests, specimens broken
(***) Ultimate load
(****) load(kN), Simulation with F.E.M program [Alluminium/CRFP] errors [0.5% / -8%].

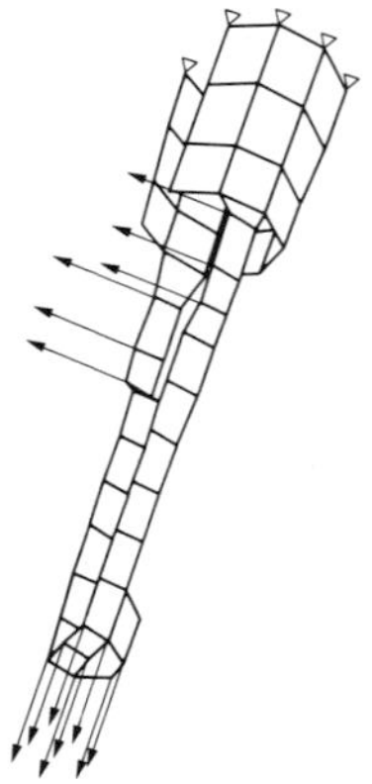

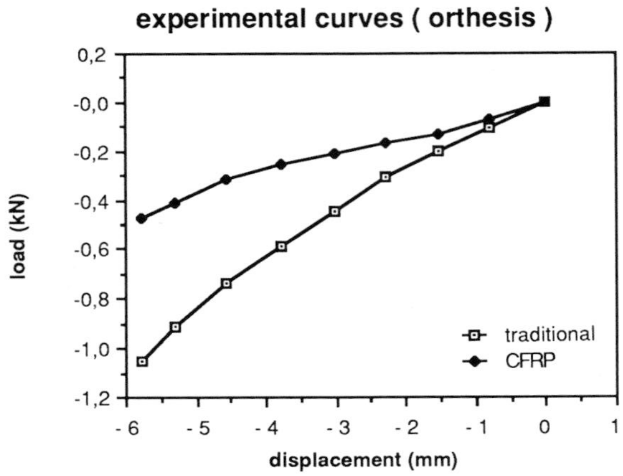

EXPERIMENTAL DATA OF ORTHESIS
(Compression test {1 cycle})

Displacement (mm)	Traditional (*)	CRFP(**)	FEM(****)
-0,474	-0.17	-0.22	-
-1,026	-0.34	-0.41	-
-2.016	-0.58	-0.65	-
-2.516	-0.69	-0.75	-
-3.016	-0.76	-0.83	-0.918 (||)/-0.77
-3.516	-0.85(*****)	-0.89	-
-3.816	-0.88(*****)	-0.94(***)	-/-0.8366

(*) load (kN), 1 test, specimen not broken
(**) load (kN), 1 test, specimen broken
(***) Ultimate load
(****) load(kN), Simulation with F.E.M program [Traditional/CRFP] errors [8% / -10%]
(*****) seriously damaged
(||) some parts work in the plastic region

Tab.3 : Summary of experimental vs. computed results

BIBLIOGRAPHY

1) S. Kodadeh, M. Mc Clelland, J.H. Patrick, Force Plate Studies of Duchene Muscular Dystrophy in: Engineering in Medicine, MEP Ltd 1987.
2) D.H. Sutherland, Gait disorders in Childhood and adolescence (Baltimore, Williams & Wilkins, 1983) chapter 11, pp. 182-183.
3) G. Kubin: Handbook of Composites (VNR Ltd 1982) chapter 28, pp. 722-723.
4) P.C. Yang, C.N. Norris, Y Stawsky, Elastic Propagation in Heterogeneous plates, in: Inter. Jour. Sol. Struct. (1966) pp. 663-648.
5) K.J. Bathe: Finite Element Procedure in Engineering Analysis (Pernetice & Hall, 1981).
6) Vinson Sierakowski, Behaviour of Structure of Composite Materials (M. Njhoff Pub., 1986).
7) T.J.R. Hughes, R.L. Taylor, W.Kanoknukulchal, a simple and efficinet finite element for bending, in: Int. Jour. Num. Meth. Eng. (1977) Vol. 11 pp. 1529-1543).
8) E.D.L. Pugh, E. Hinton, O. Wienkiewicz, a study of a quadrilateral plate element with "reduced" integration, in: Int. Jour. Meth. Engr. (1978) Vol. 12, pp. 1059-1079.

SYMBOLS

u	Movement in direction X
v	Movement in direction Y
w	Movement in direction W
σ_1	Normal stress in direction 1 (parallel to fibre)
σ_2	Normal stress in direction 2 (at right-angles to fibre)
σ_3	Normal stress in direction 3 (at right-angles to plane 1-2)
τ_{ij}	Tangential stress on J lying on normal plane 1
γ_{ij}	Tangential deformation on J lying on normal plane 1
R_1	Ultimate strength in direction 1 (parallel to fibre)
R_2	Ultimate strength in direction 2 (at right-angles to fibre)
χ_x	Curvature in direction X
χ_y	Curvature in direction Y
χ_{xy}	Curvature in direction Z
ψ_x	Rotation in direction X (rad)
ψ_y	Rotation in direction Y (rad)
u_o	U referred to mean plane
v_o	V referred to mean plane
w_o	Z referred to mean plane

Materials and Processing – Move into the 90's
edited by S. Benson, T. Cook, E. Trewin and R.M. Turner
Elsevier Science Publishers B.V., Amsterdam, 1989

MANUFACTURING AND MECHANICAL PROPERTIES OF 3-D-FIBRE REINFORCED COMPOSITES

J. Brandt*, T. Preller*, K. Drechsler**

* Messerschmitt-Bölkow-Blohm GmbH, Central Laboratories, Munich
**Institute for Aircraft Design, Stuttgart

In this report, results about the manufacturing and mechanical behaviour of various 3-d monolithic and sandwich structures made of glass and carbon fibres and epoxy resins are presented.
Furthermore, carbon/PEEK 3-d composites were fabricated by coweaving and commingling.
Vacuum impregnation, autoclave and press techniques were applied to manufacture 3-d composites with less than 2 % void content.
Tensile, compressive and impact properties have been determined. The impact behaviour of carbon composites can be improved remarkably by using a reinforcement in the 3rd direction,thus, however, reducing the in-plane properties.
In addition to the improved resistance to delaminations, this class of recently developed materials has the potential to reduce composite manufacturing costs due to the relatively low preform costs and the elimination of extensive handling.

1. INTRODUCTION

Composite materials are widely used nowadays because of their high specific strength and stiffness. Due to their construction, conventional 2-d laminates exhibit relatively poor out-of-plane properties and these structures are therefore critical when subjected to impact loading.

In this respect, composites made from 3-d woven fabrics feature great improvements [1-3]. The 3-d reinforcement in particular yields superior damage tolerance, a reduced delamination crack growth and an improved peel strength. However, the fibre reinforcement in the z-direction of 3-d composites leads to a reduction of in-plane strength and stiffness.

Out-of-plane stresses are not only a problem inherent in monolithic structures. The skins are adhesively bonded to the core, in honeycomb as well as in foam core sandwich structures. Under impact load and stresses perpendicular to the core, an improvement of delamination resistance is needed, too. An approach to overcome these problems is the use of integrally woven sandwich structures [4].

It was demonstrated in a research programme that 3-d monolithic laminates as well as 3-d sandwich preforms can be realized which not only feature very interesting properties but also the potential to be fabricated cost-effectively. The manufacturing and the description of the mechanical properties of these new composite materials is the subject of this paper.

2. MATERIALS AND MANUFACTURING

Two different types of 3-d structures have been developed (fig. 1). To facilitate manufacturing these structures, modified conventional weaving machines were used.

The use of looms with more weft tables and the periodical beat-up enabled the production of a wide variety of 3-d monolithic and 3-d sandwich structures.

To date, 3-d fabrics of aramide, glass and carbon fibres can be produced with a maximum thickness of about 5 mm. Sandwich structures can be fabricated with a maximum height of about 11 mm. It should be emphasized that these preforms can be woven on fully automatic looms, allowing extremely cost-effective production.

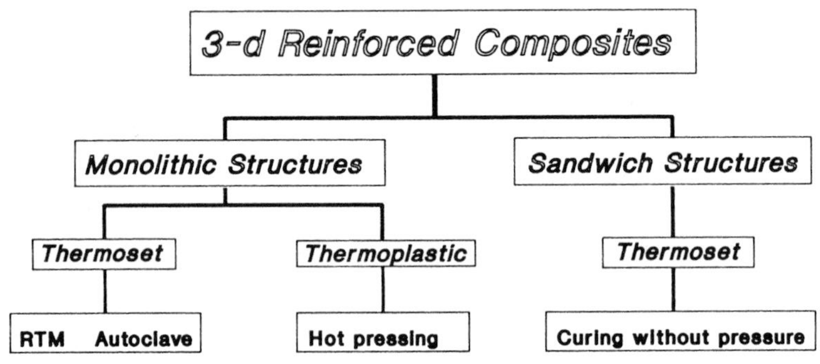

FIGURE 1

Summary of 3-d reinforced composites

2.1 Monolithic structures

Different recently developed 3-d fibre structures are shown schematically in figure 2. The fibre content of the 3-d fabrics running in z-direction was varied between 2 % and 10 %.

The fibre structure shown in figure 2d is a combination of a 3-d and interlock structure with a higher alignment of the "interlocked threads" in warp direction. For reasons of comparison, 2-d baseline fabrics were manufactured with equivalent fibres on the same weaving machine. It should be noted that many different 3-d constructions are possible which have to be optimized for the diverse applications.

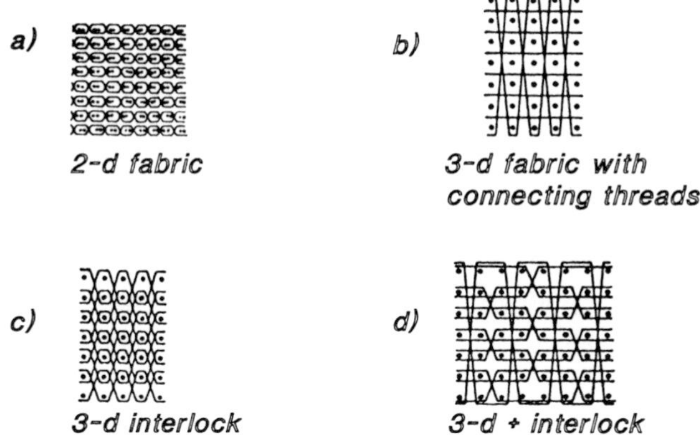

FIGURE 2

2-d and 3-d structures

Composite fabrication

The composites were manufactured by vacuum impregnation with epoxy resins in a modified test setup shown in figure 3 for the Resin Transfer Moulding (RTM) process. Of interest is the fact that due to the higher void volume in 3-d fabrics, impregnation occurred faster compared to 2-d fabrics. The time needed to impregnate a 700x250 mm² 3-d structure was only 10 minutes, in contrast to 40 minutes for conventional 2-d structures.

Experiments have shown that void-free laminates can be achieved with low-viscosity epoxy systems ($\eta \leq 1000$mPas). After impregnation, the laminates were cured in the test setup under pressure.

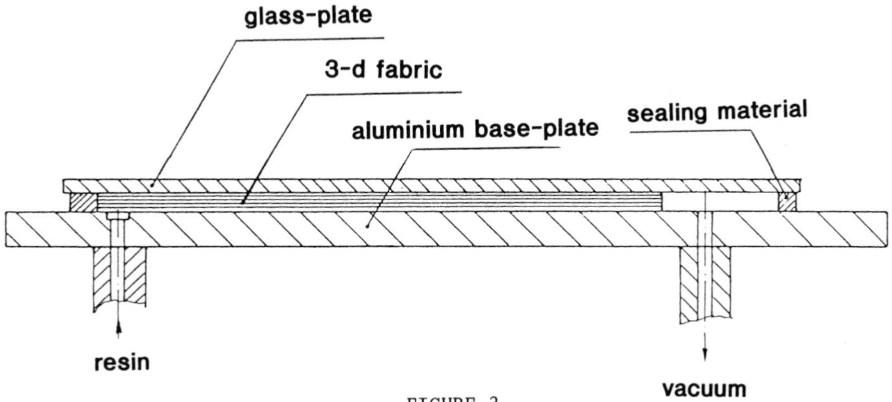

FIGURE 3

Test set-up for RTM Process

The fibre ratio realized as a function of pressure is shown in table 1. The application of autoclave pressure is needed for higher fibre volume ratios. However, in the case of higher pressure the fabric is tightened and the fibre bundles in z-direction are ondulated more or even damaged.

Due to this fact, the ILS values are reduced by a higher autoclave pressure.

Pressure [MPa]	*Fibre content* [Vol-%]	*Thickness* [mm]	*ILS-strength* *warp* [MPa]	*weft* [MPa]
0.1	46.5	3.74	56	48
0.4	53.5	3.45	34	34
0.8	55.8	3.28	31	32

TABLE 1
Relationship between autoclave pressure, fibre ratio and ILS strength of 3-d glass composites

A film impregnation technique was developed to fabricate 3-d prepregs, additionally for advanced carbon-fibre composites. For this application a prepreg resin (Shell HPT1071/1062) is cast in thin film sheets in a metal mould. Afterwards the 3-d preform is placed in the mould. This assembly is then heated in a vacuum press to the processing temperature of 130°C under full vacuum. The pressing process under vacuum removes the air from the 3-d preform. To control the resin ratio of the prepreg, spacing strips were used in the mould.

After this prepregging procedure, an autoclave process using the recommended cure parameter was applied.

3-d carbon composites produced using this technique were of consistent fibre volume fraction (50 %) with a void content of less than 2 %.

Besides composites with a thermoset matrix, 3-d hybrids of graphite and PEEK fibres were fabricated by coweaving and commingling techniques [6]. In commingling, carbon and thermoplastic PEEK filaments are located together in a single yarn to achieve a homogenous fibre/matrix distribution in the composite. These yarns were woven into 3-d fabrics as shown in figures 2a,b,d.

In coweaving, PEEK fibres are woven with separate carbon fibres to form 3-d hybrid structures.

All laminates were fabricated by compression moulding. The 3-d woven preform was placed in a metal die. The whole assembly was placed in a press preheated to the recommended process temperature.

The studies have shown (table 2) that void-free laminates could not be achieved with 3-d preforms having a 60 vol% carbon-fibre content. Much better results are obtained by using preforms with a reduced fibre volume fraction of 50 %. The optimization trials have shown that the optimum processing conditions were 1.5 MPa pressure at a temperature of 410°C and a dwell time of 30 minutes in the melt.

Moulding conditions			*Fibre content*		*Void content*
Temperature [C]	Pressure [MPa]	Time [min]	Preform [Vol-%]	Laminate [Vol-%]	Laminate [Vol-%]
400	1.0	30	60	55.0	7.0
400	1.0	60	60	57.6	6.1
410	1.5	30	60	58.2	4.9
400	1.0	60	50	46.4	2.9
410	1.5	30	50	47.8	1.8
410	1.5	60	50	48.2	1.9

TABLE 2
Moulding conditions of 3-d hybrids of graphite and PEEK fibres

3.2 Sandwich structures

Figure 4 shows schematically some of the developed 3-d sandwich preforms. The length of fibres in thickness direction, the height of the sandwich as well as the arrangement of the linking threads can be varied in a wide range by the weaving technique. Through this, both stiffness and strength properties as well as the core density can be adjusted to the respective requirements.

Up to now, glass and carbon fibres were used for the fabrication of 3-d sandwich preforms.

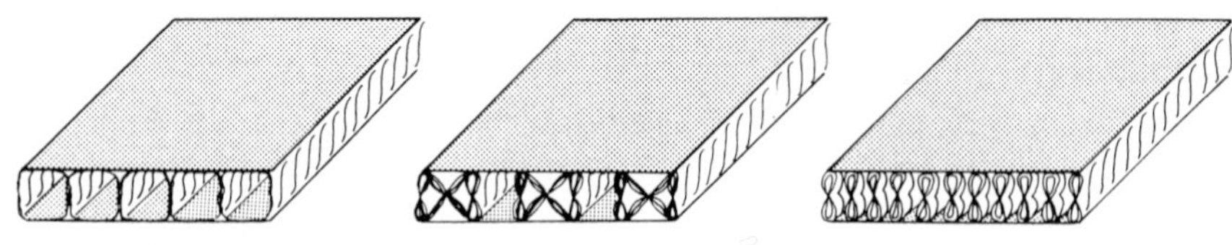

FIGURE 4
3-d sandwich preforms

The manufacturing of 3-d sandwich composites is very easy (figs. 5.1-5.3). The fabric is impregnated, for example, with a low-viscosity epoxy resin by a wet laminating process and the excess resin is squeezed out. The linking fibres between the fabric layers have such a high stiffness that they stand up by themselves after the squeezing process. After curing, this yields a hollow structure with two impregnated fabric layers and impregnated linking yarns. Because no pressure is used during the curing process, only the bottom fabric has a smooth surface.

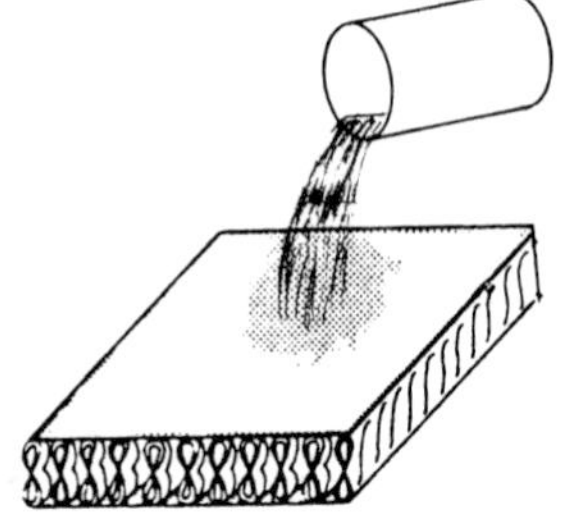

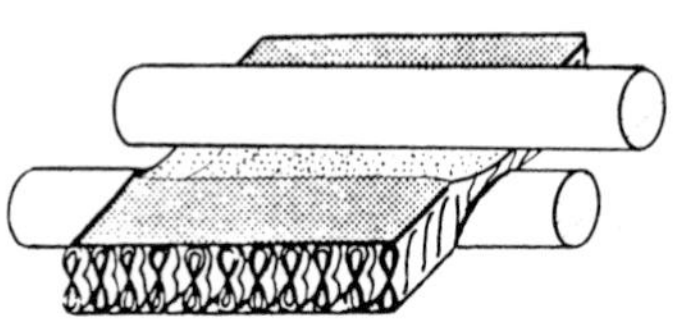

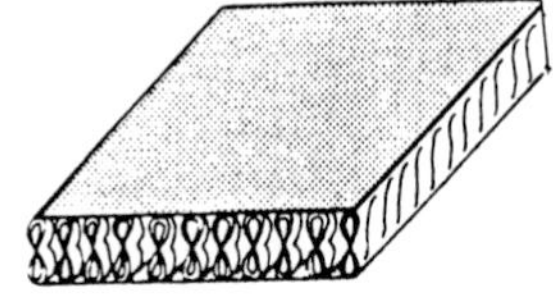

1st step: resin application

2nd step: impregnation and control of resin content

3rd step: free-standing curing

FIGURE 5
Manufacturing of 3-d sandwich composites

After impregnation (steps 1 and 2) 3-d sandwich structures can also be formed in a mould. In this way, sandwich parts of fabrics with different thicknesses, for example profiled tail spoilers for automobiles, can be manufactured in one step without preforming the core.

4. MECHANICAL PROPERTIES

The 3-d composite characterization was performed by using tensile compression and flexural test methods. Impact tests were carried out on a drop-weight instrumented impact machine for compression after impact (CAI) and through-penetration tests [3].

4.1 Monolithic structures

Figure 6 shows the tensile strength of 3-d glass composites with fibre ratios in z-direction varied between 2 % and 8.6 %. The results show that the in-plane tensile strength properties are reduced as the reinforcing fibres increase in the 3rd direction. The retention in warp direction is more significant because a certain number of warp fibres is orientated in z-direction by the weaving process [1, 2].

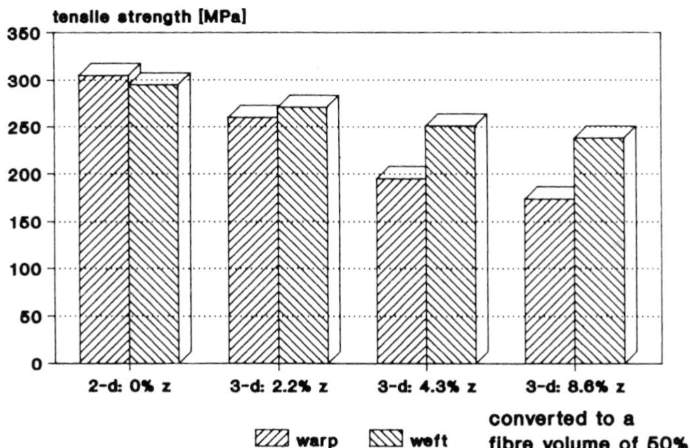

FIGURE 6
3-d glass composites with varied fibre ratios in z-direction

The improved fracture toughness crash behaviour of 3-d GFRP is reflected in the dynamic response curves of the through-penetration impact test according to DIN 53373 (fig. 7). 3-d composites show a significantly different response pattern from 2-d materials. After a gradual increase in impact load, the damage initiation point is almost coincidental with the maximum load plateau [5].

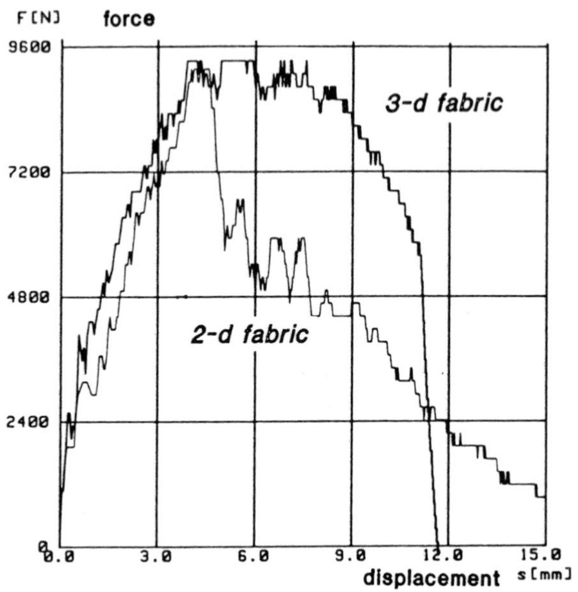

FIGURE 7
Output of through-penetration test of 2-d and 3-d glass composites

Figure 8 shows the basic relationship between ultimate and residual compressive strength after 6.7 J/mm impact energy of 2-d and 3-d glass fabric composites with the same epoxy resin. Compared to 2-d composites the 3-d material exhibited no significant advantage in respect to residual properties. This result confirms the good damage tolerance behaviour of glass fibres and the effect of the 3-d reinforcement is thus limited.

In contrast to glass fibres, a significant improvement in residual compressive strength after impact was measured for 3-d carbon laminates.

Figure 9 shows the well-known reduction of compressive strength between 60 % and 75 % after impact for 2-d fabrics, whereas the reduction of 3-d composites is only between 10 % and 50 %, depending on the ratio of fibres in the 3rd direction.

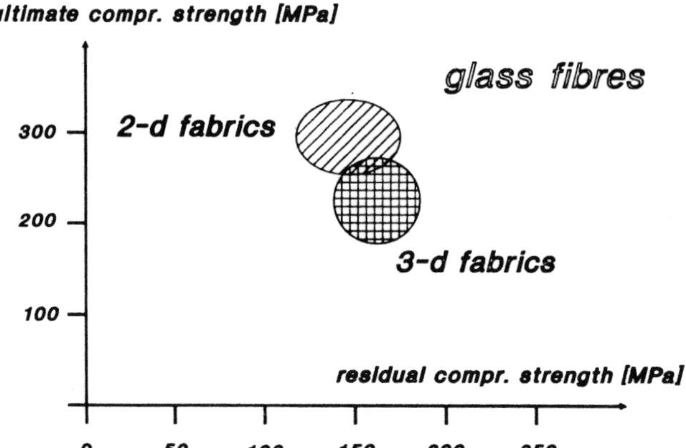

FIGURE 8
Relationship between ultimate and residual compressive strength after impact for 2-d and 3-d glass composites

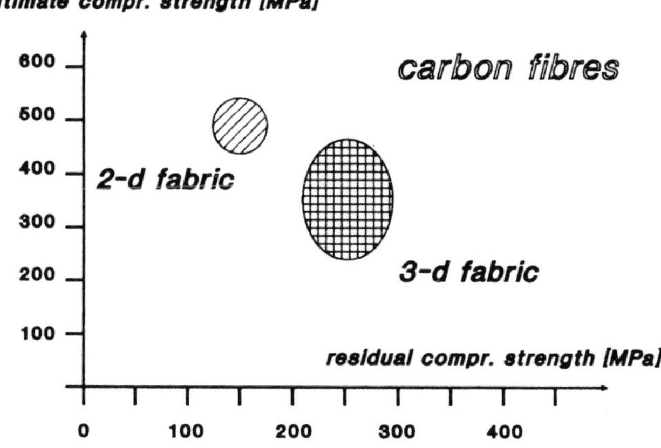

FIGURE 9
Relationship between ultimate and residual compressive strength after impact for 2-d and 3-d carbon composites

4.2 Sandwich structures

For reasons of comparison, appropriate experiments were carried out using conventional sandwich structures with honeycombs (NOMEX ECA 3.2-48) and foam (ROHAZELL 51A) cores. The core height was chosen according to the respective integrally woven 3-d structures (5 mm, 7.5 mm, 10 mm).

Glass-fibre satin fabrics (163 g/cm²) bonded to the core were used for the skins. The 3-d sandwich structures tested consisted of glass fibres, too.

Compressive and shear results of mechanical tests are summarized in figure 10. The mechanical properties of the 3-d sandwich structures are partly in the same range of foam or even honeycomb core materials. Moreover, the damage behaviour is of special interest as delamination of the skins from the "core" is impossible.

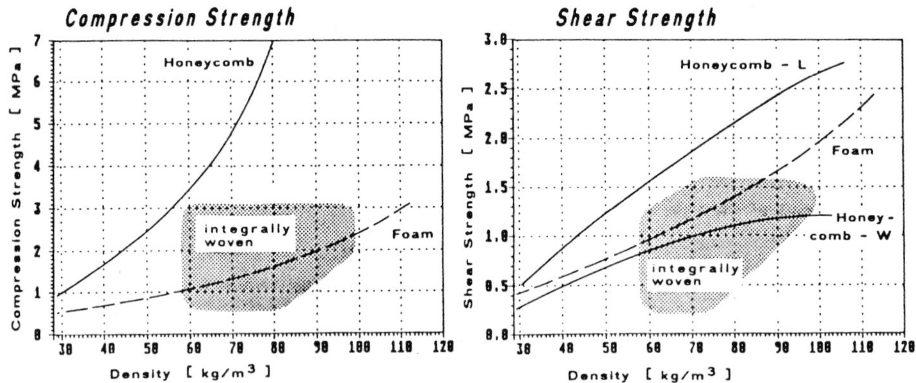

FIGURE 10
Compression and shear strength of 3-d sandwich weavings compared to conventional sandwich structures [4]

4. CONCLUSIONS

Monolithic and sandwich 3-d preforms were manufactured on modified weaving machines. 3-d fabrics made of glass and carbon fibres were realized having various constructions and fibre ratios in the 3rd direction. Furthermore, 3-d hybrids of graphite and PEEK fibres were fabricated by coweaving and commingling.

Vacuum impregnation, vacuum film, pressing and autoclave techniques using epoxy resins and prepregs were applied to impregnate and process 3-d composites with a specific fibre volume ratio and a void content of less than 2 %.

Thermoplastic laminates were fabricated by compression moulding in a heated metal die.

It could be demonstrated that the mechanical properties of 3-d structures can be adjusted to the respective requirements over a wide range by the weaving process.

It was found that the impact behaviour of carbon composite structures can be remarkably improved by 3-d reinforcing, whereas the 3-d effect in the case of glass fibres is rather limited.

ACKNOWLEDGEMENTS

This work is the result of a cooperation between the Central Laboratories of MBB GmbH, Munich, Vorwerk & Co, Kulmbach and the Institute for Aircraft-Design, University of Stuttgart and was supported by the German Ministry of Research and Technology.

REFERENCES

1) H.F. Siegling, K. Drechsler, Anwendungen von dreidimensionalen Faserstrukturen bei Faserverbundwerkstoffen mit Polymermatrix: Symposium Materialforschung, Sept. 1988, Hamm.

2) F.J. Arendts, K. Drechsler, Verbesserung der Eigenschaften faserverstärkter Kunststoffe durch den Einsatz neuer Textiltechnologien: Proceeding Verbundwerk 1988, Frankfurt.

3) F.J. Arendts, K. Drechsler, Entwicklung von Testmethoden für Faserverbundwerkstoffe mit verschiedenen Verstärkungsgeometrien: DGLR-Jahrestagung, 1988, Dortmund.

4) F.J. Arendts, K. Drechsler, J. Brandt, Integrally woven sandwich structures: Third European Conference on Composite Materials, 1989, Bordeaux.

5) F.K. Ko, Impact behaviour of 2-d and 3-d glass/epoxy composites: SAMPE Journal, July/Aug., 1986.

6) F.K. Ko, H. Chu,E. Ying, Damage tolerance of 3-d braided intermingled carbon/PEEK composites: The Advanced Composites Conference, Nov. 86, Dearborn, Michigan.

Materials and Processing – Move into the 90's
edited by S. Benson, T. Cook, E. Trewin and R.M. Turner
Elsevier Science Publishers B.V., Amsterdam, 1989

AUTOMATED R.T.M. FOR AN AIRFRAME COMPONENT

Ian **MARCHBANK**

British Aerospace, Reinforced and Microwave Plastics Group, Stevenage, Herts, U.K.

ABSTRACT

Resin Transfer Moulding, has been used for many years within British Aerospace for the production of low volume, high integrity aircraft and missile radomes. Following a recent initiative, an exercise was undertaken to assess the practicality of designing and manufacturing a high volume airframe component in composites within tight cost and weight constraints.

This paper discusses the changes required to the 'traditional' injection process and follows the evolution of both design and manufacture through an eighteen month exercise culminating in the production of full size prototype components. These components, of hybrid construction and incorporating both low density core materials and load bearing inserts, have been successfully tested and work is continuing toward flight trials.

The likely course of development is outlined and the factors expected to affect production are highlighted. The paper concludes with a description of an automated manufacturing facility of the kind envisaged for volume production of composite components.

1. INTRODUCTION

Some three years ago, at the 7th S.A.M.P.E. European Chapter Conference[1], the half-shell concept for production of missile airframes was introduced. In this concept an airframe is produced in two longitudinally split 'Half-Shells' rather than as a series of short, nominally cylindrical sections requiring subsequent assembly. This concept was considered an ideal vehicle for assessing the viability of replacing the 'traditional' airframe material, aluminium, with a reinforced plastic.

Cost reduction remained the driving force with lower mass as a secondary target. It was hoped to achieve this cost reduction in two ways; firstly by reducing the basic manufacturing cost and secondly by reducing the need for subsequent processes such as painting or plating.

The component chosen was not a direct replacement for a current unit but was a new design incorporating features and dimensions expected to be typical of the next generation of transonic missiles. In general terms the design could be described as cylindrical shell of 1.1 m length and 150 mm outside diameter with location features on both the inner and outer surfaces and simple interfaces at both ends.

The programme was seen to fall into four phases:

PHASE ONE: The assessment of the various composite materials and processes available and the selection of one or more for practical trials.

PHASE TWO: Generation of a suitable design. This phase was expected to interact and run in parallel with phase three.

PHASE THREE: Practical trials aimed at both process and design evolution.

PHASE FOUR: Manufacture and testing of a representative batch of full size half-shells.

The clearly defined aims of the programme were to produce a structurally sound composite airframe design and to demonstrate a manufacturing capability suitable for subsequent productionisation.

2. ALTERNATIVES COMPOSITE OPTIONS

The options considered together with their major advantages and disadvantages are shown in Figure 1.

FIG 1 ALTERNATIVE COMPOSITE OPTIONS

Method	Reinforcement	Comments
Injection Moulding (thermoplastic)	Short fibres	Machine size, poor control over fibre orientation, bonding problems
Compression (D.M.C./S.M.C.)	Short/medium fibres	Poor control over fibre orientation, excessive variability of properties
Contact Moulding (hand lay-up)	Long fibres	Slow, labour intensive (hence high cost), poor surface finish
Resin Transfer Moulding (R.T.M.)	Long fibres	Good mechanical properties, traditionally slow with expensive tooling
L.P. Compression Moulding (prepreg)	Long fibres	Good mechanical properties, labour intensive, poor surface finish

At this point in time, early 1987, half-shells had been produced in aluminium, using traditional manufacturing techniques and in SMC/DMC using the compression moulding process. A short section of half-shell incorporating solid wings had also been produced by injection moulding in 40% glass filled Ryton (P.P.S.).

Assessment of the other options was carried out on the basis of experience gained with other products and materials property data. It became evident that the short fibre materials were unable to meet the structural requirements of an airframe and this effectively ruled out the 'high volume ' processes of injection moulding and compression moulding.

Of the available processes using long fibre reinforcement only resin transfer moulding (R.T.M.) would produce a good finish on both surfaces as moulded and in addition was seen to offer the possibility of automation for future production.

A novel approach, best described as low pressure compression moulding of prepreg, was also investigated as a back up using the component design and tooling developed for the R.T.M. process.

BAe already had a great deal of experience in RTM for the production of high integrity, high value structures such as radomes[2] . This paper indicates how this existing technology and expertise was extended to address the high production rate low unit cost situation.

The next section of the paper details the technology available in mid 1987 and gives the aims of the development programme. The work carried out over the next eighteen months is detailed and is followed by a description of the current position and an outline of the future work.

3. PROGRAMME AIMS

3.1. DESIGN

The design at the commencement of the programme, shown in Figure 2, was a solid composite, thin skin, shell structure with local thickening in the area of highest load.

FIG 2 INITIAL DESIGN

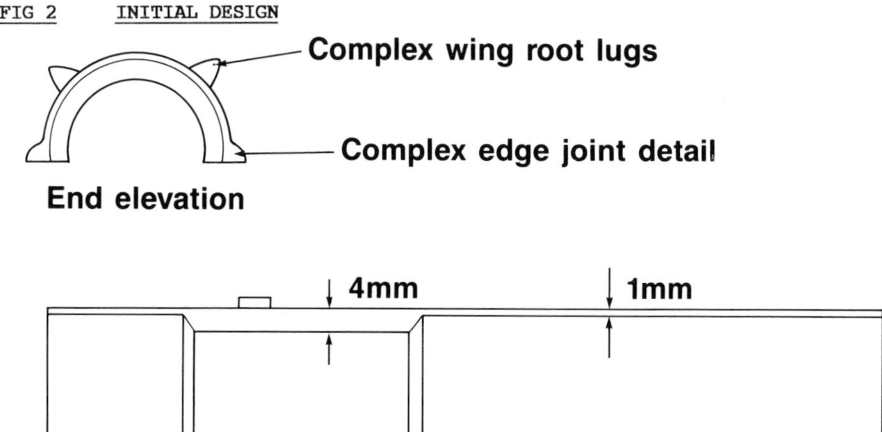

It featured complex details for both the longitudinal joint and the wing root lugs and was predicted to be lacking in stiffness. It had no internal location features and would require post-moulding operations to provide E.M.I. screening and an acceptable external finish.

The design aims for the programme were thus to simplify the existing edge and wing locations, incorporate internal location features and produce a stiffer structure, ideally with inherent screening properties and an acceptable cosmetic finish as moulded.

3.2 PROCESS

The resin transfer moulding process involves injecting a liquid thermosetting resin into a tool cavity previously loaded with dry reinforcement. The cutting and placement of the reinforcement is traditionally done by hand as is the mould clamping. The actual injection of the pre-mixed resin is done by either pressure pot or reciprocating pump and the overall cycle time is long; the production rate of trial half-shells manufactured at this time being at most one per day.

To meet the production rate and cost targets envisaged (eighty per day at a reduced cost compared to a comparable metallic structure) considerably improved productivity would be required at all stages of the process.

3.3. MATERIALS

The matrix systems commonly used in radome manufacture, both polyester and epoxy, are chosen on the basis of electrical and mechanical properties but frequently feature high cost and disadvantageous processing parameters, such as high viscosity and long cure cycles.

Typical material properties being viscosities of 200 to 400 cps and 200 to 1000 cps and cure times of 15 mins + and 30 min + (frequently accompanied by a post-cure cycle) for polyesters and epoxies respectively.

For this programme low cost and processability would be equally as important as ultimate properties. As a result experience would have to be gained rapidly with low viscosity, rapid cure materials new to R.M.P.G. and to the industry in general. The geometry of the half-shell permits the use of 'off the roll' reinforcements rather than the expensive custom made reinforcements which are often necessary in radome manufacture.

Any advances in materials properties would however have to be balanced against the other aspects under consideration; for instance a reduction in cure cycle obtained by increasing the process temperature would be counterproductive in view of the requirements it would impose on the tooling.

3.4. TOOLING

Much experience is available within BAe in the production of RTM moulds, predominantly in the field of the large matched metal tools required to achieve the tight tolerances required for radome manufacture.

The aims of this part of the programme were to investigate tooling with a reduced initial cost suitable for low cycle time automated production; i.e, low mass, fast heat up rate tooling ideally independant of ovens or bulky fluid heating equipment.

4. THE DEVELOPMENT EXERCISE

Due to the interaction of the design development and the processing trials it is both impractical and unrealistic to attempt to report them separately. Instead the exercise will be covered, as far as possible, in chronological order.

4.1 THE INITIAL STAGES

Based on the lessons already learnt a tentative design and process/tooling philosophy was agreed. The basic assumptions were that the component would have a uniform wall thickness (to ease jointing and ensure adequate stiffness) and that it would be produced on lightweight composite tooling which was to incorporate electrical resistance heater mats. It was decided to test these assumptions on a simple section of component and a tool was ordered. This testing was designed to produce a component some 600 mm long with a 4 mm wall thickness. Whilst the basic section was semi-circular a joggle was introduced along one edge with flanges on both edges to allow a number of possible jointing methods to be tried. Electrical heating was incorporated in one half of the tool. This 'short' tool was used extensively throughout the programme to evaluate new concepts or materials.

The initial components produced in the short tool were of solid wall composite construction utilising 0/90% woven glass in a polyester matrix. As experience was gained sandwich structures incorporating P.U. foam cores, bands of unidirectional fibres and various non-axial ply orientations were successfully attempted.

It was at this stage that I.C.I.'s Modar methacrylate based rapid cure resins were introduced to the programme. Designed specifically for automated processing they possessed suitable viscosities and cure cycles at acceptable temperatures (typically 60 to 230 cps and 3 minutes at 20° C). After some initial problems with air inhibition in cored samples a suitable grade, Modar 836S, was employed and rapidly became a baseline against which other contending rapid cure systems were judged.

Following testing of a batch of twenty sandwich construction samples the decision was taken to purchase a full size tool.

4.2 FULL SIZE TRIALS

The component produced from the full size trial tool was to include representative wing root lugs and, due to their orientation, required a split female tool. All tool faces were to be heated and provision made for later fitting of hydraulic tool clamps on both the horizontal and vertical tool split lines.

The design of the wing root lugs had by this stage been greatly simplified such that both fore and aft lugs could utilise the same extruded aluminium inserts. These inserts were placed into the tool together with the dry reinforcement, located on pins and encapsulated in the one shot injection process. A number of quite severe process problems were experienced such as washing and rippling of the reinforcement, leading respectively to circumferential and longitudinal resin richness, and non-uniform resin flow resulting in dry patches. These problems were largely overcome by closer control of the size and placement of both the reinforcement and the core and the introduction at strategic points of 'Tygamesh' binder produced by Fothergill and Harvey.

Whilst further developments were still being pursued it was decided at this stage to produce a batch of full length samples for mechanical testing. This would allow both design verification and the demonstration of process repeatability. The components produced in this batch were to the design shown in Figure 3.

The samples were tested as pairs with the joint being achieved by means of extruded aluminium 'H' sections to which the half-shells were bonded. Testing was carried out in three stages; simple three point bend testing, loading of the wing root lugs and a combination of the two to provide a closer approximation to 'in flight' conditions. In all cases the target loads were exceeded with generous reserve factors whilst the variation was considered well within a satisfactory range. The test results also confirmed the predicted natural frequency of the assembled structure as being undesirably low. It was felt that replacement of at least some of the woven glass with woven carbon was required. This had previously been avoided both on cost grounds and because the translucent nature of the G.R.P. composite allowed easier quality assessment and observation of the effects of process or lay-up changes.

FIG 3 TEST BATCH SAMPLES

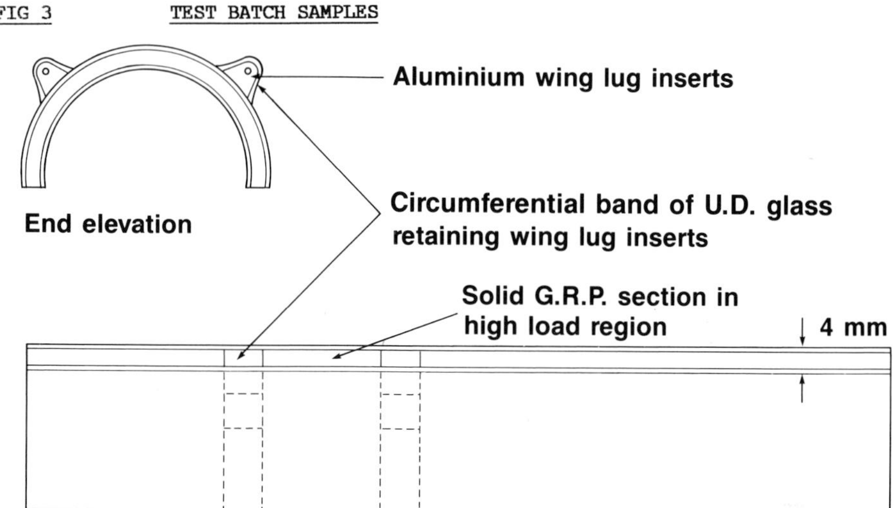

4.3 SUPPORT INVESTIGATIONS

In parallel with the work on the main tool a number of investigations were been carried out using both the short tool and flat plaque injection tools. The most important of these looked at the E.M.I. screening effects of aluminium mesh, silver coated nylon mesh and woven carbon fibre. When incorporated into the build as a single ply woven carbon fibre proved superior in terms of both screening performance and cost. The incorporation of one or more plies of woven carbon into the build thus serves the dual role of providing the required screening effect and producing the necessary mechanical properties.

The programme also investigated the effect of pigmentation and the use of alternative materials such as vinyl-ester and modified polyester resins, alternative types of fabric reinforcement and 'coremat' as a replacement for rigid foam cores. The latter was especially advantageous as it removes the need for a pre-forming operation and allows the core to be treated as 'just another ply' in the build which has obvious appeal from the point of view of productionisation.

The short tool has also been modified to incorporate representative rib details. The manufacture in-situ of these ribs by pushing pre-moulded cores through the inner skin reinforcement prior to tool closure has been successfully demonstrated. The ease with which the tool was modified for this and earlier trials with a reduced component thickness indicates one of the major advantages of composite tooling for development work.

With the exception of the very first trials using the metal tool, all injection has been carried out using a 'Hypaject' resin transfer machine produced by Plastech T.T. of Gunnislake, Cornwall. This is an ideal development tool since it is simple to use and clean, is suitable for virtually all resin systems and operates with a minimum of material wastage.

5. RESULTS OF THE PROGRAMME

5.1. COMPONENT DESIGN

The basic layout of the final design is much the same as that shown in figure 3; that is to say a 1.1 m long, 4 mm thick semi-cylindrical half tube with continuous matched skins on each surface.

The wing lugs are sections of aluminium extrusion retained by transverse bands of unidirectional fibres which run over the lugs and under the outer skin. The area between these bands is formed of solid composite whilst fore and aft of the bands the skins are separated by a low density core. Internal details (ribs) are formed by running the inner skin over pre-moulded inserts.

Jointing of half-shells is by bonding to an 'H' section extruded aluminium strip. The components are self coloured (limited range of dark colours) and incorporate carbon fibre E.M.I. shielding as part of their basic structure. This design is also applicable to the prepreg compression moulding process should suitable prepreg systems and automated lay-up techniques become available.

5.2 PROCESS

Pre-cut shapes of dry reinforcement and core material together with inserts, both metallic and non-metallic, are pre-loaded to the tool, which is set at 60+ 5 °C. Alignment and retention of the wing lug inserts is by pneumatically operated pins.

After manual alignment final tool closure and clamping is carried out hydraulically. Injection of pre-mixed resin takes approximately two minutes using a 'Hypaject' resin transfer machine. Back pressure is maintained by a controlled bleed system. The tool is cycled to 75 + 5 °C and maximum tool surface temperature following peak exotherm is in the order of 85 °C. The tool is hydraulically split; the cycle time from the beginning of tool closure to component ejection is in the order of fifteen minutes.

5.3. **MATERIALS**

The reinforcement in the outer skins comprises plies of woven carbon and glass fabrics whilst the central core section utilises woven glass rovings. The low density core used is 'coremat' and the wing lugs are retained by unidirectional glass bands stabilised with 'Tygamesh' binder. 'Tygamesh' is also utilised at strategic points to prevent relative movement of the various sections of build. Both wing lugs and 'H' section jointing pieces are lengths cut from continuously formed aluminium extrusions. The pre-moulded 'cores' of the internal ribs are currently manufactured from a polyester based filler compound.

The matrix resin currently considered most suitable is I.C.I's Modar 836S methacrylate based resin system.

5.4. **TOOLING**

Low mass composite shelled tooling with integral electrical heating is currently in use as shown in Figure 4.

FIG 4 LIGHTWEIGHT COMPOSITE TOOLING

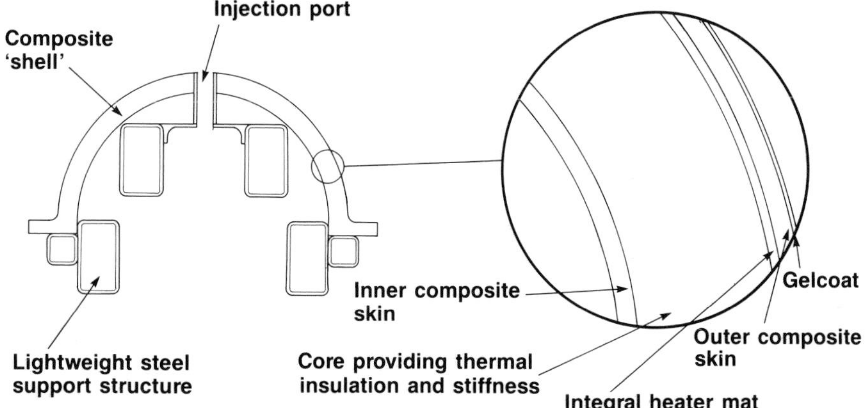

The heater mat is placed behind the first, gel coated skin and is backed by a thermally insulating core. This in its turn is backed by a further skin to form a balanced sandwich. This is grouted onto a backing structure made of hollow section steel tubing to which are attached the various stops and pins required to provide cavity thickness control and the hydraulic and pneumatic systems used for location and clamping. Injection ports are set centrally in the one piece male tool whilst the female tool has a longitudinal split line to facilitate component removal.

6. AUTOMATION

To meet the rates of manufacture and cost targets envisaged for this type of product will inevitably mean a high degree of automation within the production process.

The R.T.M. manufacturing cycle breaks down into ten basic stages. These are shown in Figure 5 which highlights the factors which will affect volumne production. It can be seen that most of these stages are amenable to automation using adaptations of current technology.

FIG 5 THE MANUFACTURING CYCLE

1)	Material Storage	-	Automated cutting/Kitting equipment available.
2)	Tool Preparation	-	Minimised by suitable choice of release agent(s).
3)	Lay-up	-	Potential area of greatest difficulty for automation.
4)	Tool Closure	-	Relatively simple to automate using existing equipment and techniques.
5)	Injection	-	Several automated systems available on the market.
6)	Cure	-	Must be short to allow high throughput.
7)	Component Ejection	-	Ease of automation dependant on both tool and component design.
8)	Post Cure	-	Requirement depends on material choice and service conditions.
9)	Component Trim	-	All edges are simple, single plane cuts.
10)	Bonding of Joint Strip	-	Metered adhesive dispensing equipment already exists.

The main risk area is the actual lay-up process and considerable interest is being shown, both within BAe and throughout the industry, in preform technology and the type of cut, pick and place technology used within the clothing industry. These offer possible alternative, or even complementary, solutions to the problems in this area.

For any high volume manufacturing process of high quality components the Q.A. aspects must be fully integrated into the manufacturing cycle hence they do not appear as separate operations in the cycle breakdown. This philosophy and the requirements it imposes are amplified in Figure 6.

FIG 6 QUALITY ASSURANCE ASPECTS

Philosophy - To minimise 'hands on'inspection and batch testing by emphasising process control/monitoring and N.D.T.

Requirements - Active process control of all major parameters (clamping force, temperatures etc).
- Interactive material/process monitoring for shelf life/pot life control.
- On line N.D.T. checking for moulding and bond quality (probably by thermography and/or ultrasonics).
- Automatic gauging for dimensional checks.
- Full electronic monitoring and logging of all measured values.
- Random sampling for destructive testing.

Whilst actual component details or future materials or process breakthroughs make forecasting a risky business it is possible to envisage the almost complete automation of the production cycle with five or six tools being served by a number of robot and automated process cells. Using the resin systems available today the desired cost and production rate targets could be achieved by running two cure cycles in parallel as shown in figure 7.

FIG 7 PROPOSED PRODUCTION FACILITY

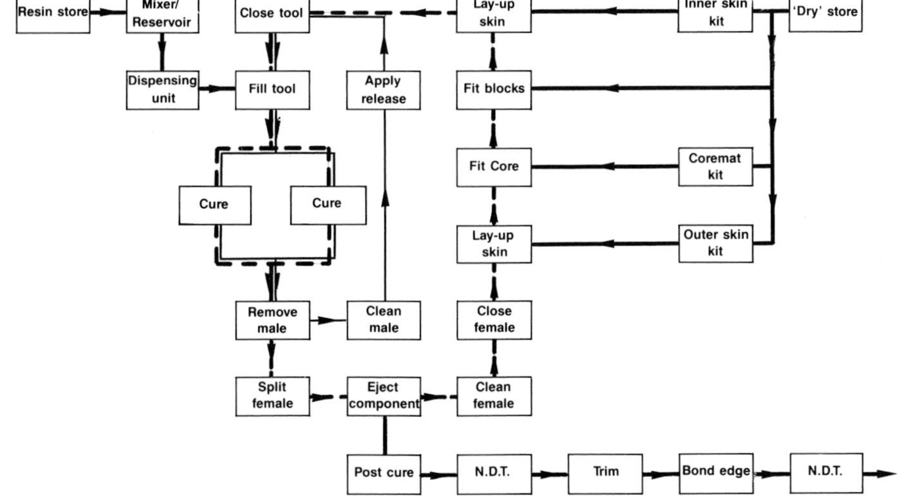

7. REVIEW

The work to date has resulted in a composite design with properties equal to or better than a metallic equivalent and has shown the feasibility of economically manufacturing such a component by demonstrating the following points:-

(1) Cycle times can be reduced sufficiently to allow high production from a small number of tools.

(2) Tooling can be used which is light, cheap to manufacture and cheap to use.

(3) High productivity can be achieved by utilising both designs and processes suitable for automation.

(4) High integrity products can be produced to a uniformly high quality standard.

However before a production facility can be set up considerable detail work remains to be done, especially in the field of process automation.

ACKNOWLEDGEMENT

The author wishes to acknowledge the vital contributions to the programme of many people within the Reinforced and Microwave Plastics Group, the Mechanical and Technology Group and the Missile Department of British Aerospace Dynamics Limited.

REFERENCES

1) C.K. Hall and T.Cook, Designs in Composites: Are They Cost Effective?, in: Materials Science Monographs, Vol 35, eds K.Brunsch, H.D. Golden and C.M. Herkert (Elsevier, 1886) pp 237 - 250.

2) T. Cook and M.C. Cray, Supersonic Radomes in Composite Materials, in proceedings of the third technology conference of the European Chapter (SAMPE) 1983. Volume 1, Page 4.

Materials and Processing – Move into the 90's
edited by S. Benson, T. Cook, E. Trewin and R.M. Turner
Elsevier Science Publishers B.V., Amsterdam, 1989

A new Braiding Process
- Robotised Braiding Mechanism.

Atsushi YOKOYAMA, Akihiro FUJITA, Hideteru KOBAYAHSI, Hiroyuki HAMADA and Zenichiro MAEKAWA.

Kyoto Institute of Technology
Matsugasaki, Sakyo-ku, Kyoto 606 Japan

It is well known that braided fabric constructions provide useful reinforcement configurations in three dimensional composites. This paper describes work carried out to modify a conventional braiding machine for the fabrication of complex shaped braided fabrics, and goes on to discuss automation of the system (self-driven system) for which computer simulation programmes were developed.

1. INTRODUCTION.

Composite materials are used widely in both primary and secondary load bearing structural components. However conventional 2-D laminates have very poor through thickness strength and as a result are limited in their application and utilisation for complex shaped structural parts. In order to overcome this limitation, methods of fabricating 3-D reinforced structures are being studied. One technique of particular interest is braiding. Braiding is an established textile process for the formation of a wide range of shaped fabrics by combining bundles of bias fibers delivered from sets of moving carriers or spindles to form biaxial and triaxial fabrics.

KO[1,2] and Brown[3] have developed several 3-D braiding processes; Li[4] and Florentine[5] have also investigated 3-D braided composites. Papers have been previously presented discussing the properties of braided tubes and braided flat bar, [6-8]. In addition modelling techniques for basic flat and tubular braiding mechanisms have been developed[9].

In the study, described in this paper, the modification of a conventional braiding machine was carried out and various shaped fabrics were produced. A new more flexible fabrication process is proposed based on an automated (self-driven) braiding system.

A computer model to simulate the new system was set up to support the further development of the process.

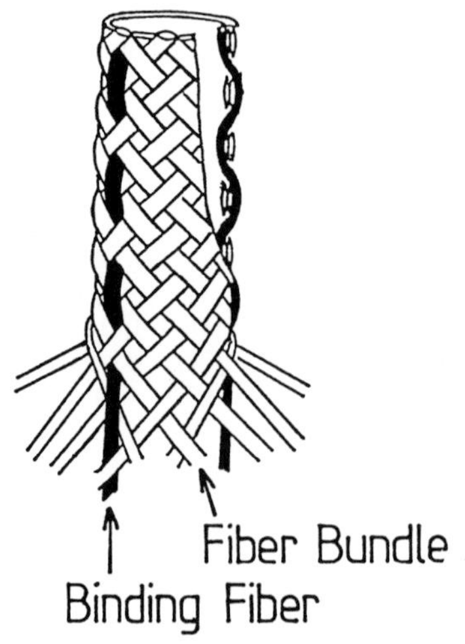

Fig.1 Schematic diagram of braided pipe with binding fibers.

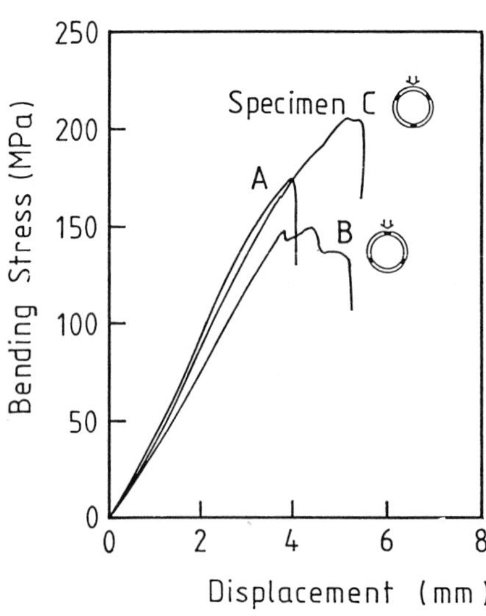

Fig.2 Relation between bending stress and displacement.

2. MECHANICAL PROPERTIES.

Four point bend tests were carried out to evaluate the mechanical properties of braided tube composites. The specimens tested, were made from 1600 tex glass rovings in an epoxy resin. Three types of specimen were evaluated :

Specimen A - standard braided tube

Specimen B - } braided tube with longitudinal reinforcement

specimen C - } braided tube with longitudinal reinforcement

A schematic diagram, of the braided tube construction is shown in Figure 1. A standard braid angle of 26° was used, and the position of the "Binding fibers" - in this case longitudinal reinforcements - is illustrated. Specimens B + C were classified by the different positions of the longitudinal fibers. In both cases three longitudinal reinforcing fibers were introduced, however in specimen B the longitudinal fibers were located on the compression side during testing and in Specimen C on the tension side. Figure 2 shows the relationship between bending stress and deflection of the three specimens. Specimen C demonstrates a higher flexural strength than Specimen A - the standard braided tube - whereas the strength of specimen B is lower. Examination of the fracture surfaces of the specimens showed compression failure of the longitudinal fibers in Specimen B, however in Specimen C no failure occurred in the longitudinal reinforcement.

The position of the longitudinal binding fibers clearly determines the flexural strength and accordingly the position and amount of longitudinal reinforcement can be selected to meet the load considerations of the required structure. The fact that the longitudinal fiber can be introduced in any position simultaneously during the braiding process is a further useful advantage.

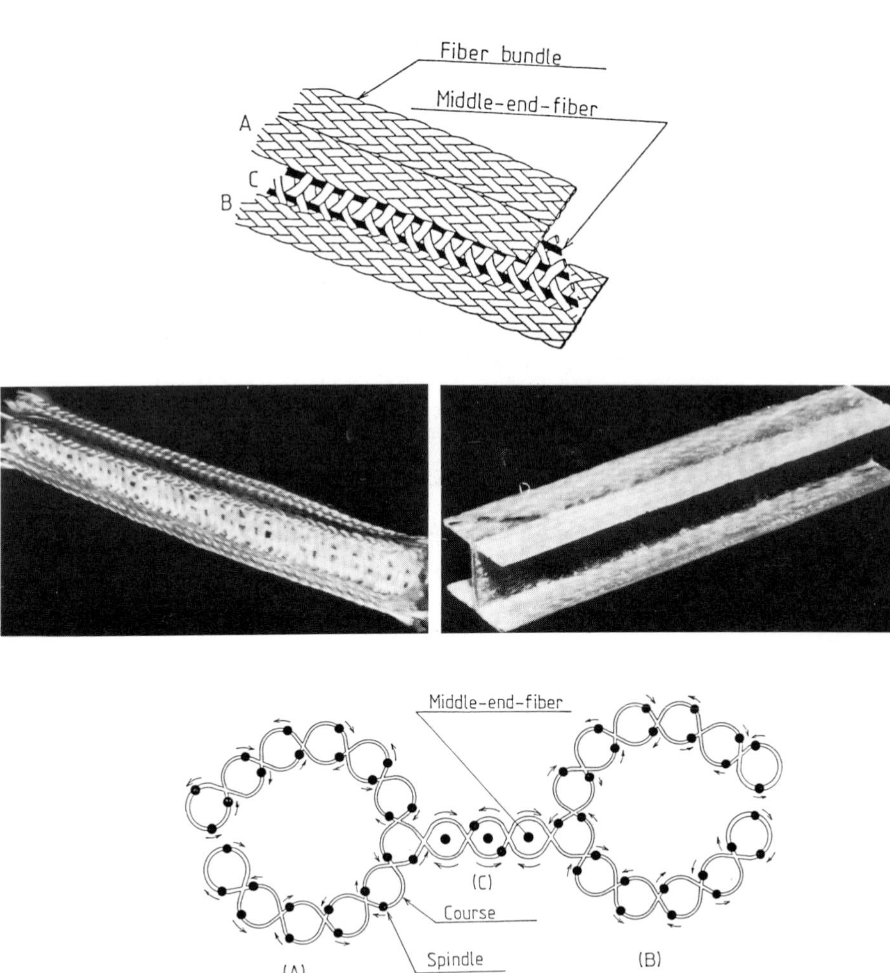

Fig.3 Outside view of braiding fabric and composite with I-Beam shape and the configuration of the course.

3. Modification of a conventional Braiding Machine.

A conventional braiding machine was modified to achieve the fabrication of various complex shaped fabrics. Figure 3 shows the construction of the braided reinforcement and the resultant composite for an I-Beam structure. The configuration of the spindle track (or course) is also illustrated. The flanges of the I-Beam braid, parts A and B, are made by flat braiding and the shear web (C) of the beam is also built up at the same time by the sequential motion of the spindles through the centre section (c) of the track. The I-Beam braided structure clearly has a single-fabric construction.

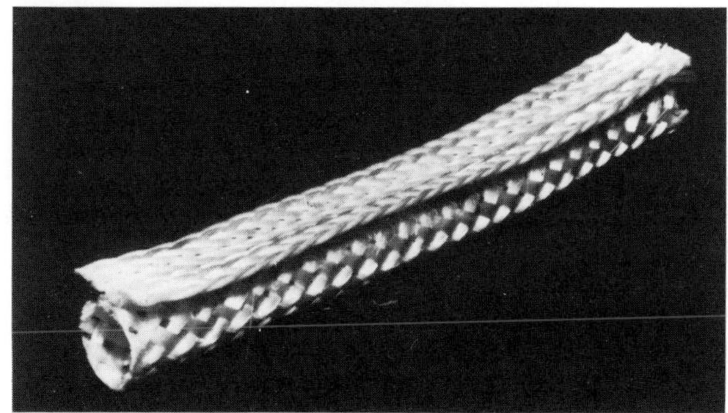

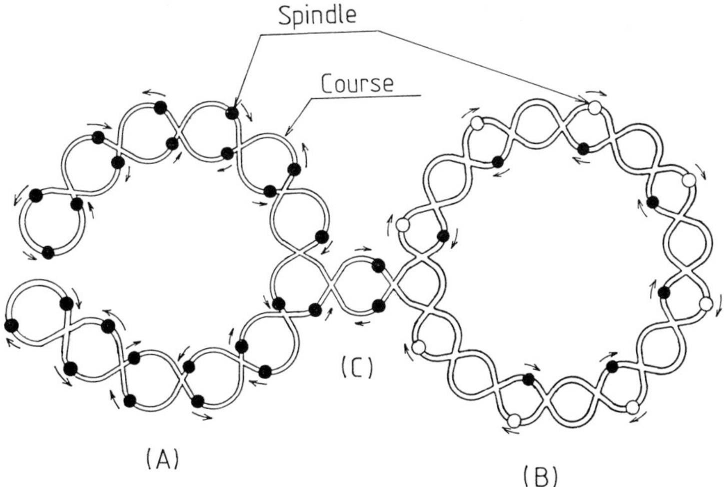

Fig.4 Outside view of braiding fabric and the configuration of the course. (Flat-Cord braiding fabric)

Figure 4 shows the construction of a braided fabric, combining flat and tubular elements. The configuration of the spindle track is also illustrated in the lower part of the diagram. Parts A and B are fabricated by flat and tubular braiding mechanisms respectively. The open circles indicate the spindles moving only in part B of the track and the solid circles represent the spindles moving sequentially through both parts of the track to form the combination of the flange and tube.

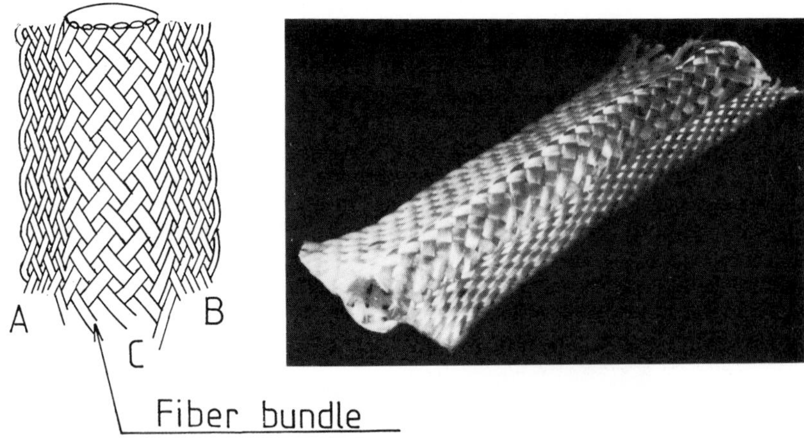

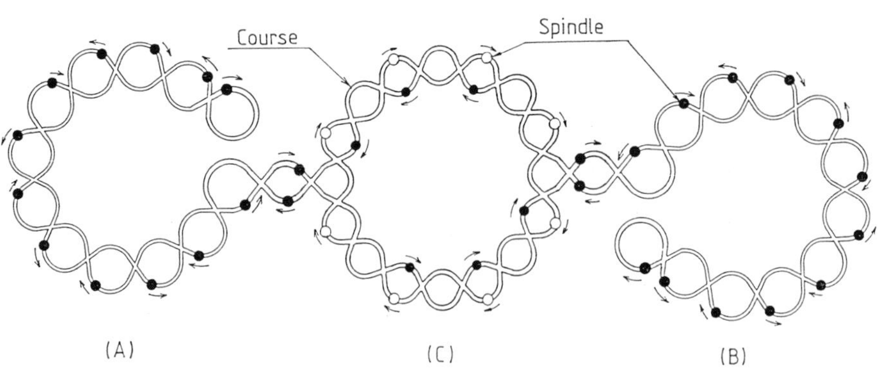

Fig.5 Outside view of braiding fabric and the configuration of the course. (Flat-Cord-Flat braiding fabric)

Figure 5 shows a braided fabric construction similar to that in Figure 4 combining flat and tubular braids. The configuration of the spindle track is illustrated in the lower half of the diagram. Parts A and B are fabricated by flat braiding, and part C by a tubular braiding mechanism. The spindles indicated by the open circles move on section C of the track and spindles represented by solid circles move through all sections of the track in the following sequence (A)-(C)-(B)-(C)-(A).

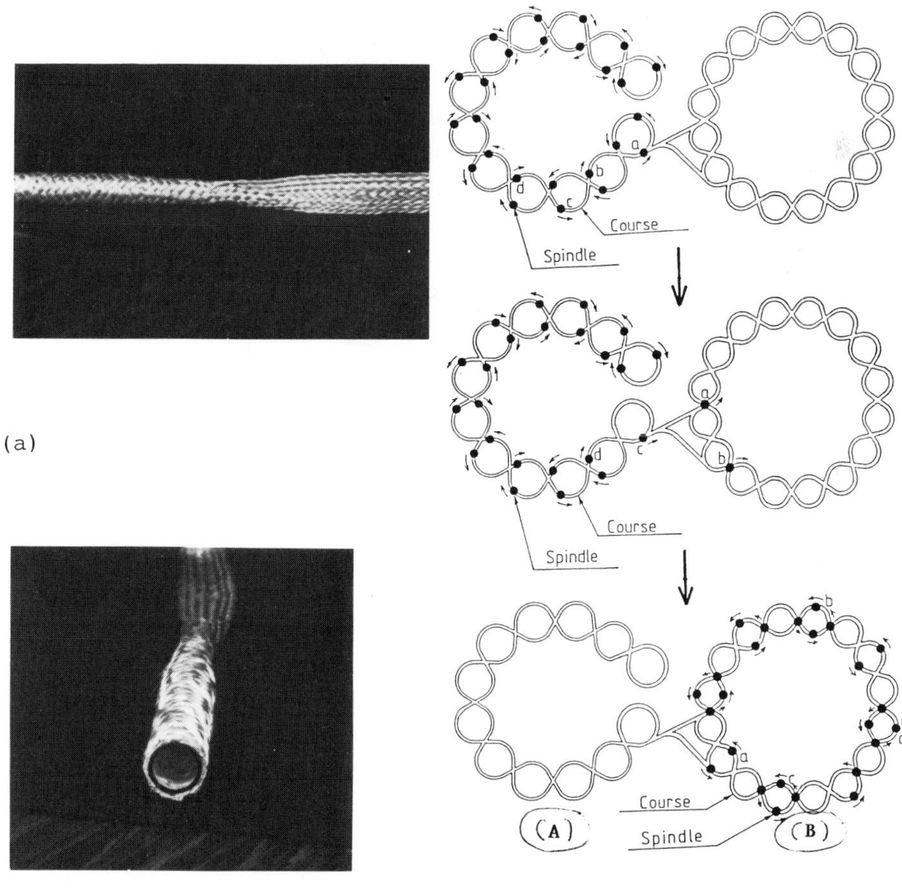

Fig.6 Outside view of braiding fabric with two different cross section and the configuration of the course.

A further braided shape is illustrated in Figure 6 together with the spindle track configurations. This particular braid is characterised by variable cross section - changing from flat to circular. Part A is fabricated by flat braiding and Part B by tubular braiding. The transformation of cross sectional shape is achieved by progressively transferring spindles from Section A of the track to Section B.

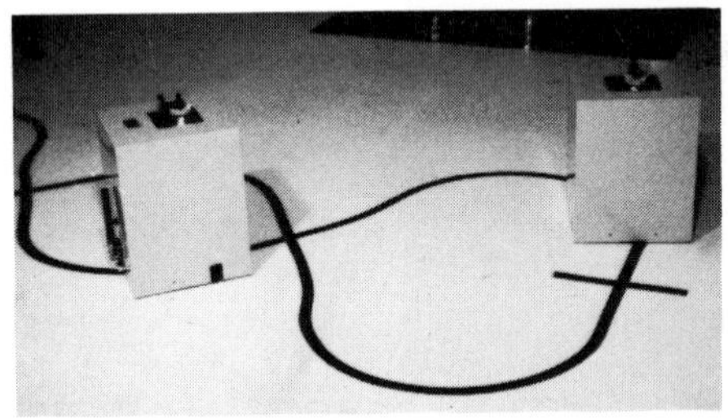

Fig.7 Outside view of the robot.

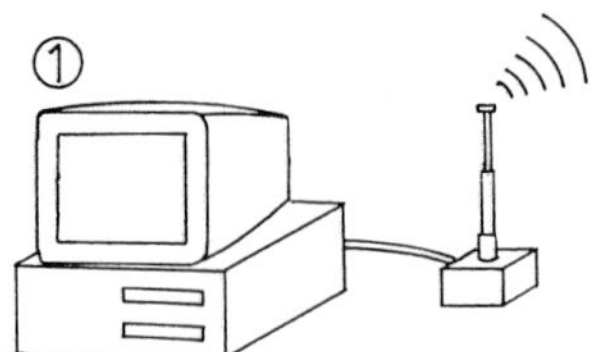

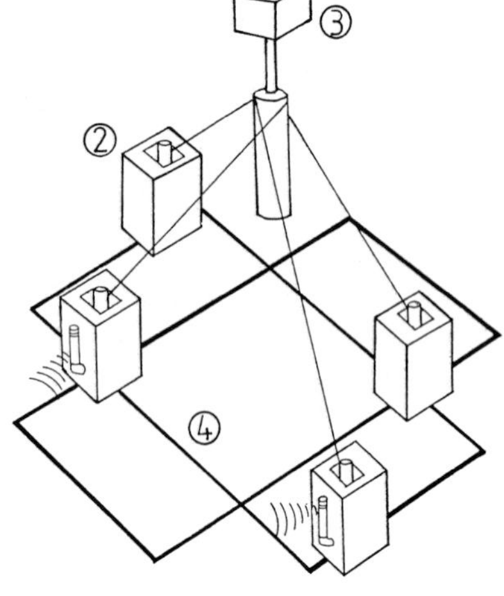

Fig.8 Self-driven system.

4. SELF DRIVEN SYSTEM

In order to fabricate the full range of braided shapes, illustrated above in Section 3, using a single braiding machine, it is necessary to control the motion of each individual spindle and to have the facility to change easily the configuration of the spindle track. By meeting these requirements the freedom of the spindle is extended allowing a wider range of fabrics to be produced from a single braider. A robotically driven system has been developed for moving the spindles, controlled by a personal computer.

The system developed is illustrated in Figure 7, which shows the method for creating and modifying the spindle track by the use of a guide tape placed on the floor. The robot responds to optical sensors which enable the system to follow the track laid down by the guide tape, as illustrated in Figure 8.
In this way, creation and modification of the spindle track are extremely easy to achieve.

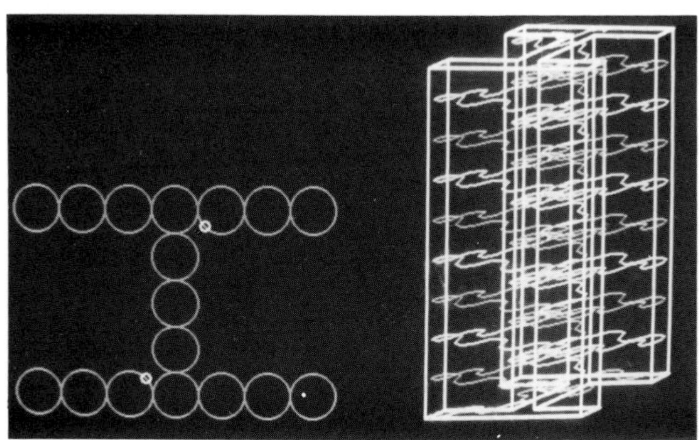

Fig.9 Simulation result. (I-Beam braiding fabric)

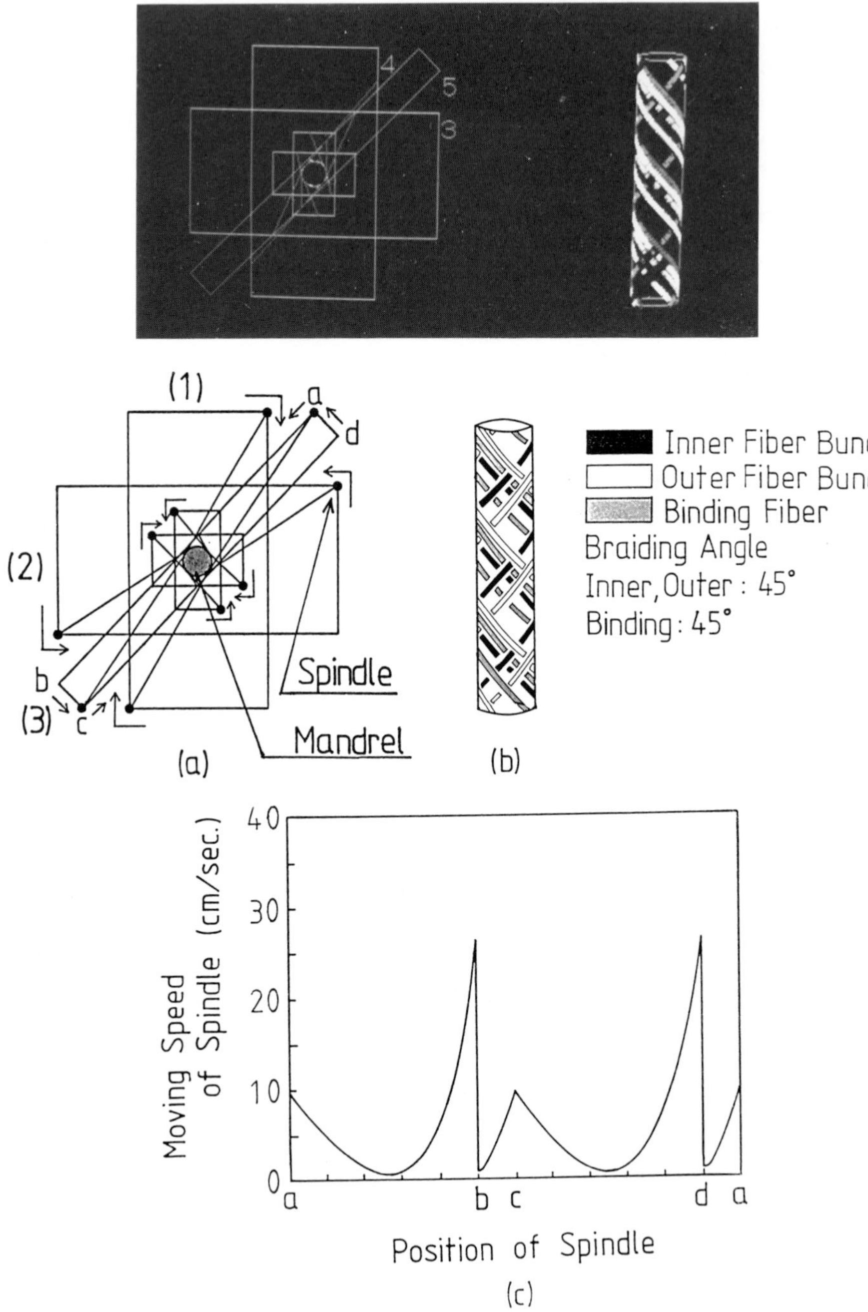

Fig.10 Simulation result. (Cord braiding fabric with binding fiber)

5. COMPUTER SIMULATION.

In order to illustrate the relationship between spindle motion and fiber path, a computer programme has been developed which generates a graphic display. Figure 9 shows the computer simulation of the woven geometry of the I-Beam braid. Figure 10 shows a similar display for the tubular braid containing reinforcements. The fibers delivered from spindles moving in section (3) of the track, illustrated in Figure 10(a), represent the additional reinforcements (Binding fibers). Figure 10(c) shows the relationship between spindle speed and position in the track for this construction. If the spindle speed is controlled according to the pattern in Figure 10(c), braided fabrics can be produced automatically with a some braid angle for both the main braided structure and the additional reinforcement.

Figure 11 shows the spindle track configuration for the automated system (self-driven system); and the computer simulation illustrates the geometry of a tubular fabric with 3 longitudinal reinforcing fibers positioned on one side of the construction. From the results obtained for mechanical performance in Section 2, it is clear that superior flexural strengths are obtained from braided tubes containing longitudinal reinforcement on the tension side. The computer simulation demonstrates how to fabricate this type of tubular braids in order to achieve improved flexural properties.

6. CONCLUSION.

This study has examined a range of fabrics capable of being produced on conventional braiding equipment. A system has been developed for automating the motion of the spindles by the use of robots, which facilitates the manufacture of braids with automatically controlled configurations.

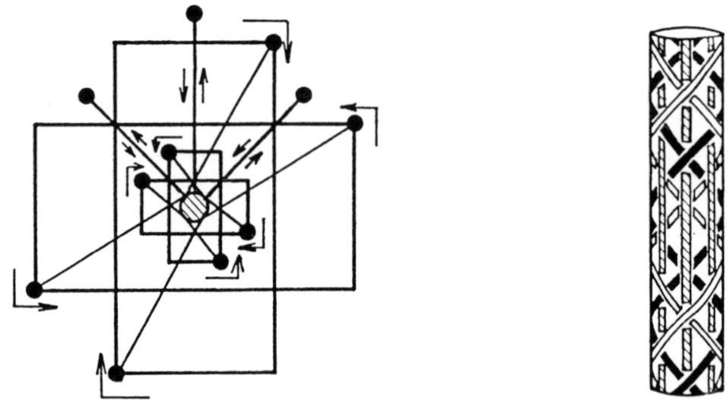

Fig.11 Simulation result.
(Cord braiding fabric with 3 longitudinal binding fibers)

REFERENCES

1) Frank K. Ko, Progress in Science and Engineering of COmposites (1982) 1609.

2) Frank K. Ko, H. Benny Soebroto and Charles Lei, 33rd International SAMPE Symposium (1988) 912.

3) R. T. Brown, 30th National Sample Symposium (1985) 1509.

4) Wei Li and Aly El Shiekh, 33rd International SAMPE Symposium (1988) 104.

5) Robert A Florentine, 33rd International SAMPE Symposium (1988) 922.

6) Z. Maekawa, H. Hamada, T. Horino, A. Yokoyama and Y. Iwasaki, Composite Structures 4, Proceedings of 4th International Conference on Composite Structures 2 (1987) 192.

7) Z. Maekawa, H. Hamada, T. Horino, A. Yokoyama and Y. Iwasaki, Journal of the Textile Machinery Society of Japan 41 (1988) 103.

8) Z. Maekawa, H. Hamada, A. Yokoyama and S. Ueda, Journal of the Japan Society for Composite Materials 14 (1988) 116.

9) A. Yokoyama, Z. Maekawa, H. Hamada, K. Yamaki and Y. Iwasaki, Transactions of the Japan Society of Mechanical Engineers 502 (1988) 1181.

Materials and Processing – Move into the 90's
edited by S. Benson, T. Cook, E. Trewin and R.M. Turner
Elsevier Science Publishers B.V., Amsterdam, 1989

THERMOPLASTIC COMPOSITES, PAST, PRESENT AND FUTURE

G R GRIFFITHS

Business & Technology Manager, Westland Helicopters Ltd, Yeovil, Somerset, BA20 2YB, United Kingdom

1 INTRODUCTION

Thermoplastic composite materials have been readily available from many of the major material suppliers for the past five years. The materials boast enhanced properties, due largely to the toughness of the matrix, together with the ability to be converted into structural components by simple and rapid processes, the latter offering the potential to reduce cost.

This paper describes the work that has been done in the past, reviews the current state of manufacturing technology and indicates the work that needs to be completed in order to launch thermoplastic composites into large-scale economic production.

2 THE ADVANTAGES AND DISADVANTAGES OF THERMOPLASTIC COMPOSITES

The advantages have been well recorded in the past[1] and can be summarised as:-

1 A tough matrix leading to increased component robustness and reduced damage propagation rates
2 Excellent resistance to moisture and solvents by many, but not all, matrix types
3 Freedom from volatiles leading to good quality laminates and good performance in space applications by many thermoplastic materials
4 Very rapid forming of components is possible taking advantage of the thermoplastic nature of the matrix
5 Multiple heating cycles possible, leading to advantages in manufacturing complex shapes, joining methods, scrap rate reduction and repairability.

[1] A C Duthie, Engineering Substantiation of Fibre Reinforced Thermoplastics for Aerospace Primary Structure, SAMPE, March 1988

No new development is without its drawbacks and for these materials, the following seem to be major challenges for the future:-

1 Material costs have come down this year significantly but some of the better materials still command a premium that makes them unlikely to replace thermosets on cost-sensitive applications
2 The need for a cost-effective high-speed, automated method of laminating well-consolidated sheets of complex lay-up is important for making both components and preforms for subsequent processing
3 Joining methods and the associated NDT are in their infancy. Well-developed processes for which consistent data, on large "aircraft-sized" joints, have yet to be demonstrated in a production environment.

3 POTENTIAL APPLICATION AREAS

The range of applications for which thermoplastic composites are now being considered far exceeds that which was anticipated a few years ago. Such applications include:-

1 Aircraft applications, usually where toughness is critical, but cost, EMC and other factors are invariably also important
2 Underwater weapon systems, driven by the need for robustness in a wet environment
3 Space where the lack of outgassing and environmental resistance is essential. Some projects requiring significant production quantities dictate cost-effective processing
4 Missile applications are potentially very promising due to the high stresses, and the need for cost-effective manufacturing in large numbers.

4 COMPONENT MANUFACTURE

4.1 The Past

In the past few years, Westland have manufactured a large number of test specimens for themselves and other companies. These include:-

1 A Lynx Access door in APC-1* (flew in 1984)

2 A Tail plane in APC-2* and PEI+ carbon, bonded with epoxy adhesives[2]

3 A Tetrahedron truss for a Space Application.

These were manufactured by hand lay-up of laminates that were subsequently formed to shape by a variety of methods, rapid press-forming, shown schematically in Figure 1, being the method of first choice wherever applicable. This method is preferred because of the cost saving, due both to the short manufacturing times and the possible use of low-temperature tooling. Joining methods have in the past been either adhesive bonding, with elaborate pretreatments, or mechanical fastening.

4.2 The Present

Two recent developments have brought a new reality to the manufacture of components in these materials at Westland. Firstly, thermoplastic composites have gone into full-scale production for a European aircraft manufacturer. Secondly, joining techniques have developed in which thermoplastic materials are fused directly to one another. These have been demonstrated by a number of organisations using a variety of processes.

The production item mentioned above is an ice protection panel fitted to the side of the Fokker 50 aircraft to protect the fuselage from damage if ice is thrown off the propeller. The panel made from PEI-Kevlar measures some 1.2 m x 1.7 m. The net-shape moulded panel exhibits features such as an edge seal and reinforcements for the window apertures.

[2] Griffiths, Hillier and Whiting, Thermoplastic Composite Manufacturing Technology for a Flight Standard Tail Plane, SAMPE, March 1988

* Carbon/PEEK (Poly-Ether-Ether-Ketone)

+ Poly-Ether-Imide

Much work has been done on joining methods in which the use of adhesives has been eliminated. Instead, thermoplastic materials are fused to one another with the use of films of thermoplastic material or in the case of some amorphous materials by a process analagous to the diffusion bonding of titanium. This process has been further developed to make thermoplastic (PEI) skins bond to nomex honeycomb without the use of an adhesive. The use of ultra-high modulus fibres, such as P75 in PEEK, has been successfully demonstrated in order to build test specimens which feature a zero coefficient of expansion, for space applications.

The other recent innovation has been the development of reliable filament winding, which produces high quality tubes which are impervious to gases and liquids. Tubes of between 50mm and 200mm have made with fibres at angles of 0°, ±45° and 90°. These can now be made directly on the mandrel without the need for post-consolidation in order to remove voids.

4.3 The Future

It was, perhaps naively, believed until recently that many applications would justify a small premium for thermoplastic composite components because of their improved performance. This is only rarely the case and therefore most of the future effort must be directed at cost-effective component production. Factors affecting economic manufacture are: Raw Material; Labour content; Process. More recently, price levels have fallen and now approach those of epoxy-based materials, which is a welcome development.

Figure 2 shows the three stages leading to the production of a structure and the main options available. Rapid forming techniques have been described by many authors including Westland[1]. These offer cheap manufacturing routes since they occupy capital equipment for minimal time (a few minutes) and tooling does not have to withstand high temperatures.

[1] A C Duthie, Engineering Substantiation of Fibre Reinforced Thermoplastics for Aerospace Primary Structure, SAMPE, March 1988

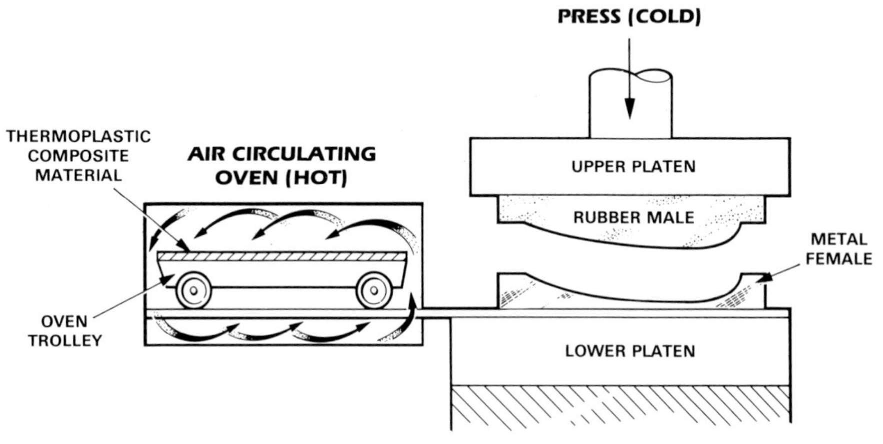

Figure 1
PRESS FORMING OF PRE-CONSOLIDATED LAMINATES

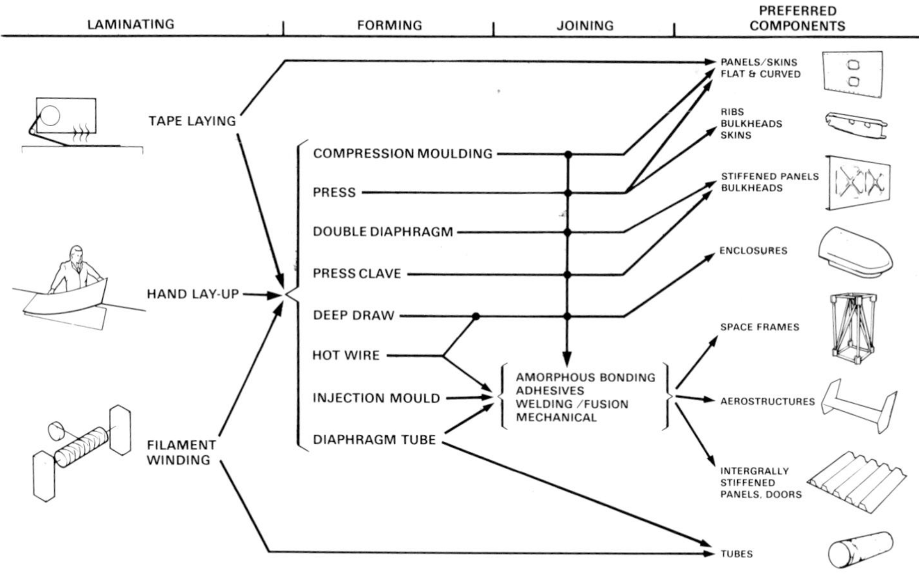

Figure 2
THERMOPLASTIC COMPOSITE MANUFACTURING CAPABILITY

The two missing links which are now receiving the attention of Westland and other R & D facilities, are the mechanised manufacture of well-consolidated and complex laminates and secondly the joining of components either post-forming or during the forming process itself.

The first step towards automatically laminating a preform has been achieved by filament winding. Well-consolidated cylinders have been wound, cut into segments, reheated and then formed into ribs as shown in Figure 3. However, for the economic preform manufacture with ply drop-offs and complex lay-ups, of the type invariably required by aircraft designers, a flat bed tape layer is needed. By the time this paper is presented it is anticipated that such laminates will be laid at up to 6 meters/minute using full width tapes (200mm). Figure 4 shows a schematic of a machine currently being developed at Westland.

Figure 5 shows how the Thermoplastic Press-forming manufacturing route, even with hand-laminating and consolidation, produces components in less time and using capital equipment for a shorter period than with the use of epoxy materials[3]. However, a significant cost saving is dependent on there being only a small premium for the material, over epoxy, and the costing conventions used to value direct labour and the utilisation of capital equipment. No new technology can flourish if it is dependent on the whims of accountants and the pricing policies of the materials suppliers.

However, very significant cost savings will be achieved when tape laying can be used to remove the manual lay-up and press consolidation phase. Today, a twin head filament winder goes some way to meeting this objective but without the ability to insert plys in the middle of a laminate. A considerable amount of effort is being made at Westland and elsewhere to produce a suitable tape layer.

A further stage may be direct lamination of curved components, such as wing skins and perhaps even fuselage sections, although initial design studies indicate that this will not be easy except for very gentle curves.

[3] J A S Whiting, Automated Manufacturing of Thermoplastic Composites

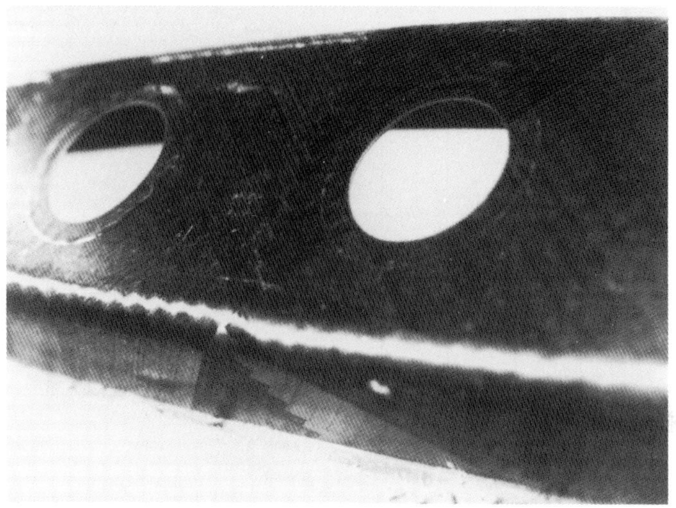

Figure 3
THERMOPLASTIC RIB PRESS FORMED FROM FILAMENT WOUND FEEDSTOCK

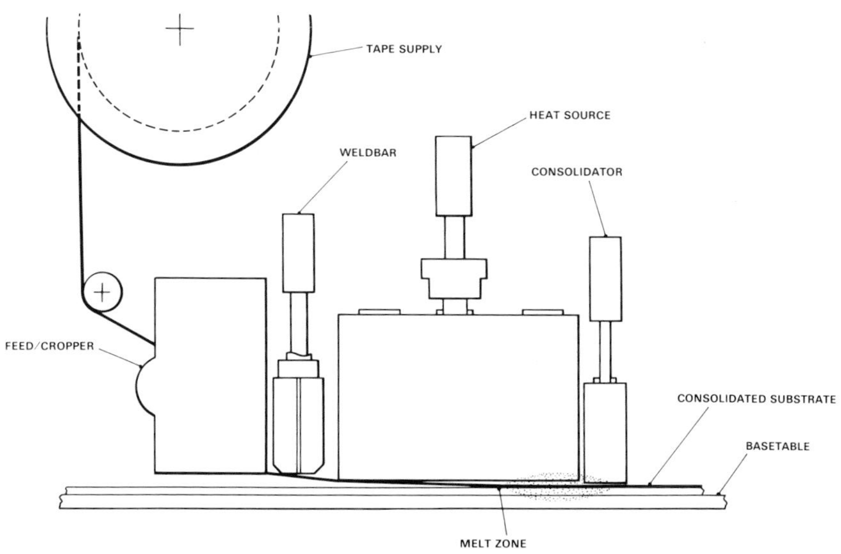

Figure 4
THERMOPLASTIC TAPELAYER

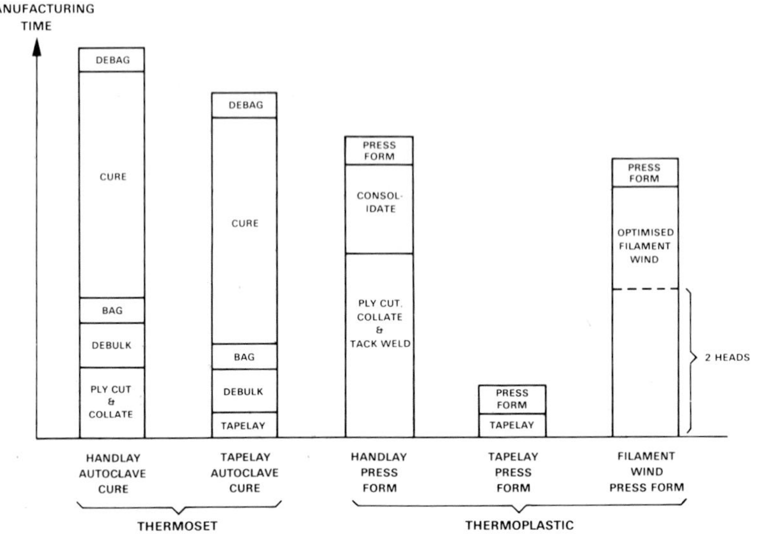

Figure 5
MANUFACTURING TIMES FOR THIN SECTION COMPONENTS

CONCLUSIONS

1 Thermoplastic composites are finding wide-spread applications. In addition to those areas where traditional composites have been used, opportunities to replace metals have occurred because of their increased toughness and moisture resistance.

2 The majority of applications are very cost-sensitive and prices have to be no more than the cost of epoxy components, of many aircraft applicatons.

3 Using manual lay-up methods cost-competitative prices can be achieved for a selected range of components by the use of rapid forming techniques.

4 Major cost savings will be achieved when well-consolidated preforms can be automatically laid at high speed. This technique is very close to being available.

5 The need to design for thermoplastic composite processes is paramount in reducing costs. Consideration of joining methods is a critical area in which further development is necessary.

REFERENCES

1) A C Duthie, Engineering Substantiation of Fibre Reinforced Thermoplastics for Aerospace Primary Structure, SAMPE, March 1988

2) Griffiths, Hillier and Whiting, Thermoplastic Composite Manufacturing Technology for a Flight Standard Tail Plane, SAMPE, March 1988

3) J A S Whiting, Automated Manufacturing of Thermoplastic Composites Automated Composites '88, PRI, September 1988

Materials and Processing – Move into the 90's
edited by S. Benson, T. Cook, E. Trewin and R.M. Turner
Elsevier Science Publishers B.V., Amsterdam, 1989

PROCESSING AND PROPERTIES OF POLYETHERIMIDE COMPOSITES

S. Peake[*], A. Maranci[*], D. Megna[+], J. Powers[+], and W. Trzaskos[+]

*American Cyanamid Company, Chemical Research Division, PO Box 60, Stamford, CT 06904-0060, USA
+American Cyanamid Company, Engineered Materials Technology, Old Post Rd, Havre de Grace, MD 21078, USA

A new class of polyetherimides has been developed for advanced composite applications. In cooperation with General Electric Plastics, polyetherimides were screened for use as composite matrix resins with the following performance targets: ease of processing, resistance to common solvents, and service temperature up to 220°C. Two polymers were selected for further development: one for service up to 150°C and the second up to 180°C. Both are true thermoplastics which lend themselves to rapid forming techniques. Laminates have been formed using pressures as low as 400 kPa and cycle times as low as 3 min. Consolidation temperatures are from 300 to 345°C for the 150 °C service resin or from 350 to 385°C for the 180°C service resin. The mechanical properties of both prepregs are comparable to conventional thermosets, but toughness is much greater. This paper will describe processing conditions and mechanical properties of these polyetherimide composites.

1.0 INTRODUCTION

Thermoplastic polymers are receiving increasing interest as matrix resins for advanced composites. The principal reason is that thermoplastic composites may be shaped or formed using rapid forming techniques such as thermoforming. Because no cure chemistry occurs during the forming process, processing time is dependent only on the time required to achieve complete consolidation. Furthermore, properties are independent of consolidation cycle (at least for amorphous polymers) and laminates may be reconsolidated to correct imperfections. These factors should lead to lower manufacturing costs for thermoplastic composites. A second benefit is derived from the inherent ductility of thermoplastic polymers; this leads to greater toughness by post impact compressive strength than thermoset composites.

A wide variety of engineering thermoplastics, including amorphous and semicrystalline polymers, have been examined in continuous fiber composites. In selecting a resin for aerospace composites, many of the performance criteria are the same as for thermosetting polymers: stiffness, toughness and retention of properties under hot/wet conditions. Issues of particular importance to thermoplastic polymers are consolidation temperature and pressure, solvent resistance, and extent of crystallinity.

For this work, the performance of a series of polyetherimide polymers reinforced with continuous graphite and glass fibers was investigated. Polyetherimides are a class of amorphous condensation polymers which offer a good balance of high temperature and mechanical performance, outstanding fire resistance, and commercial availability. Several grades of polyetherimide resins are sold by General Electric under the Ultem® trademark for injection molding and sheet applications.

Laminates made using standard Ultem resins can be consolidated under low pressures to high quality, void free laminates, which show good in-plane mechanical properties and toughness. In particular, the residual compression strength after impact of an Ultem 1000 resin/graphite composite was 300 MPa.[1] However, Ultem 1000 is not suitable for aircraft structure because it is attacked by a number of solvents. In cooperation with General Electric Plastics, new polyetherimide resins have been identified which are resistant to common solvents and which retain the processing advantages and mechanical properties of Ultem 1000. The processing and properties of laminates made using these novel polymers are described below.

2.0 RESULTS AND DISCUSSION

New polyetherimides were screened by dissolving in N-methylpyrrolidinone and coating the solution on swatches of AS-4 plain weave fabrics. The solvent was evaporated and the prepreg was compression molded at 100°C above the glass transition of the polymer. The resulting composites were examined microscopically to determine laminate quality. If void free laminates were obtained, the laminates were screened for solvent resistance as described below. The final test was to examine the retention of short beam shear strength vs. temperature. A composite with a minimum service temperature of 150°C was the target. Two polymers were selected for further development: a polyetherimide with a glass transition of 217°C for service up to 150°C and a second polymer with a glass transition of 275°C for service up to 180°C. Large scale evaluation was carried out on prepreg containing no solvent which was impregnated by a proprietary process.

2.1 Resin Properties

Resin pellets, obtained from General Electric Company, were compression molded at approximately 100°C above the glass transition. The higher Tg polymer was used in a prepreg designated CYPAC® X7156-1 and the lower Tg polymer in CYPAC 7005 prepreg. Modulus and strength of these resins are comparable; the tensile elongation of the two resins used in CYPAC are less than Ultem 1000. The mechanical properties are compared to Ultem 1000 in Table 1.

2.2 Laminate Fabrication

Both compression molding and autoclave consolidation techniques were used to consolidate prepregs made from the above resins. Temperature was the most important variable in obtaining well consolidated laminates. A temperature at least 70°C above the glass transition is required to consolidate the dry prepreg
and assure adequate resin flow. The temperatures required were 300 to 375°C for CYPAC 7005 and 345 to 385°C for CYPAC 7156-1. For either resin,

TABLE 1 - Neat Resin Properties

Property	Ultem 1000	Resin from CYPAC 7005	Resin from CYPAC 7156-1
Tg by DMA, °C	217	230	275
Tensile Strength, Ult., MPa	105	95	102
Tensile Elongation, %	60	20	15
Flex Modulus, GPa	3.3	3.0	3.1
Flex Strength, MPa	145	130	128
Izod Impact			
Notched, 3.2mm, J/m	50	50	
Unnotched, 3.2mm, J/m	1300	1300	
Specific Gravity, g/cm^3	1.27	1.28	

temperatures above 385°C had a deleterious effect on laminate quality and mechanical performance. Pressure was a less important factor. Good quality laminates were obtained using as little as 400 kPa, but the most reproducible results were obtained at 1000 kPa. Since no cure was occurring during molding, time at the forming temperature could be quite short. Laminates held at the molding temperature from 3-120 min have equivalent quality and mechanical properties.

Because of the relatively high molding temperatures, mismatch in the coefficient of thermal expansion between the laminate and the mold (or caul plate) had a dramatic effect on laminate quality. This was more critical for laminates made from unidirectional tape than for fabrics. When an aluminum caul plate (α=7X10^{-6}/°C) was used, fiber kinking and warpage of the laminate was observed. These problems were absent when a borosilicate glass caul plate (α=3X10^{-6}/°C) was used. In addition, it was critical that the press platens be flat and parallel to achieve void free laminates.

CYPAC 7005 laminates made in this fashion were amenable to thermoforming or thermostamping. Using conditions optimized at General Electric's Plastic Application Center, flat laminates were heated at 285°C then transferred to a mold heated at 120°C. The press was closed at 25 cm/min and 4.8 MPa pressure was applied for 5 sec. Thermostamping was successful using blank temperatures from 285 to 345°C, mold temperatures from 120 to 180°C and molding pressures from 4.6 to 13.8 MPa. The parts thus formed showed smooth surfaces with little evidence of fiber buckling or delaminations in the interior curvatures.

While low cost processing is the primary driver behind thermoplastic composites, autoclave processing may be advantageous in some circumstances. In the initial stages of development for example, existing tooling and equipment can be used rather than invest the time and money necessary to develop thermoforming processes. The tacky prepreg is easier to handle and lay-up with accurate fiber alignment, particularly for complex shapes. For this reason a prepreg was developed containing a solvent to lend tack and drape. This prepreg, designated CYPAC 7000, is based on the same polymer as CYPAC 7005, but it is in the from of a polyamic-acid in N-methylpyrrolidinone solution. The volatile level required to produce tack in the prepreg is 25-28% by weight. The cure cycle is similar to other condensation polyimides (see Figure 1). Mechanical properties of laminates derived from CYPAC 7000 or CYPAC 7005 prepregs are indistinguishable; this suggests that essentially the same polymer is derived from either material.

2.3 Solvent Resistance

The effect of solvent and environmental exposure has been a particular concern for amorphous thermoplastics. Unfortunately, there is no generally accepted method for assessing solvent resistance. We examined the following properties during and after solvent exposure: weight gain, flex creep, and retention of short beam shear strength (see Figure 2). The CYPAC 7005 laminates showed little or no effect when exposed to jet fuel (JP-4) or Skydrol* (hydraulic fluid). A slight amount of flex creep was noted for exposure to 2-butanone (MEK). Severe plasticization was observed for CYPAC 7005 samples exposed to dichloromethane. Nelson and Seferis[2] found that dichloromethane induced crystallization in CYPAC 7005 resin. The higher Tg

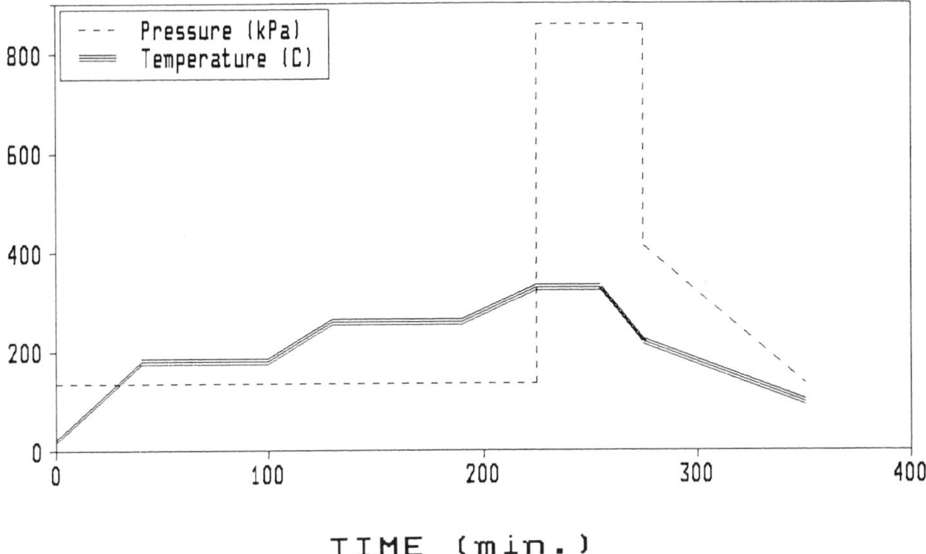

FIGURE 1 - CYPAC 7000 Autoclave Cure Cycle

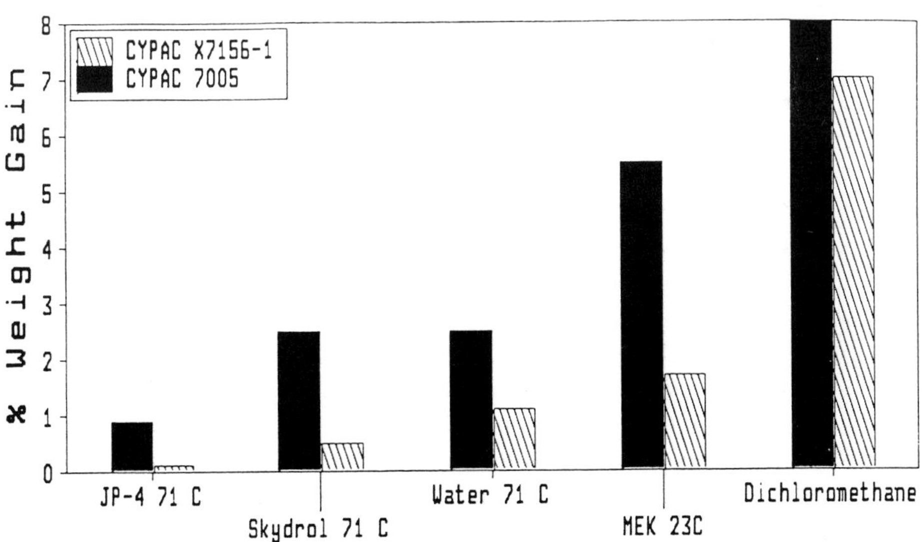

FIGURE 2 - Thirty Day Weight Gain of CYPAC Composites in Solvents

resin, CYPAC X7156-1, does not show any evidence of solvent induced crystallization, and showed improved solvent resistance over 7005 across the board.

2.4 Mechanical Properties

Mechanical properties of CYPAC laminates have been measured for a variety of product forms and reinforcements. The compressive properties for CYPAC 7005/graphite laminates in fabric, uni-fabric and uni-tape are shown in Figure 3. Room temperature properties were comparable to those seen in thermosets, and greater than 50% of the room temperature values are retained up to 150°C.

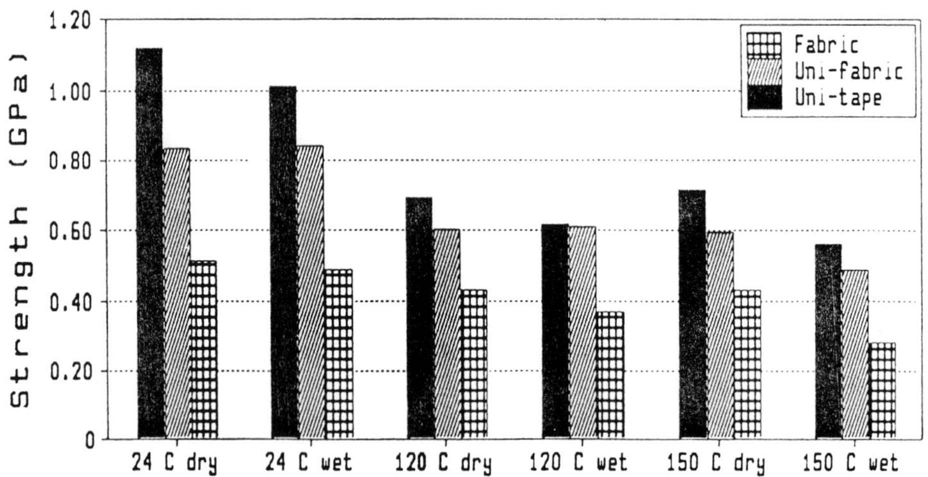

FIGURE 3 - CYPAC 7005 Laminate Compressive Strength vs. Temperature

The toughness as measured by post impact compression strength was 320 MPa for fabric laminates and 290 MPa for tape laminates (impact energy 6.67 J/mm thickness). Fracture toughness, G_{Ic} measures by double cantilever beam, was 4.0 kJ/m^2 for fabric or 2.6 kJ/m^2 tape laminates. Composites were also fabricated using Kevlar, glass, silicon carbide, and quartz fabric reinforcement (Table 3).

The compressive strengths of CYPAC 7156-1 tape and fabric laminates are comparable to CYPAC 7005 at room temperature, but service is extended to 180°C. Retention of compressive strength vs. temperature for CYPAC 7156-1/graphite is summarized in Figure 4. The room temperature compressive strengths of graphite composites are 1240 MPa for IM-8 unidirectional tape and 550 MPa for AS-4 plain weave fabric.

TABLE 2. CYPAC 7005 Fabric Laminate Mechanical Properties

	Graphite AS-4/3K70P		E-Glass 7781 Style		Kevlar* 49	
Laminate Tg, °C	225		225			
Fiber Volume, %	54		52			
Compression Strength,	MPa	(ksi)				
23 °C Dry	503	(73)	554	(80.4)	207	(30)
23 °C Wet	434	(63)	555	(80.5)		
120 °C Dry	400	(58)				
120 °C Wet	365	(53)				
150 °C Dry	414	(60)	355	(51.5)	131	(19)
150 °C Wet	289	(42)	332	(48.1)	97	(14)
177 °C Dry	322	(46)	--	--		
177 °C Wet	228	(33)	--	--		
Compression Modulus,	GPa	(Msi)				
23 °C Dry	56	(8.0)	23.5	(3.41)		
150 °C Dry	54	(7.8)	19.0	(2.75)		
150 °C Wet	54	(7.7)	16.5	(2.39)		
Tensile Strength,	MPa	(ksi)				
23 °C Dry	733	(105)	408	(59.2)	455	(66)
150 °C Dry	768	(100)	330	(47.9)	365	(53)
Tensile Modulus,	GPa	(Msi)				
23 °C Dry	57	(8.1)	22.5	(3.26)	30	(4.3)
150 °C Dry	57	(8.2)	21.8	(3.16)	24	(3.5)
Short Beam Shear Strength,	MPa	(ksi)				
23 °C Dry	66	(9.5)	70.3	(10.2)	33	(4.8)
23 °C Wet	60	(8.6)	70.3	(10.2)		
150 °C Dry	41	(5.8)	38.7	(5.6)		
150 °C Wet	35	(5.0)	28.8	(4.2)	19	(2.7)
Flexural Strength,	MPa	(ksi)				
23 °C Dry	838	(120)	595	(86.3)		
150 °C Dry	559	(80)	418	(60.6)		
150 °C Wet	405	(58)	352	(51.1)		
Flexural Modulus,	GPa	(Msi)				
23 °C Dry	56	(8.0)	21.5	(3.1)		
150 °C Dry	54	(7.9)	20.1	(2.9)		
150 °C Wet	53	(7.7)	18.6	(2.7)		

* Kevlar fiber of E. I. Dupont de Nemours and Co.

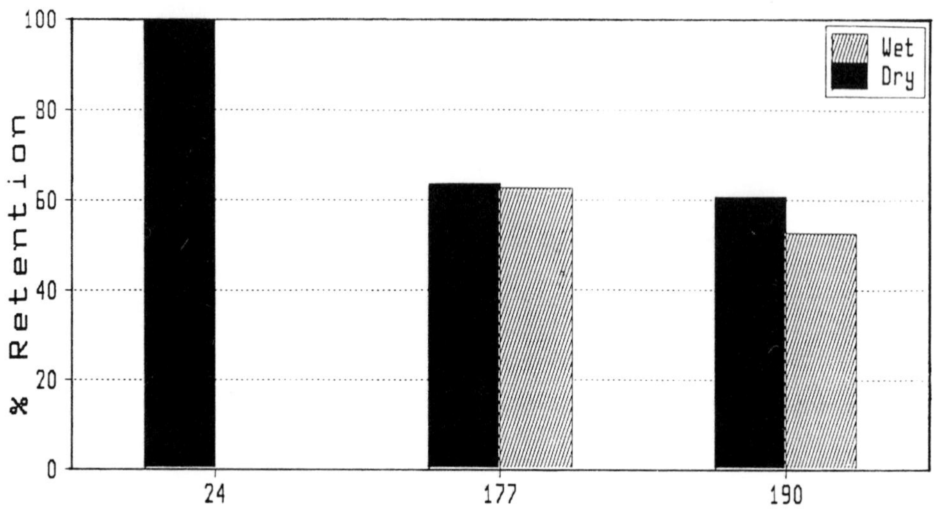

FIGURE 4 - CYPAC X7156-1 Laminate Compressive Strength vs. Temperature

3.0 SUMMARY

Two polyetherimides were developed as matrix resins for advanced composites. The polymers' thermoplastic character permitted rapid and repeated consolidation at temperatures from 300 to 375 C. The composites display adequate solvent resistance and excellent mechanical properties with service up to 180°C. Toughness as measured by post-impact compressive strength was much greater than seen for typical thermoset composites. Thermostamping of consolidated polyetherimide laminates was demonstrated.

ACKNOWLEDGMENT

The authors would like to thank the General Electric Company and Dr. D. C. Bookbinder in particular for their assistance. Mechanical testing and data analysis by A. J. Cronin and S. Kaminski is gratefully acknowledged.

REFERENCES

1. S. L. Peake and A. Maranci, 32nd Int. SAMPE Symp., (1987) 420.

2. K. M. Nelson, J. C. Seferis, and H. G. Zachmann, 34th Int. SAMPE Symp., in print, 1989.

3. D. M. Maguire, SAMPE Journal, **25**(1) (1989), 11.

4. A. Benatar and T. G. Gutowski, SAMPE Quarterly, **18**(1) (1986) 34.

5. W. Trzaskos and J. Powers, 34th Int. SAMPE Symp., in print, 1989; D. Kohli, Proceeding of the 10th International SAMPE Conference, in print July, 1989.

CYPAC and FM are registered trademarks of American Cyanamid Company

Materials and Processing – Move into the 90's
edited by S. Benson, T. Cook, E. Trewin and R.M. Turner
Elsevier Science Publishers B.V., Amsterdam, 1989

THE INFLUENCE OF PROCESSING CONDITIONS ON THE PROPERTIES OF PEEK MATRIX COMPOSITES

Erinann Corrigan David Leach Tim McDaniels

ICI Composites Inc., Fiberite Composite Materials,
2055 E. Technology Circle, Tempe, Arizona 85284, USA

The effect of cooling rate on cystallinity and properties of APC-2 poly(ether-ether-ketone), PEEK/carbon fibre composites is examined. The first two phases of the program are described covering the literature review and processing of the laminates. The cooling rates used in the program were selected as 0.3, 5, 39, and 247°C/min on the basis of previous work and processing experiments.

1. INTRODUCTION

Continuous fibre reinforced, thermoplastic matrix composites are now being used in aerospace and other applications. One such material is APC-2, poly(ether-ether-ketone)/carbon fibre (PEEK/CF) composite[1]. An attractive feature of thermoplastic composites is the opportunity to use a wide range of fabrication techniques in order to manufacture parts more economically[2]. It is possible to use conventional techniques such as autoclave forming, and rapid processing techniques such as hydroforming and sheet stamping.[2] Consequently, parts may have significantly different thermal histories especially in terms of cooling rate.

PEEK is a semi-crystalline polymer and therefore the microstructure and properties may be dependent on processing history[3-5]. A four phase program has been initiated to characterize these effects. These four phases are:

1. Literature Review
2. Identification of processing conditions and manufacture of laminates
3. Assessment of mechanical properties, crystallinity and microstructure
4. Interpretation and further evaluations

Due to the size of the overall program only the first two phases will be discussed in this paper.

There have been several studies of crystallinity and properties in PEEK[3-5] but relatively few studies relating processing to properties in the composite. It has been noted that the composite behaves differently to the neat resin due to differences in thermal properties and nucleation sites[6]. In order to simplify the work, this program will only consider PEEK composites reinforced with Hercules AS4 carbon fiber.

2. EFFECT OF COOLING RATE AND ANNEALING ON CRYSTALLINITY

2.1. Crystallinity vs. Cooling Rate.

Several authors have examined the effect of cooling rate on the crystallinity of the PEEK/CF composite[4,7-13]. A number of techniques may be used to assess crystallinity including Differential Scanning Calorimetry (DSC), Wide Angle X-Ray Scattering (WAXS) and Fourier Transform Infra-red Reflection Spectroscopy (FTIR). DSC is the most common technique but has the lowest reproducibility and accuracy. It has generally been agreed that WAXS is the most accurate technique for assessment of crystallinity[7,8].

The effects of cooling rate on crystallinity from the various studies[7-13] are plotted in Figure 1. Crystallinity decreases from 40% to 30% as cooling rate increases from 0.3°C/minute to 10°C/minute. The crystallinity is between 25-30% for cooling rates in the range 10°C/minute to 600°C/minute. At cooling rates of greater than 600°C/minute the crystallinity decreases though there is little agreement between various authors. Possible reasons for the differences include the difficulty of accurately measuring cooling rate under these conditions, and problems in determining crystallinity from the DSC due to crystallization during the DSC run. These extremely high cooling rates may not be relevant in practice as they can only occur in extremely thin laminates pressed between cold tool surfaces of high thermal conductivity.

2.2. Crystallinity vs. Annealing Conditions

Under certain circumstances it may be necessary to anneal parts. The effect of annealing has been examined in two studies[7,4]. Blundell et al[7] quenched samples into the amorphous state and then annealed for 30 minutes at various temperatures. Annealing temperatures of 200-300°C gave crystallinites of 22-27%. Annealing below 200°C gave very low crystallinity and annealing above 300°C gave crystallinities in excess of 35%. Therefore, there is a broad temperature window for annealing.

Berglund[4] annealed a sample which already had 31% crystallinity, at 310°C for 2 hours and 50 hours. The crystallinities increased to 38% and 41% respectively. Therefore irrespective of the previous processing history it is possible to anneal the composite to achieve a high level of crystallinity.

3. EFFECT OF COOLING RATE AND CRYSTALLINITY ON PROPERTIES

3.1 Glass-Rubber Transition Temperature (Tg).

Only the amorphous regions of the polymer undergo relaxation during the glass-rubber transition. The Tg may still be affected by crystallinity as the crystalline regions impose a constraint on the surrounding amorphous polymer.

The effect of cooling rate on Tg has been examined by Curtis et al[10], Tung and Dynes[11] and Sichina and Gill[15]. The results are summarized in Table 1.

Table 1. Effect of Cooling Rate on Tg of PEEK/CF, APC-2/AS4

COOLING RATE °C/min	GLASS-RUBBER TRANSITION TEMPERATURE (Tg)°C Curtis et al [10]	Tung & Dynes [11]	Sichina & Gill [15]
1	152 ± 2	--	--
1.5	--	163	--
2	--	--	166
3	160 ± 2	--	--
5	--	159	166
19	160 ± 2	--	
20	--	--	163
70	--	164	--
2500	--	171	--
Quenched	--	--	161

Tg is not significantly affected by cooling rate. Within one set of data the largest variation in Tg is 12°C. The results do not show any consistent variation in Tg as a function of cooling rate. Therefore, it can be concluded that the service temperature is not affected by cooling rate.

3.2 Toughness

The effect of cooling rate and crystallinity on toughness has been discussed by Curtis et al[10], Davies et al[13], Berglund[14], and Talbott et al[17]. Mode I interlaminar fracture toughness (G_{1c}) was examined using a double cantilever beam test[10,13,16], a notched three-point bend test[14] and a center notched tensile test[13]. Mode II interlaminar fracture toughness was examined using the end-notched flexure test[10,13,16].

The effect of cooling rate on G_{1C} is shown in Figure 2. There is a tendency for decreasing toughness with decreasing cooling rate, although there is some variation in the absolute values of toughness. The G_{1C} values of Talbott et al[16] on the center-notched specimen are somewhat lower than the values from the DCB tests [10,13]. It should also be noted that the center-notched values were calculated from K_{IC} values and were not measured directly[16]. The two earlier studies[10,16] show a dependency of toughness on cooling rate but the recent study[13] does not show this effect.

PEEK matrix composites exhibit a mixture of stable and unstable (or 'stick-slip') propagation in the DCB test[10,13]. Davies et al[13] reported that the fracture was stable at cooling rates of 1°C/minute and greater, and a mixture of stable and unstable at 0.3°C/minute[13]. This is an indication of slightly reduced toughness at very low cooling rates.

The Mode II interlaminar fracture toughness follows a similar trend to the Mode I values.

3.3 In-Plane Shear Properties.

In-plane shear properties have been examined in two papers[10,12]. Tensile failure strain was reported to decrease from 10.5% at cooling rate of 19°C/minute to 2.5% at a cooling rate of 1°C/minute[10]. This failure strain exceeds the requirements of most practical applications. In-plane shear modulus was not reported but stress-strain curves were identical up to 2% strain over the range 0.3-19°C/minute[10].

In another study the in-plane shear modulus was examined as a function of cooling rate[12] and these results are reproduced in Figure 3. In-plane shear modulus was independent of cooling rate in the range 0.5-50°C/minute, but at higher cooling rates reduced slightly.

3.4 Other Mechanical Properties

Three properties of practical interest are open-hole tension (OHT), open-hole compression (OHC) and post-impact compression (PIC). The effect of crystallinity on these properties has been reported[17] and the results are summarized in Table 2. The OHT and OHC strengths are independent of crystallinity over the range examined. PIC strength decreases with increasing crystallinity. The controlling failure mechanism in PIC is delamination, and the PIC strength follows a similar trend to the interlaminar fracture toughness discussed earlier.

Table 2. EFFECT OF CRYSTALLINITY ON ROOM TEMPERATURE PROPERTIES OF PEEK MATRIX COMPOSITES. Data From Reference 17

CRYSTALLINITY %	OPEN-HOLE TENSILE STRENGTH MPa	OPEN-HOLE COMPRESSIVE STRENGTH MPa	POST-IMPACT COMPRESSIVE STRENGTH* MPa
26	403	290	310
31	401	295	292
34	372	294	273
41	381	292	252

*Impact Energy 30J.

Longitudinal and transverse properties of unidirectional laminates were examined by Grossman and Amateau[12]. Strength was not affected by cooling rate. The modulus in both orientations decreased and failure strain increased at higher cooling rates (>50°C/min). The reasons for these changes were not fully discussed but it was suggested that they were due to variations in fiber microbuckling.

3.5 Discussion

From the previous sections it is apparent that the mechanical properties of PEEK/CF composites are affected by cooling rate and crystallinity under some circumstances. Most of the effects can be explained qualitatively. Higher crystallinity gives increased matrix modulus but reduced matrix ductility. Conversely low crystallinity cause a reduction in matrix modulus and increase in ductility.

4. IDENTIFICATION OF PROCESSING PARAMETERS AND MANUFACTURE OF LAMINATES.

Cooling rates were selected to represent a range of fabrication techniques used to process PEEK/CF. From the previous work, it was anticipated that these cooling rates would result in different levels of crystallinity. An extremely slow rate of 0.3°C/minute was selected to represent the slowest cooling rate experienced during autoclave processing. The intermediate cooling rates (5°C/min and

39°C/min) represent manufacturing techniques such as diaphragm forming, press molding or stamping. A fast cooling rate was chosen to represent the most severe cooling possible in manufacturing environments and still produce a high quality laminate. Some initial experiments were needed to determine the maximum cooling rate achievable with our equipment.

4.1 Experimental

Cooling rates were determined by fabricating a 32-ply, 18" x 18" APC-2/AS4 laminate. Cooling rate was monitored through the thickness and across the plane of the laminate using thermocouples embedded in the panel.

Laminates were consolidated using two 20" x 20" high temperature presses. For the three slower cooling rates heat-up, consolidation and cool-down took place in a single press. For the highest cooling rate both presses were used in a transfer technique.

Temperature was monitored continuously and cooling rate was determined as the rate during the crystallization. Blundell[17] reported that crystallization occurs between 220°C and 270°C, so these temperatures were used to determine cooling rate.

The slowest cooling rate of 0.3°C/min was achieved by turning the press heaters off after consolidation and allowing the laminate to cool by natural losses. The intermediate cooling rates were achieved by using a programmable controller with the press. The fastest cooling rate was achieved by transferring the entire laminate layup from the press at 390°C to a second press at a lower temperature. Initial experimental results indicated that a flexible graphite gasket was required to evenly distribute pressure on the laminate. At high cooling rates, however, the gasket acted as a insulator for the top surface of the laminate and the cooling rate within the panel was uneven. To ensure even cooling, the gasket was used on both sides of the layup. The maximum cooling rate possible in our press was determined by setting the temperature of the second press at 200°C, 143°C and room temperature. The cooling rates achieved in these tests were 247, 426 and 2800°C/min respectively. The laminate transferred to a 200°C press was well consolidated as shown in the photomicrograph in Figure 4. The laminates transferred to a 143°C or a room temperature press were of poor quality and contained a significant amount of voids. Poor consolidation was attributed to the slow closing rate of the press which allowed the laminate to cool and solidify before full pressure could be applied in the second press.

4.2 Manufacture and Inspection of Test Laminates

The test laminates were consolidated using the press molding techniques discussed above at four cooling rates: 0.3°C/min, 5°C/min, 39°C/min, and 247°C/min. To complete the required mechanical property evaluations laminates of the following lay-ups were fabricated for each cooling rate: $[\pm 45]_{2s}$, $[0]_{16}$, $[0]_8$, $[+45/90/-45/0]_{4s}$.

Each laminate was ultrasonically inspected with a pass criteria of -3 db. Crystallinity was determined by DSC and photomicrographs were taken at 250X to evaluate laminate quality. The results of these evaluations are shown in Figure 4. These results indicate no difference in laminate quality due to variations in cooling rate. All laminates passed ultrasonic inspection, had good fiber/resin distribution and no porosity as shown in the photomicrographs. The thickness of all laminates was within 132 ± 8μm.

Crystallinity varied less than expected from previous studies. There was no difference in the crystallinity of laminates cooled at 0.3°C/min and 5°C/min as expected from previous work[7,12,13]. At both of these rates, crystallinity was approximately 33%. At 39°C/min and 247°C/min cooling, crystallinity was 26.8% and 27.8% respectively. The difference between these results is probably due to the error in measurement; therefore, the crystallinity at these cooling rates are equal. Previous investigations have found very little reduction in crystallinity at cooling rates between 10°C and 600°C/min.

5. CONCLUSIONS

Crystallinity varied less than expected with variation in cooling rate. Although previous work indicated a marked increase in crystallinity at cooling rates lower than 0.5°C/min, our study found no significant increase in crystallinity when cooling rate decreased from 5°C/min to 0.3°C/min. Crystallinity did not significantly decrease at higher cooling rates. As expected, crystallinity at cooling rates of 39°C/min and 247°C/min was approximately 27%. Laminate quality was not affected by cooling rate. Laminates molded at 0.3°C/min, 5°C/min, 39°C/min and 247°C/min were well consolidated.

REFERENCES

1. Aromatic Polymer Composite Data Sheets, ICI Fiberite, Fiberite Europe GmbH, Erkelenzer Strasse 20, D-4050 Monchengladbach, W. Germany.

2. J.B. Cattanach, FN Cogswell, Processing with Aromatic Polymer Composites, Developments in Reinforced Plastics - 5, Ed G. Pritchhard, Applied Science Publishers, 1987.

3. D.J. Blundell, B.N. Osborn, The morphology of poly(aryl-ether-ether-ketones), Polymer, 24, (1983) p 953-958.

4. C.N. Velisaris, J.C. Seferis, Heat transfer effects on the processing-structure relationships of polyether-ether-ketone (PEEK) based composites, J. Composite Materials, in print.

5. P. Cebe, S.Y. Chung, S-D Hang, Effect of thermal history on mechanical properties of polyetheretherketone below the glass transition temperature, J. Applied Polymer Science, 33, (1987), p 487-503.

6. D.J. Blundell, R.A. Crick, B. Fife, J.A. Peacock, A. Keller, A.J. Wadden, The spherulitic morphology of the matrix of thermoplastic PEEK/carbon fiber aromatic polymer composites, International Physics Conference Series No. 89, Session 1, Institute of Physics, London, 1988.

7. D.J. Blundell, J.M. Chalmers, M.W. MacKenzie, W.F. Gaskin, Crystalline morphology of the matrix of PEEK-Carbon fiber aromatic polymer composites I. Assessment of crystallinity, SAMPE Quarterly, 16, (1985) p 22-30.

8. M.R. James, D.P. Anderson, Determination of crystallinity in graphite fiber-reinforced thermoplastic composites, Advances in X-ray Analysis, 29, (1986), p 291-303.

9. R.J. Downs, L.A. Berglund, Apparatus for preparing thermoplastic composites, J. Reinforced Plastics & Composites, 6, (1987).

10. P.T. Curtis, P. Davies, I.K. Partridge, J.P. Sainty, Cooling rate effects in PEEK and carbon fiber-PEEK composites, Proc 6th International Conference on Composite Materials, Elsevier, London (1987), p 4.401-4.412.

11. C.M. Tung, P.J. Dyens, Morphological characterization of polyetheretherketone-carbon fiber composites, J. Applied Polymer Science, 33, (1987), p 505-520.

12. S.P. Grossman, M.F. Amateau, The effect of processing on graphite fiber/polyether-ether-ketone thermoplastic composite, Proc 33rd SAMPE Symposium, (1988), p 681-692.

13. P. Davies, W. Cantwell, H. Richard, C. Mowlin, H.H. Kausch Interlaminar testing of carbon fiber/PEEK composites, Proc 3rd European Conference on Composite Materials, Bordeaux, France, March 20-23, 1989.

14. L.A. Berglund, The effect of annealing on the fracture toughness of filament wound carbon fiber/PEEK composite, J. Composites Technology and Research, in print.

15. W. Sichina, P.S. Gill, Characterization of Composites using dynamic mechanical analysis, Proc 33rd SAMPE Symposium, (1988), p 1-11.

16. M.F. Talbott, G.S. Springer, L.A. Berglund, The effects of crystallinity on the mechanical properties of PEEK polymer and graphite fiber reinforced PEEK, J. Composite Materials, 21, (1987), p 1056.

17. The place for thermoplastic composites in structural applications, National Materials Advisory Board, National Research Council, NMAB-435, 1987.

18. D.J. Blundell, B.N. Osborn, Crystalline morphology of the matrix of PEEK-carbon fiber aromatic polymer composites II. Crystallization behavior, SAMPE Quarterly, 17, (1985), p 1-6.

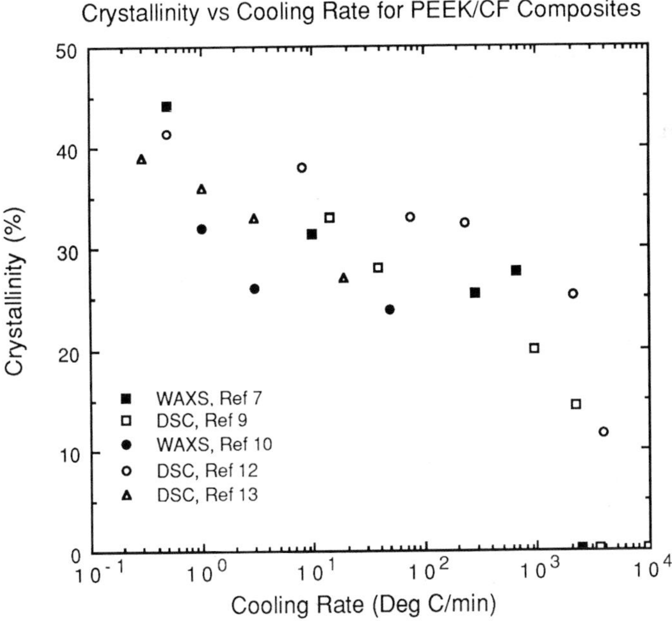

Figure 1.

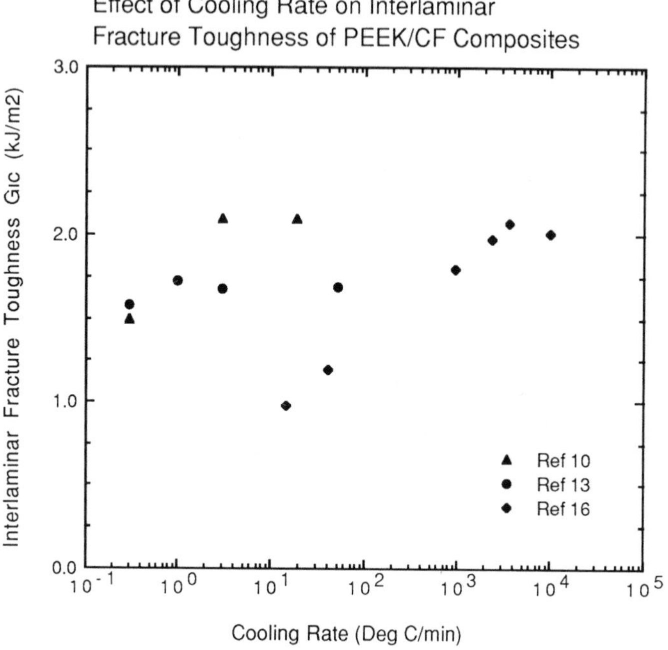

Figure 2.

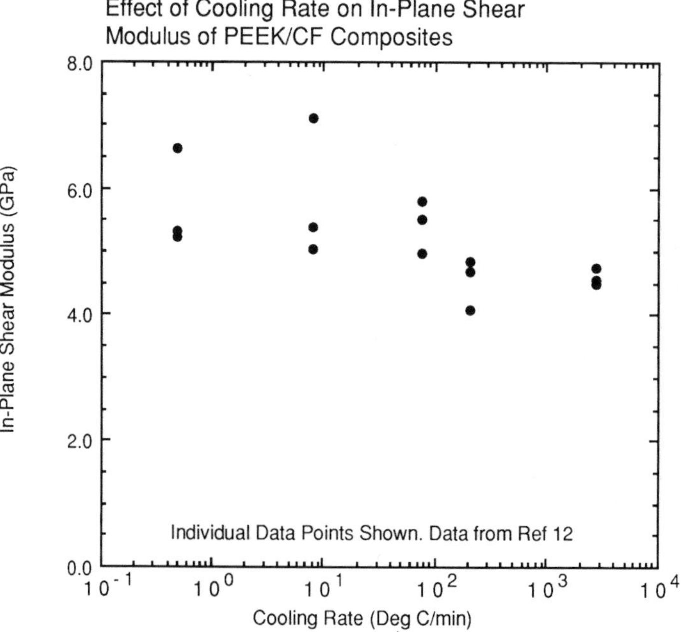

Figure 3.

0.3°C/minute Cooling

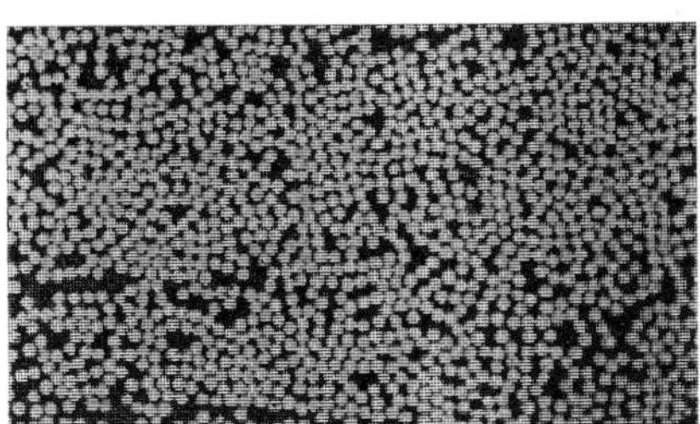

Crystallinity = 32.8%

Average Thickness = 1.03mm

Average Ply Thickness =
129μmm

5°C/minute Cooling

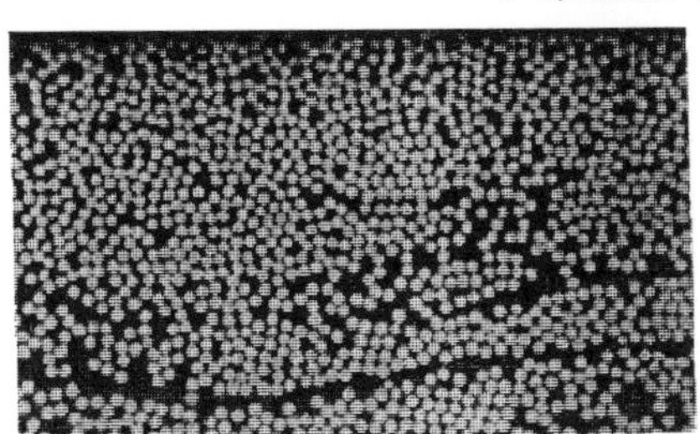

Crystallinity = 32.7%

Average Thickness = 1.02mm

Average Ply Thickness =
128μmm

39°C/minute Cooling

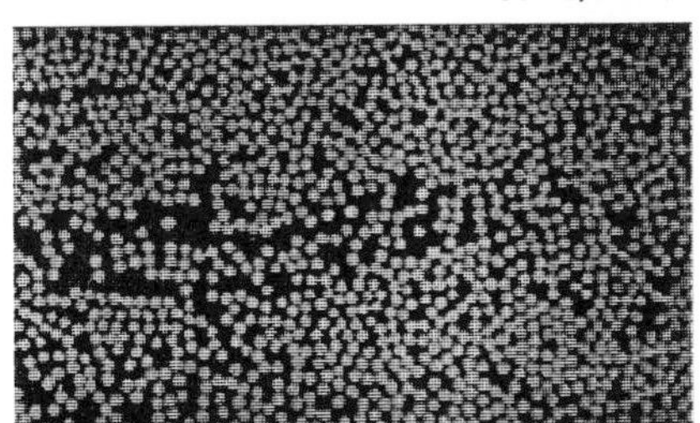

Crystallinity = 26.8%

Average Thickness = 1.04mm

Average Ply Thickness =
130μmm

247°C/minute Cooling

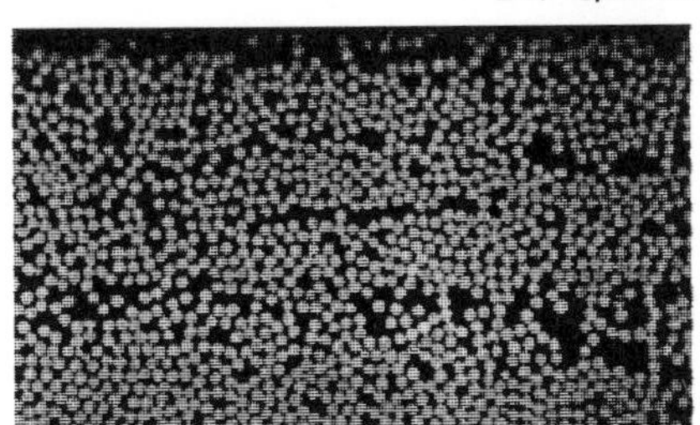

Crystallinity = 27.8%

Average Thickness = 1.03mm

Average Ply Thickness =
128μmm

Figure 4. Laminate Inspection Results

Materials and Processing – Move into the 90's
edited by S. Benson, T. Cook, E. Trewin and R.M. Turner
Elsevier Science Publishers B.V., Amsterdam, 1989

MODELLING OF THERMAL AND CRYSTALLIZATION BEHAVIOR OF THE PROCESSING OF THERMOPLASTIC MATRIX COMPOSITES

A.M. MAFFEZZOLI, J.M. KENNY, L. NICOLAIS

Department of Materials and Production Engineering, University of Naples, Piazzale Tecchio, 80125 Naples, ITALY

INTRODUCTION

The fabrication of thermoplastic based composite laminates is performed by application of heat and pressure under controlled conditions. Heat can be supply by autoclave operation, by radiation or also by circulation of electric current (Joule effect) as in the case of some welding procedures[1]. During the first part of the process the laminate is consolidated by melting the thermoplastic matrix while pressure is applied. In the last part, depending on the cooling rate, different crystalline contents and morphologies can be obtained[1-7]. Moreover, the mechanical properties of thermoplastic matrix composites are related to the crystalline characteristics[6,8,9]. Then, processing of these composites must be optimized in order to obtain adequate temperature and crystallinity profiles, short operation times and good mechanical performance of the formed parts.

A mathematical model for the description of the welding process of PEEK-carbon fiber laminates has been recently proposed[10]. In the present work, a second part of the same study, we extend the proposed model to another typical technology for the processing of PEEK-carbon fiber laminates like press thermoforming. The model accounts for heat transfer inside the composite and phase changes in the polymeric matrix, and for different boundary conditions deriving from the use of different tools. An analysis of process variables is performed in order to optimize processing times and crystallinity characteristics of the formed products.

PROCESS DESCRIPTION

Processing of PEEK-carbon fiber laminate can be performed by using different techniques, tools and process conditions.

Press thermoforming of PEEK matrix composites is usually achieved in two steps. First, the laminate is heated over the crystal melting temperature (near 400 ^{o}C), then it is placed into a mold and pressure is applied. After consolidation the system is cooled down from the molten state to a temperature below the glass transition temperature of the amorphous matrix (< 150 ^{o}C). It has been recommended to use a cooling rate in the range 10-700 ^{o}C/min[1]. In fact, there is little variation in crystallinity in this range[2-6]: The maximum level of crystallinity (near 40%) is obtained by cooling the melt at very low cooling rates (< 1 ^{o}C/min); at moderate cooling rates, between 10 and 700 ^{o}C/min, the final crystallinity level is in the range of 27% - 33%; and by quenching the melt polymer at cooling rates higher than 1000 ^{o}C/min an amorphous material is obtained. Cooling rates during processing are influenced by the use of different tool materials and temperatures and their effect must be taken into account in order to predict the final matrix crystallinity content.

Another case study presented in this work is the welding of PEEK/carbon fiber composite laminates, carried out by electrical heating[1,10]. A prepreg tape (APC-2) is introduced between the laminates and electrical power is supplied through the carbon fibers of the tape. Electrodes are fixed to the exposed fibers which must be treated to remove the polymeric matrix. The heat generated by the filaments, acting as resistance heaters, produces the in situ fusion of the matrix welding the two parts together. A density of current (J) of approximately 600 A/cm^2, considering a value of 1000 $\mu\Omega$cm for the resistivity (R) of the carbon fibers, provides sufficient heat production to obtain welding times on the order of one minute[10]. The inclusion of a thin PEEK film (0.2 mm) between the surfaces, in order to give a degree of resin richness that improves the welding performance, is recommended. During the entire process contact pressure must be maintained in order to consolidate the welding parts.

In the welding process, the electrical heating must be sufficient to melt the PEEK matrix of the single ply, the PEEK film and the PEEK matrix on the surface of the laminates. Welding time must be limited to assure the economical performance of the process. Also in this case optimum processing conditions must be determined in order to produce adequate cooling rates leading to optimum temperature profiles, crystallinity contents, short operation times and good mechanical performance of the welded parts.

CRYSTALLIZATION AND MELTING KINETICS

In order to develop a mathematical model of the processing of PEEK/carbon fiber composites it is necessary to take into account the crystallization and the crystal melting kinetics of the semicrystalline polymeric matrix.

The crystallization kinetic model reported by Velisaris and Seferis[7] has been adopted. In order to describe the non-isothermal crystallization the proposed model accounts for a dual mechanism of crystal nucleation and growth using a linear combination of two integral Avrami expressions:

$$X_{vc}/X_{vce} = W_1\ F_{vc1} + W_2\ F_{vc2} \tag{1}$$

where:

$$F_{vc1} = 1 - \exp\left[-C_{11}\int_0^t T \exp\{-[C_{21}/(T - T_g + 51.6) + C_{31}/T\ (T_{m1} - T)^2)]\}\ n_1\ t^{(n1-1)}\ dt\right] \tag{2}$$

$$F_{vc2} = 1 - \exp\left[-C_{12}\int_0^t T \exp\{-[C_{22}/(T - T_g + 51.6) + C_{32}/T\ (T_{m1} - T)^2)]\}\ n_1\ t^{(n1-1)}\ dt\right] \tag{3}$$

X_{vc} is the crystallinity volume fraction, C_{ij} are model constants, T_g is the glass transition temperature and n_1 and n_2 the Avrami exponents. A value of X_{vce}=0.37 has been assumed[4], being X_{vce} the equilibrium crystallinity volume fraction.

The weight factors W_1 and W_2 in Eq. 1 are practically constant in a broad range of the cooling rate (0.16 - 114 $^oC/s$) and are related by the following relationship:

$$W_1 + W_2 = 1 \tag{4}$$

The model constants C_{ij}, the physical properties of neat PEEK resin and APC-2, the weight factors and the onset of crystallization temperatures are reported in Table I.

TABLE I : Physical properties of neat PEEK resin

Heat of fusion : H_f =	130 J/g
Glass transition temperature : T_g =	144 ^oC
PEEK crystal density : ϱ_c =	1.4006 g/cm^3
Amorphous PEEK density : ϱ_a =	1.2626 g/cm^3
Carbon fiber resistivity : R =	1000 cm

Constants for the crystallization kinetic model[4]

n_1 =	2.5	n_2 =	1.5
T_{m1} =	593 K	T_{m2} =	615K
C_{11} =	2.08E10 $s^{-n}K^{-1}$	C_{12} =	2.08E10 $s^{-n}K^{-1}$
C_{21} =	4050 K	C_{22} =	7600
C_{31} =	1.8E7 K^3	C_{32} =	3.2E6 K^3
W_1 =	0.73 (PEEK)	W_1 =	0.61 (APC-2)

In order to describe the crystal melting kinetics, it has been assumed that the heat rate developed in dynamic DSC scans is proportional to the crystallinity content of the thermoplastic matrix[2]. Then a "degree of melting" (X_f) has been defined as:

$$X_f = 1/Q_T \int_0^t dQ/dt \; dt \tag{5}$$

Where Q_T is the heat of fusion measured in the DSC test and dQ/dt is the heat rate. The relationship between the degree of melting and the volume fraction crystallinity is given by:

$$X_f = (X_{vci} - X_{vc})/X_{vci} \tag{6}$$

Where X_{vci} is the initial crystallinity content of the PEEK matrix.

For the crystal melting a simple n-order kinetic equation has been assumed[10]:

$$dX_f/dt = K(1-X_f)^n \tag{7}$$

Where n is the kinetic order and the kinetic constant, K, is given by:

$$K = K_0 \exp(-E_a/RT) \tag{8}$$

Where T is the temperature, K_0 is the pre-exponential factor and E_a is an activation energy for the crystal melting process. The results of the kinetic analysis are reported in Table II.

TABLE II: Kinetic parameters for melting model

n =	0.5
E_a (KJ/mol) =	397
$\ln K_0$ (s^{-1}) =	73

Experimental data from DSC scans performed on APC-2 composite at different heating rates are well compared with theoretical predictions of Eq. 6 and 7 in Fig. 1.

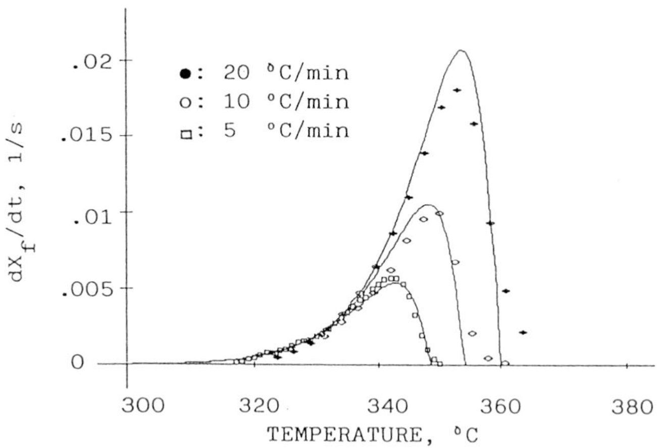

Fig 1: Dynamic DSC thermograms performed at different cooling rates on APC-2. ---- Theoretical model o,•, □ , experimental points.

HEAT TRANSFER MODEL

The final properties of PEEK-carbon fiber composite strongly depend on the crystalline content and morphology of the thermoplastic matrix. Temperature and crystallinity profiles during press forming and welding of APC-2 laminates can be computed by means of a mathematical model. The main part of the model is represented by the energy balance coupled with appropriate expressions for the crystallization and melting kinetics. The model must take also into account the system geometry, the complex thermal diffusivity of the composite and of the tool and the heat developed during crystallization and absorbed during the crystal melting.

The principal assumptions included in the mathematical model are:

- Composite laminates are studied; then only heat conduction in the transverse direction is considered. The geometries assumed for press forming and for electrical welding are shown in Fig. 2. The geometric parameters used in this work for both technologies are reported in Table III.

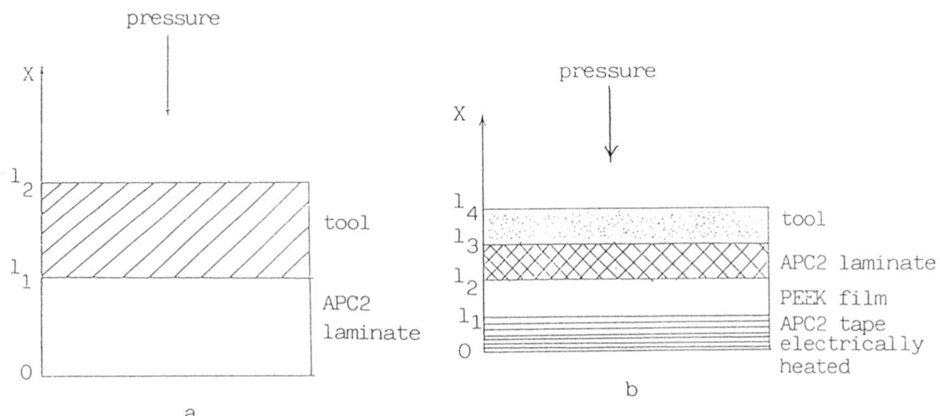

Fig 2: Geometry: A: Press forming; B: electrical welding

TABLE III: Geometric parameters of the welding system

Electrical welding		Press forming	
l_1 =	0.125 mm	l_1 =	2mm
l_2 =	0.2 mm	l_2 =	30mm
l_3 =	2 mm		
l_4 =	30 mm		

- Values of specific heat (C_p), thermal conductivity (k_x) and density (ϱ) of neat PEEK resin and APC-2 measured by Blundell et al.[4] have been adopted.

In the following the subscript a,p and t have been used to characterize the properties of APC-2, of neat Peek resin and of the material of the tool respectively.

The energy balance during the cooling stage, for both the technologies considered, can be written in the following form:

$$\varrho_i C_{pi} \ \partial T/ \ \partial t = k_{xi} \ \partial T/ \ \partial x^2 + H_f \ \varrho_c dX_{vc}/dt \qquad (9)$$

Where the subscript i indicates the layer that is being considered and H_f is the total heat of fusion of the PEEK crystals. Equation 9 must be solved coupled with the crystallization kinetic model reported by Velisaris and Seferis[4] represented by:

$$dX_{vc}/dt = X_{vce}\ d(W_1\ F_{vc1})/dt \qquad \text{for } T < 342\ ^oC \tag{10}$$

$$dX_{vc}/dt = X_{vce}\ d(W_1\ F_{vc1} + W_2\ F_{vc2})/dt \qquad \text{for } T < 320\ ^oC \tag{11}$$

The balance equation for the heat flow in the tool is given by:

$$\varrho_t C_{pt}\ \partial T/\ \partial t = k_{pt}\ \partial T/\ \partial x^2 \tag{12}$$

For the electrical welding also the heating phase of the process has been considered. Two more layers are present in the geometry reported in Fig. 2; therefore during heating the energy balance becomes:

$$\varrho_a C_{pa}\ \partial T/\ \partial t = k_{pa}\ \partial T/\ \partial x^2\ - X_{vci}\ H_f\ \varrho_c dX_{vc}/dt \tag{13}$$

Equation 13 must be solved coupled with the crystal melting kinetic equation (Eq. 7).

For the APC-2 tape electrically heated the energy balance is:

$$t<t_h \qquad \varrho_a C_{pa}\ \partial T/\ \partial t = k_{pa}\ \partial T/\ \partial x^2\ - H_f\ \varrho_c dX_{vc}/dt + P \tag{14}$$

Where t_h is the heating time and P is the heat power generated by the electrical current ($P = R\ J^2$).

The following initial and boundary conditions are assumed:

Initial conditions ($t = 0$):

For press forming:

APC-2: $T = T_0$ $\qquad X_{vci} = 0$

Tool: $T = T_0$

For electrical welding:

APC-2: $T = T_0$ $\qquad X_{vci} = 0.27$

PEEK film: $T = T_0$ $\qquad X_{vci} = 0$

Tool: $T = T_0$

Boundary conditions ($t > 0$)

For press forming:

$$x = 0: \qquad \frac{\partial T}{\partial x} = 0$$

$$x = l_1: \qquad k_{xa}\frac{\partial T}{\partial x} = k_{xt}\frac{\partial T}{\partial x}$$

$$x = l_2: \qquad T = T_0$$

For electrical welding:

$$x = 0: \qquad \frac{\partial T}{\partial x} = 0$$

$$x = l_1: \qquad k_{xa}\frac{\partial T}{\partial x} = k_{xp}\frac{\partial T}{\partial x}$$

$$x = l_2: \qquad k_{xp}\frac{\partial T}{\partial x} = k_{xa}\frac{\partial T}{\partial x}$$

$$x = l_3 : \qquad k_{xa} \frac{\partial T}{\partial x} = k_{xt} \frac{\partial T}{\partial x}$$

$$x = l_4 : \qquad T = T_0$$

RESULTS AND DISCUSSION

The results of the mathematical simulation of the cooling stage, during press forming of a 2 mm thickness APC-2 laminate, are shown in Fig. 3 and 4.

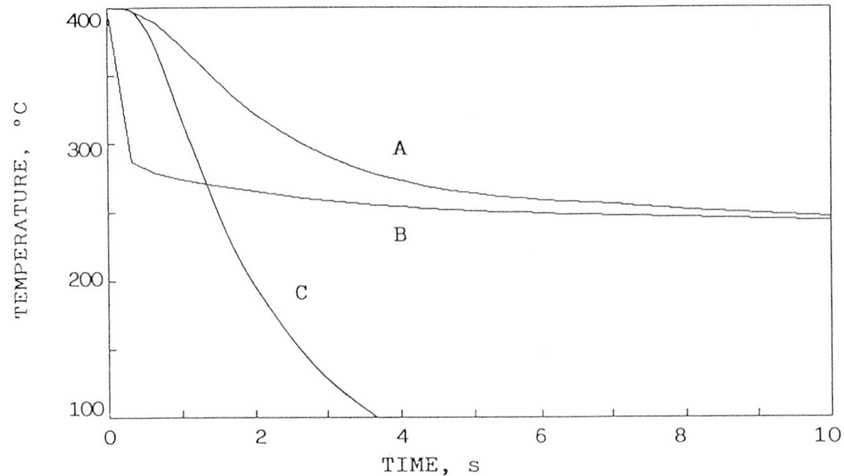

Fig. 3: Temperature vs. time for typical press forming processes and for different process conditions (see text).

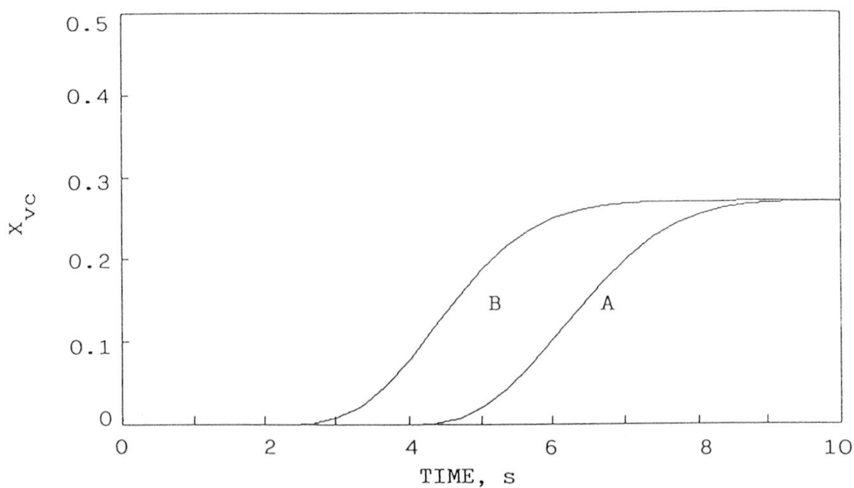

Fig.4: Crystallinity vs. time for the same conditions as Fig.3

It has been assumed that the composite is heated up to 400 ^{o}C and then it is rapidly placed into the press. The influence of different tool temperatures has been studied through the mathematical simulation. Curves A and B represent the temperature profiles calculated at the center and at the interface between the tool (steel at 230°C) and the composite. Smooth temperature profiles are obtained leading the composite the possibility to develop an adequate level of crystallinity. The effect of the tool temperature is shown in curve C. The use of a steel tool at room temperature induces very high cooling rates leading to quenching conditions. Curves A and B shown in Fig. 4 represent the crystallinity content of the composite in the same positions and for the same process conditions as curves A and B in Fig. 3. As expected, good crystallinity levels are reached by the material through the whole thickness; while no crystallinity is developed when the tool at room temperature is used.

The thermal and morphological behavior of the composite during welding process is reported in Fig. 5 and 6.

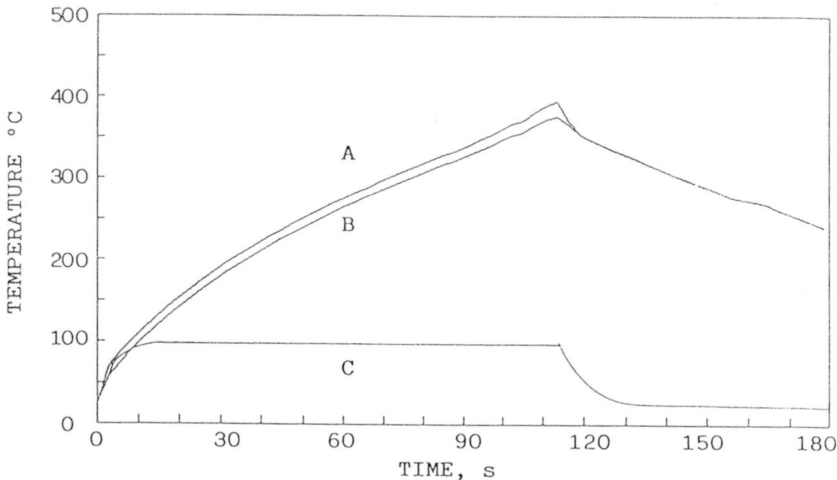

Fig. 5: Temperature vs. time for typical welding processes and for different process conditions (see text).

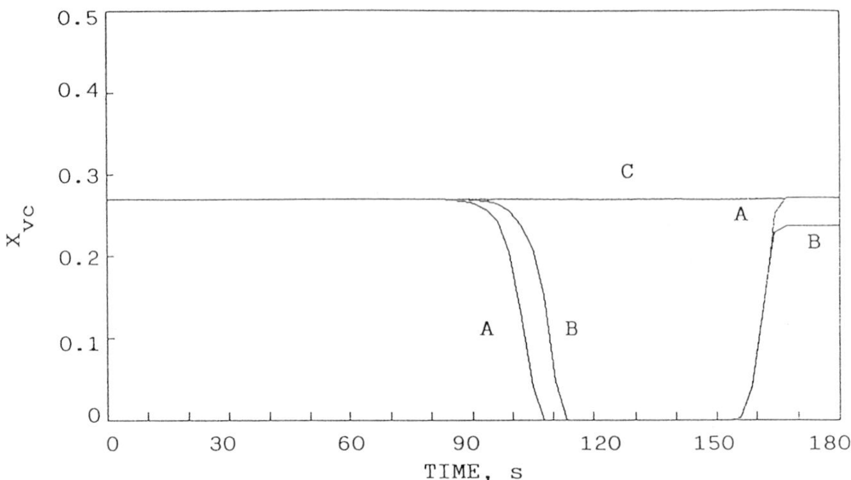

Fig. 6: Crystallinity vs. time for the same conditions as Fig.5

Curves A and B in Fig. 5 show the temperature profiles at the electrically heated APC2 tape and at the interface between the neat PEEK film and the composite laminate, using an asbestos tool at the room temperature. A current density of 600 A/cm^2 is applied for 115 s. The temperature must be kept in the welding zone in a narrow range between a maximum given by degradation conditions (approx. 400 C) and a minimum given by the melting point (350 C). Curve C shows the temperature profile calculated when a steel tool is used to achieve the welding process. Due to the higher thermal diffusivity of the steel, the temperature is not more enough to lead crystals to melt.

The crystallinity profiles obtained with the same conditions used for Fig. 5 are shown in Fig. 6. The initial crystallinity decreases from Xvci= 0.27 to Xvc=0 when the temperature reaches the melting zone (curves A and B). Once the electrical heating is interrupted the moderate cooling rate determines the crystallization of the matrix. Curve B shows a slightly lower value of the final crystallinity content due to the lower effect of the fibers on the crystallinity development at the interface. Curve C corresponds to the crystallinity of the center tape when steel is used as contact tool. In this case, as already noted in Fig. 5 the melting zone is not reached, the crystallinity content remains at his initial value and the welding process is not performed.

CONCLUSIONS

A mathematical model for the description of typical processing technologies of thermoplastic matrix laminates has been presented. The model, accounting for heat transfer inside the composite and phase changes in the polymeric matrix, has been applied to the description of different processing operations like press thermoforming and welding performed by electrical heating. The effect of boundary conditions deriving from the use of different tools has been pointed out. Tool materials and temperatures can be properly chosen in order to obtain adequate crystalline properties.

ACKNOWLEDGEMENT

The financial support from Aeritalia Saipa for the research concerning this work is gratefully acknowledged.

REFERENCES

1. "Fabricating with aromatic polymer composite, APC-2", Data Sheet 5, Fiberite corp., (1986).

2. D.J. Blundell and B.N. Osborn, Polymer, 24 (1983) 953.

3. D.J. Blundell and B.N. Osborn, SAMPE Quarterly, 17 (1985) 1.

4. D.J. Blundell and F.M. Willmouth, SAMPE Quarterly, 17 (1986) 50.

5. P.Cebe and S.D.Hong, Polymer, 27 (1986) 1183.

6. J.C Seferis, Polym. Compos., 7 (1986) 158.

7. C.N. Velisaris and J.C. Seferis, Polym. Eng. Sci., 26 (1986) 1574.

8. S. Saiello J.M. Kenny L. Nicolais J. Mater. Science in print.

9 T. Murayama, Dynamic Mechanical analysis of Polymeric Material, (Elsevier Sci. Publ., N.Y., 1978)

10. A.M. Maffezzoli, J.M. Kenny and L. Nicolais Sampe J., 25 (1989) 35.

Materials and Processing – Move into the 90's
edited by S. Benson, T. Cook, E. Trewin and R.M. Turner
Elsevier Science Publishers B.V., Amsterdam, 1989

FATIGUE TRANSVERSE PLY CRACK PROPAGATION IN FIBER REINFORCED COMPOSITE LAMINATES

Catherine HENAFF-GARDIN and Marie-Christine LAFARIE-FRENOT

Laboratoire de Mécanique et de Physique des Matériaux, URA 863 CNRS
ENSMA, rue Guillaume VII, 86034 POITIERS Cedex, FRANCE.

Two carbon/epoxy laminates ($[0_3/90/0_4]_S$ (A) and $[0_7/90]_S$ (B)) are studied in tension-tension fatigue. The influences of the stacking sequence and the loading amplitude on the transverse ply cracking are emphasized. Transverse cracks are found more numerous but far much shorter in A than in B. The total measured cracked surface rate decreases throughout test. A cracked surface saturation value is reached : it is about twice lower in A than in B, but it is independent of the loading amplitude. Moreover, the initiation cycle number seems to determine the transverse damage state.

1. INTRODUCTION

Fatigue damage in composite laminates consists in the development and accumulation of three main kinds of defects : matrix cracks along the fibers, generally in the off-axis plies, delaminations between plies, and fiber breakage which usually leads to the ultimate failure [1]. During a fatigue test, the first damage observed *on the edge* of the specimen is an increase in the number of transverse ply cracks with the fatigue cycle number. A saturation value of the spacing has been first observed by Reifsnider in 1977[2]. Models of this " Characteristic Damage State" have been proposed [2,3], using a shear lag analysis. This crack spacing saturation value is a characteristic of the laminate, depending on the constituents and on the stacking sequence [4,5], and it was first considered as independent of the loading history. More recently, L. Boniface et al (1987)[7], M.C. Lafarie-Frenot and J.M. Rivière (1988)[8] have shown, in GFRP and CFRP cross ply laminates respectively, that the number of transverse ply cracks measured *on the edge* of the specimen is greater in fatigue than in quasi-static tests, and depends on the fatigue loading levels.

Moreover, the cycle number at which the first crack initiates depends on the loading level[9]. E.Petitpas and col. (1988)[10] have confirmed this property in T300 and T400 carbon/epoxy laminates ; they have proposed to represent it by a "fatigue curve" in which the applied maximum stress is plotted against the cycle number pertaining to the first ply failure. They have shown, too, that the transverse crack initiation rate during a fatigue test depended on the loading level. Finally, the initiation rate and the cycle number necessary to reach the CDS both depend on the test frequency[11].

As transverse cracks initiate and develop, the longitudinal stiffness decreases. Many authors, considering that, at the beginning of a fatigue test, this decrease is only due to transverse ply cracking, have proposed to model it with one microscopic parameter e.g. crack spacing [2,12,13]. When the CDS is reached, and in most of the studied laminates, this model seems to be experimentally confirmed.

Recent observations in GFRP laminates on the initiation and growth of transverse ply cracks during fatigue[14], have shown that cracks nucleate at random throughout the specimen and grow stably as fatigue cracks. Several attempts have been carried out in order to predict crack growth rates in transverse plies. These models have been based on fracture mechanics. So, S.L. Ogin et al[15] have shown that the transverse crack growth rates can be related to a stress intensity factor (depending on the applied stress, on the spacing between cracks and on the ply thickness), by an expression in the form of a Paris law. Some other models have been proposed[16,17] based on the strain energy release rate related to the crack growth. The crack saturation could be explained by such an approach.

The purpose of this *experimental* study is to observe transverse ply cracking in fatigue, and particularly the growing of the cracks in the *width* of the specimens. Cross-ply laminates have been used, and the influence of the stacking sequence and of the loading level has been emphasized.

2. MATERIALS AND EXPERIMENTAL METHODS

Fatigue tests were performed on carbon/epoxy T300/914 laminates which have a nominal ply thickness of 0.12 mm. Two stacking sequences have been studied : $[0_3/90/0_4]_S$ (A) and $[0_7/90]_S$ (B). Both of them have the same monotonic mechanical properties, but the thickness of the 90° ply is either single or double.

Coupons are 180 mm long, 15 mm wide and about 2 mm thick. Glass/epoxy end-tabs 65 mm long have been used for gripping in the Instron servo-hydraulic machine. The specimens were loaded under tension-tension fatigue in a sinusoïdal load controlled testing mode, with a 0.1 load ratio and a 10 Hz frequency. Two fatigue loading levels were studied : 0.8 F_R and 0.6 F_R, F_R being the static failure load.

A microscope (X100) mounted on the testing machine has permitted to count the crack number on the polished edge of the specimen. Complementary microscopic observations have been achieved after unloading at a higher magnification (X 250).

Moreover, the widthwise growth of the transverse ply cracks in the specimen was investigated throughout fatigue tests using an X-ray penetrant technique. Zinc iodide penetrant was applied to both edges of the coupons. A Pantak industrial X-ray unit produced photographs of the specimens (exposure conditions : 13 kV, 35 mA, 42 s).

Note that : - fatigue tests have never led to specimen failure,

- after 10^4 cycles, some edge delaminations have been observed in B laminates but never in A.

3. EXPERIMENTAL RESULTS AND DISCUSSION

3.1. Transverse ply crack spacing on the specimen edge

For each specimen, the edge crack spacing has been measured in each off-axis ply on a 10 mm distance. The evolutions of crack spacing are plotted against the cycle number in figures 1A

($[0_3/90/0_4]_s$) and 1B ($[0_7/90]_s$). The cross-symbols represent the values obtained in situ at 0.8 F_R at a low magnification. The observed scattering is analogous for the two stacking sequences. The other symbols are used to represent the measurements obtained at higher maginfication : the following comments will be based on these values.

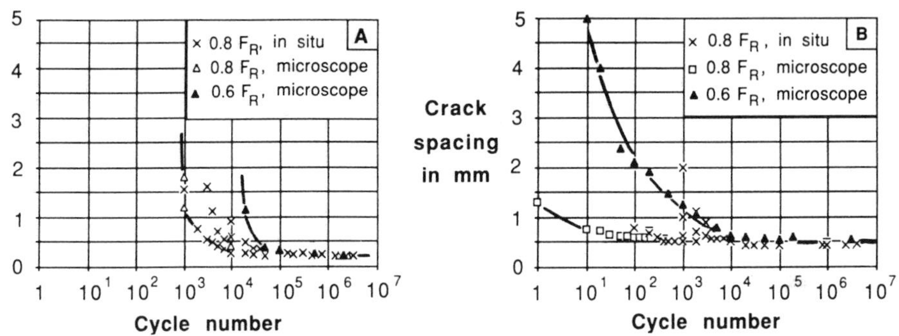

FIGURES 1 A and 1B
Edge crack spacing in mm vs cycle number ; A : $[0_3/90/0_4]_s$;B :$[0_7/90]_s$

For the two stacking sequences and the two loading levels studied, crack spacing evolutions are similar: after initial crack formation, the crack spacing decreases rapidly and finally tends to a saturation value. For both laminates, at a given cycle number, the spacing is lower at 0.8 F_R than at 0.6 F_R. But although the saturation value is independent of the loading level, it is reached earlier when the loading level is higher. These observations are similar to those obtained by Daniel and Charewicz[18], and by Petitpas et al[10] for various graphite/epoxy laminates.

Yet, first cracks appear later in the $[0_3/90/0_4]_s$ laminate than in the $[0_7/90]_s$ laminate.The crack spacing saturation value depends on the stacking sequence and it is proportional to the 90° ply thickness (A : 0.22 mm ; B : 0,44 mm). Moreover, the saturation is reached later in A than in B. In the litterature, many experimental and theoretical studies concerning the influence of the 90° ply thickness can be found : but, in most of them, the loading was a quasi-static tensile one. Our fatigue results reveal a similar tendency :

- the onset of transverse cracking is more difficult when the thickness decreases [19,20,21],
- the larger the 90° ply thickness, the higher the crack spacing saturation value[22].

Following Charewicz and Daniel[23], the results presented here point out the need of improvements in modelling the "fatigue effect".

3.2. Crack cartographies

X-ray radiographs (see for example figure 2) allow the visualization of the 90° ply crack (2 plies in A, 1 in B) along the specimen width. At the beginning of fatigue tests, all the cracks initiate on the specimen edge. Cracks have been classified according to their length. Various length classes have been defined : 1, 3, 5, 7, 9, 11, 13, 15 mm (the specimen being 15 mm wide).

Figure 3 shows the influence on crack distribution of the stacking sequence and the fatigue loading amplitude, for some cycle numbers. In each histogram, the cracks visualized and measured on the radiographs along a specimen 10 mm distance have been reported.The total crack number can be obtained by adding up the x-dimensions of the various rectangles and their total length corresponds to the grey area.

For a given laminate and loading level, the cracks are growing in number and length, with the cycle number, according to previous observations in GFRP[14].

At 0.8 F_R, there are many differences between the two stacking sequences : in A laminate ($[0_3/90/0_4]_s$), there are no cracks at 1 cycle, where as in B laminate ($[0_7/90]_s$), long cracks already exist and even some of them span the entire specimen width. As a consequence, the cartographies are completely different. At the end of fatigue tests (near a few millions of cycles), cracks are more numerous in A laminates than in B ones, but these cracks are far much shorter ! (ultimate average crack length ; A, 0.8 F_R : ~ 2 mm ; B , 0.8 F_R : ~ 8 mm).

In B laminates, the evolution with the cycle number is similar for both loading levels. Yet, this evolution in crack number and length is faster at 0.8 F_R than at 0.6 F_R (ultimate average crack length ; B, 0.6 F_R : ~7mm).

If only specimen edge observations are taken into account, the transverse damage (characterized by edge crack spacing) is more important in $[0_3/90/0_4]_s$ laminate than in $[0_7/90]_s$ laminate. But it is the contrary when the average crack length is considered. As a consequence, it is interesting to define a parameter which depends on both crack number and length. The cracked surface measured in a properly chosen elementary volume seems to us as adequate.

3.3. Cracked surface

The cracked surface has been calculated in a fictitious specimen 1 mm long, i.e. in an elementary volume 1 mm long, 15 mm wide and about 2 mm thick. Note that the obtained values depend on the volume dimensions. This parameter takes into account the number and the thickness of the 90° plies.

Many fatigue tests have been achieved, but very little dispersion has been observed throughout tests, as can be seen in figures 4 and 5 where all the results have been plotted.

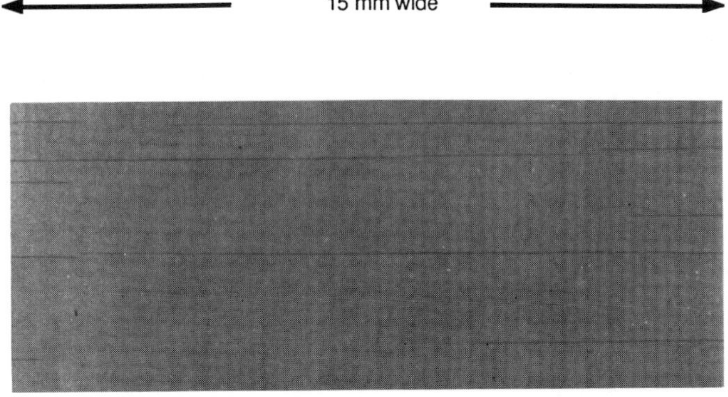

FIGURE 2
X-Ray-radiograph ($[0_7/90]_S$, 0.8 F_R, 1 cycle)

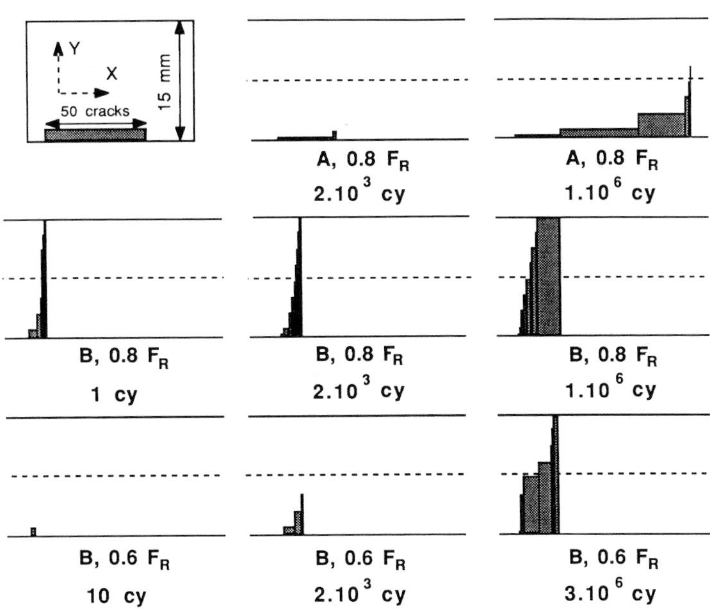

FIGURE 3
Evolution of transverse crack cartographies with the cycle number for various laminates and loading levels.

In figure 4, cracked surface S is plotted against cycle number using linear scales. The cracked surface reaches a saturation value which doesn't depend on the loading level but which is strongly influenced by the stacking sequence. This saturation value is approximately proportional to the 90° ply thickness. Considering the cracked surface values, the increase of the 90° ply thickness is harmful to the specimen, as far as transverse cracking is predominant. Complementary experiments are in progress in order to confirm these primary observations. For both stacking sequences, the cycle number necessary to reach the cracked surface saturation value is lower at 0.8 F_R than at 0.6 F_R.

Figure 5 is a lin-log representation of the same values than in figure 4 ; that way, we focus our attention on the test beginning. It can be seen that, at one cycle, no cracked surface is observed in all the studied cases but one : the B laminate cycled at 0.8 F_R for which the cracked surface value is already high due to the presence of few cracks that instantaneously span the entire specimen width. This is in accordance with Boniface et al' observations[7] in CFRP laminates tested in fatigue. These authors attributed this fast fracture to a cyclic loading above the monotonic load threshold corresponding to the onset of transverse cracking. This convenient interpretation leads us to presume that this threshold depends on the laminate : it would be located between 0.6 F_R and 0.8 F_R for the B laminate, and above 0.8 F_R for the A one.

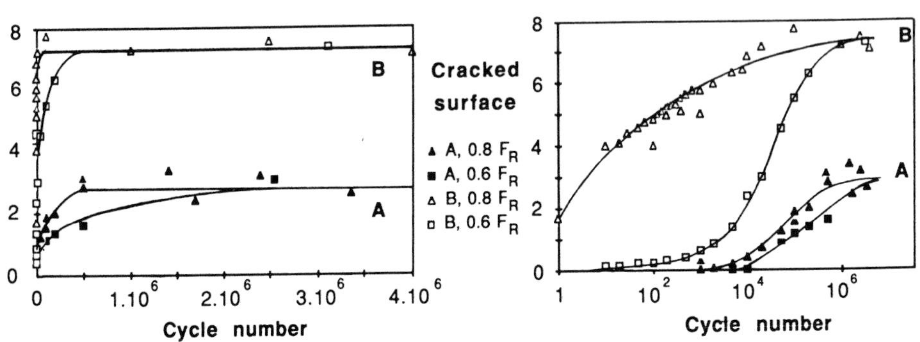

FIGURE 4
Cracked surface (S) vs cycle number (N)
lin-lin representation

FIGURE 5
Cracked surface (S) vs cycle number (N)
lin-log representation

3.4. Total cracked surface growth rate in transverse plies

In order to characterize the cracked surface evolution with the fatigue cycle number, we have calculated the cracked surface growth rate in transverse plies dS/dN defined as follows :

$$\frac{dS}{dN} = \frac{S_{i+1} - S_i}{N_{i+1} - N_i}$$

where S_i is the cracked surface corresponding to the cycle number N_i ;

this value is associated with the $(\frac{N_{i+1}+N_i}{2})$ th cycle . So, it represents the approximate derivative of the growing cracked surface as plotted in figure 4. The results obtained for all the tests are shown in a log dS/dN versus log N diagram (see fig. 6).

In each studied case, the measured cracked surface growth rates belong to a proper scatter band. The main difference between the various cases consists in a very different *initiation* cycle number, represented by a vertical line in figure 6 : this value especially depends on the laminate stacking sequence. So, it is one thousand times higher for A laminates than for B ones, and only ten times higher at 0.6 F_R than at 0.8 F_R. This initiation stage highly determines the consecutive transverse ply damage state ; indeed, as soon as transverse cracks are observed, the cracked surface growth rate can be expressed by a single relationship, as follows :

$\frac{dS}{dN} = \alpha N^{-\beta}$ for the two laminates and the two loading amplitudes.

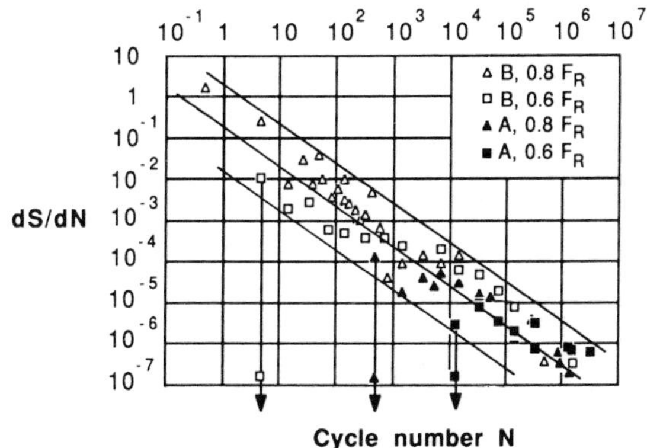

FIGURE 6
Total cracked surface growth rate in transverse plies (dS/dN) vs cycle number (N)

As a consequence :

- the global damage parameter is continuously decreasing throughout the fatigue test. A more detailed analysis of the behaviour of each transverse ply cracks during test permits to explain such a decrease (to be published). In accordance with most of Ogin's assumptions[14], it may be due to (a) no more crack nucleation on the edge, (b) crack annihilation and (c) decrease of individual crack growth rate because of interaction.

- the later transverse plys cracks appear, the slower they grow. Work is in progress in order to verify and explain this result in other cases.

4. CONCLUSIONS

The main results of these fatigue experiments are the following :

i/ when the saturation is reached, the transverse ply cracks in the whole laminate are four times more numerous but very much shorter when the 90° ply thickness is single ($[0_3/90/0_4]_s$), than when it is double ($[0_7/90]_s$).

ii/ considering the total cracked surface, a transverse damage saturation is observed which is higher when the ply thickness is double, but which doesn't depend on the fatigue loading amplitude.

iii/ the total cracked surface growth rate is continuously decreasing throughout a fatigue test.

iV/ the transverse damage stage highly depends on the cycle number at which the first cracks initiate.

ACKNOWLEDGEMENT

We would like to thank here the Aerospatiale Laboratory of Suresnes (France) which provided us with the specimens.

REFERENCES

1) W.W. Stinchcomb, Composites Science and Tech. 25 (1986) 103.

2) K.L. Reifsnider, Some fondamental aspects of the fatigue and fracture response of composite materials, in : Proc. of the 14 th annual meeting of Society of Engineering Science (USA, Bethlehem, 1977) pp.373-384.

3) K.W. Garret and J.E. Bailey, J. of Materials Science 12 (1977) 157.

4) W.W. Stinchcomb and K.L. Reifsnider, in : Fatigue mechanisms, ASTM STP 675 (1979) pp. 762-787.

5) F.W. Crossman and A.S.D. Wang, in : Damage in composite materials, ASTM STP 775, ed K.L. Reifsnider (1982) pp. 118-139.

6) J.E. Masters and K.L. Reifsnider, this volume, pp.40-62.

7) L. Boniface et al, in : Proc. of 6th Int. Conf. on composite materials, ICCM and ECCM, vol. 3, eds. F.L. Matthews et al. (U.K. , London, 1987) pp. 156-165.

8) M.C. Lafarie-Frenot and J.M. Rivière, in : Proc. of JNC6, eds. J.P. Favre and D. Valentin (Paris, 1988) pp.635-647.

9) J. Odorico et al., in : Proc. of JNC5, eds. Bathias and Menkès (Paris, 1986) pp.191-208.

10) E. Petitpas et al, in : Proc. of JNC6, eds. J.P. Favre and D. Valentin (Paris, 1988) pp. 621-633.

11) D.O. Stalnaker and W.W. Stinchcomb , in : Composite materials, testing and design (5th Conf.) ASTM STP 674, eds S.W. Tsai (1979) pp. 620-641.

12) R. Talreja, Fatigue of composite materials : damage mechanisms and fatigue life diagrams, in Proc. of Royal Society of London A 378 (1981) pp. 461-475.

13) A. L. Highsmith and K.L. Reifsnider, in : Damage in composite materials, ASTM STP 775, ed. K.L. Reifsnider(1982) pp. 103-117.

14) S.L. Ogin, in : Proc. of Eur. Symp. on damage development and failure processes in composite materials, eds. I. Verpoest and M. Wevers, (Belgium, Leuven, may 1987) pp. 56-61.

15) S.L. Ogin, P.A. Smith and P.W.R. Beaumont, Transverse ply crack growth and associated stiffness reduction during the fatigue on a simple cross-ply laminate, in : CUED/C/MATS/TR 105, Sept. 1984, Report from Cambridge University , Eng. Dept., Trumpington Street, Cambridge CB2 1PZ, U.K.

16) A.S.D. Wang et al, Composites science and tech. 24 (1985) 1.

17) A. Poursatip, Aspects of damage growth in fatigue of composites, PhD Thesis, Cambridge University Eng. Dept. (1983).

18) I.M. Daniel and A. Charewicz, Eng. Fracture mechanics, 25, (1986) 793.

19) D.L. Flaggs, M.H. Kural, J. Composite materials 16 (1982) 103.

20) A. Parvizi et al., J. Materials science 13 (1978) 195.

21) S.L. Ogin and P.A. Smith, Scripta metallurgica 19 (1985) 779.

22) A. Parvizi and J.E. Bailey, J. Materials science 13 (1978) 2131.

23) A. Charewicz and I.M. Daniel, in : Damage mechanisms and accumulation in graphite/epoxy laminates ASTM STP 907, ed. H.T. Hahn (1986) pp. 274-297.

Materials and Processing – Move into the 90's
edited by S. Benson, T. Cook, E. Trewin and R.M. Turner
Elsevier Science Publishers B.V., Amsterdam, 1989

DAMAGE TOLERANT CARBON FIBER EPOXY COMPOSITES FOR AEROSPACE APPLICATIONS

H. G. Recker,

BASF Structural Materials Inc., Narmco Materials, 1440 N. Kraemer Blvd.,Anaheim, CA 92806

T. Allspach, V. Altstädt, T.Folda, W. Heckmann, P. Ittemann, G. Linden, H.Tesch, T. Weber,

Polymer Research Laboratory, BASF AG, D-6700 Ludwigshafen, Federal Republic of Germany

Based on fundamental studies on thermoplastic modified epoxy neat resins, prepreg resin systems with a significant improvement in fracture toughness and damage tolerance have been achieved. Key investigations focused on understanding and control of morphology within phase separated resin systems. A special morphology was observed that correlates to a distinct maximum in neat resin fracture toughness. Additionally an improved fatigue crack propagation resistance was observed in the case of this morphology. Further understanding of structure-property-relationships within laminates was used to optimize mechanical properties. Highly damage tolerant systems with excellent hot/wet properties and solvent resistance were obtained.

1. INTRODUCTION

Low impact resistance of carbon fiber reinforced thermoset composites is one of the major restrictions for designing load-bearing primary aircraft structures from those materials. To obtain the utmost weight saving potential of advanced composites in aircraft structures, new high-strain carbon fibers and improved toughened polymer matrices for a better delamination resistance of the composites have to be used.

For commercial aircraft applications the service temperature requirements usually do not exceed 120 °C, so that 180 °C- curable epoxy resin systems are prime candidates.

New epoxy prepreg resins for commercial aircraft applications have to be significantly improved in toughness, with sufficiently high values in modulus and glass transition temperature, as well as little sensitivity of these values to moisture.

Numerous attempts have been made to increase the toughness of epoxy matrices for composites. Two frequently used approaches are modification with liquid rubber or reduction of the crosslink density of the thermoset network. In both cases the increase in toughness can often only be achieved at the expense of high temperature performance of the material.

Recently excellent results were obtained by modifying epoxies with engineering thermoplastics[1-5]. However this approach seems to imply various shortcomings such as solvent sensitivity and handling characteristics.

The goal of the present study was to understand the fundamental structure-property-relationships within thermoplastic modified epoxy resin systems.

This understanding was successfully used for developing matrix systems with improved property profiles. The overall performance of these modified thermoset resins meets the requirements for primary aircraft structural applications.

2. STATUS ON THERMOPLASTIC MODIFICATION OF EPOXY PREPREG RESINS

The incorporation of thermoplastics into epoxy resins has already been described in the late 1960s / early 1970s [6]. Various thermoplastics like polyethersulfone and polyetherimide have been suggested, and in most cases no, in other cases some improvements in toughness by addition of a thermoplastic modifier were reported[7,8,9]. But until recently there have been no studies showing that a synergistic combination of the advantageous processing characteristics of thermosets and the toughness of thermoplastics can be achieved.

The modification of epoxy resins with specially synthesized thermoplastic compounds is reported to create significant improved toughness without high temperature performance being sacrificed. Mc Grath et al. found functionally terminated low molecular weight thermoplastics, e.g. polysulfones to be most effective for increasing toughness[1].

Those thermoplastic modifiers, chemically linked to the epoxy network, are superior to those simply physically blended. The toughening effect is reported to be strongly dependent on the molecular weight of the thermoplastic. The morphologies are phase-separated with polysulfone inclusions dispersed in the continuous epoxy matrix. At higher polysulfone loading phase inversion was observed, and the thermoplastic becomes a continuous matrix with dispersed spheres of epoxy [2].

Phase-inversion at high concentrations of a thermoplastic oligomer in an epoxy resin is also reported by Kim and Brown [10].

Sefton et al.[3,4] describe a thermoplastic modified epoxy system which shows optimum toughness properties because of its spinodal, a two-continuous phased morphology. It is shown that the excellent resin toughness translates into a composite exceeding the goal of 0.6 % residual strain-to-failure for primary aircraft applications considerably without compromising hot/wet properties.

In a most recent report Odagiri et al.[5] also described highly damage tolerant composites, which have been achieved by a not exactly defined thermoplastic modification of epoxy resins.

3. EXPERIMENTAL

3.1. Materials

Resin chemistry

The resin systems selected for this study were based on various epoxies including di-, tri- and tetrafunctional epoxies with different backbones. The thermoplastic modifiers with functional endgroups of different molecular weight (Mn), determined by titration of the functional endgroups, and backbone structure were synthesized according to proprietary procedures. In all cases the epoxy was prereacted quantitatively with the thermoplastic modifier. After this procedure the curing agent diaminodiphenylsulfone (DDS) was added.

For basic neat resin mechanical investigations different formulations were prepared by systematically varying the molecular weight of the thermoplastic, and by varying the thermoplastic/ epoxy ratio. Neat resin plaques of 4 mm thickness were prepared by casting degassed resin mixtures between parallel polished steel plates. The mixtures were heated to and held at 180 °C for 2 hours for curing and then post cured for another 2 hours at 200 °C.

In the case of prepreg resin systems, resin viscosities were adjusted to guarantee sufficient tack and drape at room temperature even under normal out time conditions.

Fibers

Two intermediate modulus carbon fibers, Celion G 40-800-12K (BASF Structural Materials, Inc., USA) and Hercules IM7-12K (Hercules Corp., USA) were selected for prepreg fabrication.

Prepreg and laminate fabrication

UD-prepregs were produced by a hot melt process. The fiber areal weight was 145 ± 5 g/m^2, the resin content 33 ± 2 %. Laminates were laid up and cured in an autoclave for 3 h at 180 °C with a heating and cooling rate of 2.5 °C/min. The fiber volume content in the cured panels was 60±3 %.

3.2. Test methods

Neat resin testing. Specimens of the compact-type (CT) configuration were used. Fracture toughness experiments to determine the quasi static value for G_{IC} were performed with a loading rate of 5 mm/min following procedures described previously[11]. Fatigue crack propagation tests were conducted at 10 Hz using identical CT-specimens as with monotonic tests. The applied waveform was sinusoidal with a constant load amplitude and a minimum-to-maximum load ratio, R, of 0.1. The tests were performed under computer control and crack length determination was based upon front face crack opening displacement (COD) measurements combined with analytical solutions for specimen compliance described in the literature[12,13].

Composite testing. All mechanical testing of laminates was done as outlined in Table 1. For further detailed information see references [11,14,15].

Optical and scanning electron microscopy. For morphological investigations thin epoxy films were made by curing the neat resin between two glass carriers.

Hot/wet performance. To evaluate the hot / wet performance, laminate specimens of 1 mm thickness were stored in hot water at a temperature of 70 °C for 14 days. The moisture uptake was monitored gravimetrically. Softening temperatures T_S associated with the onset of the glass transition region of the resin were determined with a DuPont Thermal Analyzer. A test frequency of 1 Hz and a heat-up rate of 4 K/min was applied. The softening temperatures were determined in accordance to DAN 432.

Solvent resistance. Solvent uptake of neat resin systems was determined gravimetrically by immersing samples for 30 days in methylethylketone (MEK). The solvent resistance of laminates was tested by interlaminar shear strength after soaking of specimens for 21 days in MEK, acetone or methylene chloride at room temperature.

Test method	Lay-up sequence	Test procedure
Compression 0°	$[0]_8$	SACMA SRS 1-88
Fract. Toughn. G_{IC}	$[0]_{24}$	Ref. 16
Fract. Toughn. G_{IIC}	$[0]_{24}$	Ref. 17
CAI	$[+45/0/-45/90]_{4s}$	SACMA SRS 2-88
ILSS	$[0]_8$	SACMA SRS 8-88
DMA	$[0]_8$	DAN 432

Table 1: Mechanical test methods for laminates, lay-up sequences, and applied test procedures.

4. RESULTS AND DISCUSSION

4.1. Neat resin fracture energy and morphological investigations

It has been shown for 180 °C-curable single phase as well as for 125 °C-curable single phase and phase-separated CTBN rubber modified systems that there is a strong correlation between composite interlaminar fracture energies under Mode I and the fracture energies of corresponding neat resins[11,14].

Consequently neat resin toughness of thermoplastic modified systems was evaluated by fracture energy measurements under Mode I loading conditions. Additional fatigue crack growth experiments were conducted. Although it is not yet fully understood what the controlling neat resin parameters for interlaminar crack resistance and composite damage tolerance in multiphase epoxy / thermoplastic blends are.

4.1.1 Influence of thermoplastic loading on monotonic neat resin fracture toughness

Different thermoplastic modified epoxy systems with a thermoplastic modifier of optimized molecular weight[18] were prepared for testing the influence of thermoplastic content on neat resin fracture energy. The influence of the thermoplastic content on Mode I neat resin fracture energy is shown in Figure 1. G_{IC} increases with the increase of thermoplastic modifier up to 300 J/m^2 at a concentration of about 23 % thermoplastic. Surprisingly, a pronounced and reproducible G_{IC} maximum of more than 700 J/m^2 is observed at a thermoplastic concentration of about 25 %. Beyond this maximum G_{IC} drops back to its initial level.

4.1.2. Morphological investigations

The G_{IC} maximum in Figure 1 corresponds to a special morphology, in which epoxy domaines from below 1 to about 100 µm are dispersed throughout a thermoplastic continuous phase. The large epoxy domaines themselves are of discontinuous structure with thermoplastic

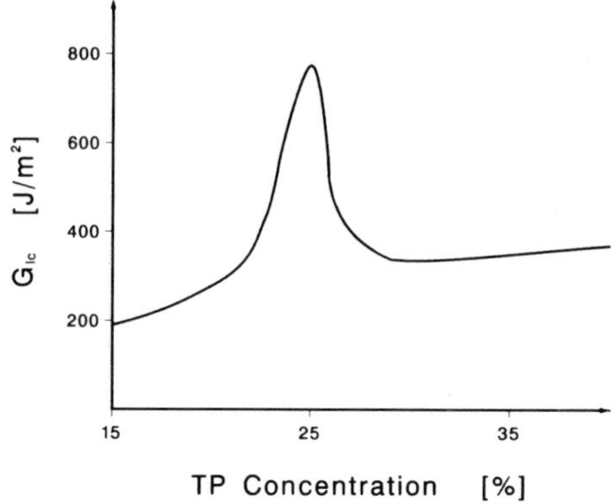

Figure 1: Neat resin G_{IC} of epoxy systems with different concentrations of thermoplastic modifier.

inclusions from about 0.1 to 3µm in diameter, which is clearly shown in a scanning electron microscope photograph of the fracture surface of a K_{IC} test specimen in Figure 2.

The morphologies beside the distinct G_{IC} maximum are characterized by a continuous epoxy phase with discontinuous thermoplastic inclusions for formulations with low loadings of thermoplastic, and by phase-inverted structures with the thermoplastic being the continuous phase and small dispersed epoxy domaines for formulations containing higher loadings of thermoplastic (Figure 3).

4.1.3. Fatigue crack growth experiments

The growth of cracks from preexisting defects under fatigue loading is a major concern in the application of advanced composites in primary structures. Therefore improvement of the fatigue behaviour is an important issue. In Ref.[13] the quantitative interrelationships between characteristic fatigue properties of neat resins and the fatigue delamination behaviour of corresponding composites have been investigated. Significant variations in fatigue crack growth resistance of neat resins and corresponding composites were observed for matrix systems with different neat resin fracture toughness.

As with other isotropic materials, fatigue crack propagation (FCP) rates, da/dN (a is the crack length, N is the number of cycles), in neat polymers are usually related to the cyclic stress intensity factor range, ΔK. From the standpoint of evaluating a material´s fatigue resistance, any decrease in crack growth rates at a given value of ΔK or, alternately, any increase in ΔK to drive a crack at a given speed is, of course, beneficial. The threshold value ΔK_{th} determines whether the existing cracks or defects grow to final fracture. As illustrated in Figure 4, both ΔK_{th} and ΔK_c must be shifted towards higher values to indicate improved fatigue behaviour.

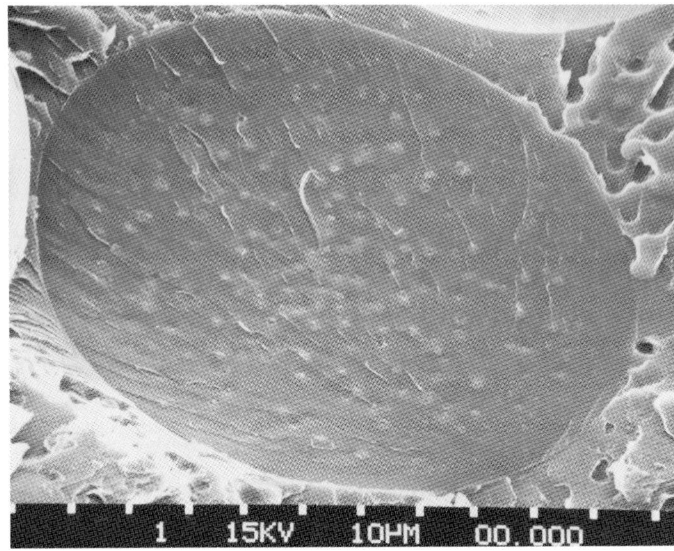

Figure 2: Epoxy domain of about 100 µm in diameter with thermoplastic inclusions of below 3 µm, dispersed in a continuous thermoplastic phase. (Crack surface of a thermoplastic modified (25 %) epoxy system by SEM)

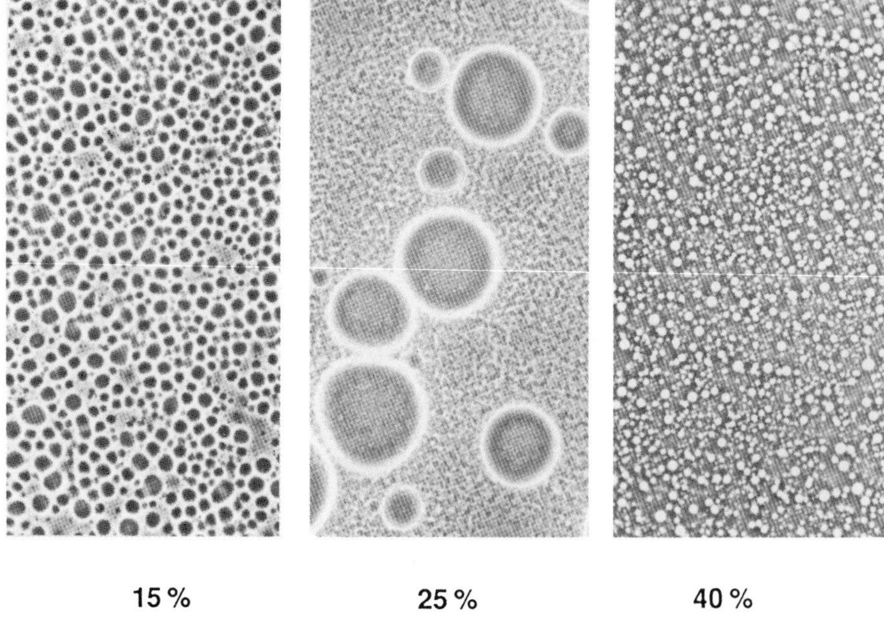

Figure 3: Morphology of thermoplastic modified (15 %, 25 %, 40 %) epoxy resins (Phase contrast light microscopy). [50 µm ├──┤]

FCP rates for three selected resin systems with different content of thermoplastic are plotted in Figure 5 (log da/dN as a function of ΔK). For comparison the FCP curve of Rigidite 5208, a brittle, first generation epoxy system, is added as a reference.

At high crack velocities the ranking of the three resin systems is in direct agreement with the results from monotonic tests. In fact, values of K_{IC} calculated from ΔK_{max} under fatigue loading conditions, are very close to the K_{IC} values determined for monotonic loading. The overall shape of the FCP-curve of the system without thermoplastic is very similar to Rigidite 5208. Once a crack is initiated in these more brittle systems, it will accelerate and grow very rapidly to a critical size at which unstable fracture occurs. The sigmoidal shape of the FCP curves of the systems with thermoplastic shows a clear separation of a threshold region and a stable crack growth range. But in contrast to the system with 25 % thermoplastic, the threshold value of the system with 40 % thermoplastic is as low as with the systems without thermoplastic. Thus, in terms of fatigue crack propagation resistance, there is a clear advantage of the system with the outstanding morphology over the entire crack growth rate range.

4.2. Evaluation of laminate properties

Based on previous neat resin investigations and additional understanding of structure-property-relationships and morphologies several optimized toughened epoxy prepreg resin systems were developed. The following data outline the basic features of two developmental prepreg systems.

4.2.1. Fracture toughness and damage tolerance

Interlaminar fracture toughness of developmental systems, designated A and B, on IM7 carbon fiber are shown in Figure 6. Clear advantages in toughness for the new experimental systems over Rigidite 5245C/IM7 are reflected in improved Mode I as well as Mode II crack growth resistance. For example the system A shows a nearly 50 % improvement in both G_{IC} and G_{IIC} over Rigidite 5245C.

Of special importance for damage tolerant applications is the increase in compression after impact (CAI), which is in good agreement with improved G_{IC} and G_{IIC} fracture toughness. Figure 7 shows the comparison of CAI-data for first and second generation prepreg systems with newly developed systems A and B on the same IM-fiber. The new developmental system A shows a CAI of 310 MPa (45 ksi) at an impact energy of 6.7 kJ/m (1500 inlb/in) versus 200 MPa (29 ksi) for Rigidite 5245C and 150 MPa (22 ksi) for Rigidite 5208. This increase in damage tolerance is also reflected in a substantial damage area reduction for our new system. System B developed for improved hot/wet performance (see 4.2.3.) also turns out to be significantly improved in damage tolerance over Rigidite 5245C.

4.2.2. Comparison of intermediate modulus carbon fibers

In Table 2 interlaminar fracture toughness and damage tolerance performance of system A on two different intermediate modulus carbon fibers (Hercules IM7 and Celion G40-800) are compared. The G40-800 laminate shows a lower G_{IC} fracture toughness than the IM7 laminate, while G_{IIC} and CAI data are comparable in the range of accuracy of measurement.

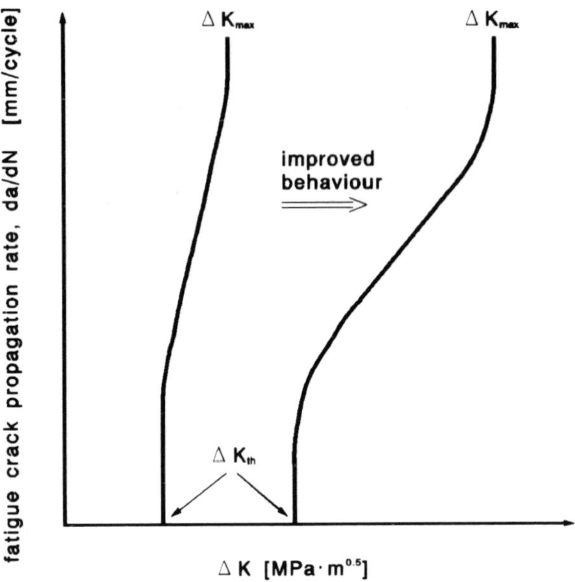

Figure 4: Schematic diagram showing different fatigue crack propagation behaviour.

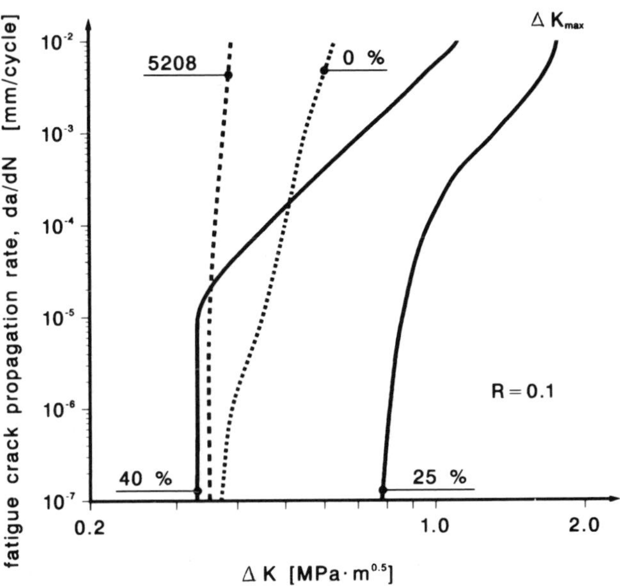

Figure 5: Fatigue crack propagation behaviour of resin systems with different thermoplastics content (0%, 25%, 40%) and Rigidite 5208 as reference.

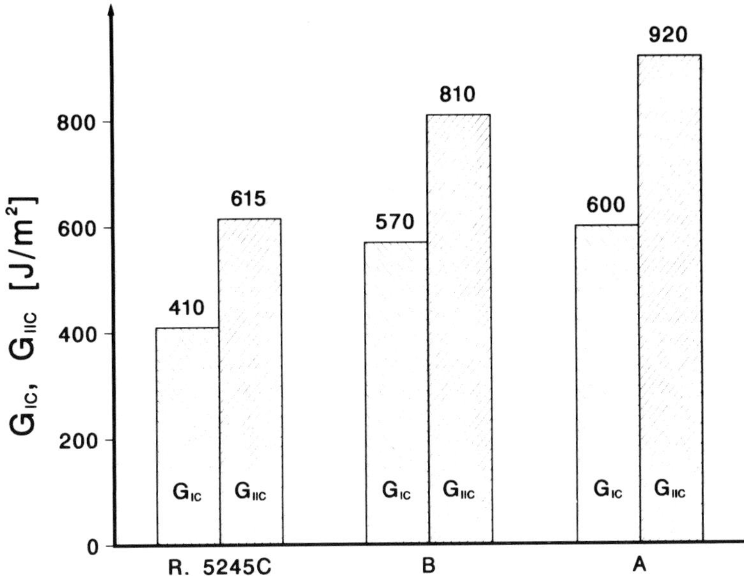

Figure 6: Comparison of interlaminar fracture toughness data G_{IC} and G_{IIC} of developmental systems A and B with Rigidite 5245C on IM7 carbon fiber.

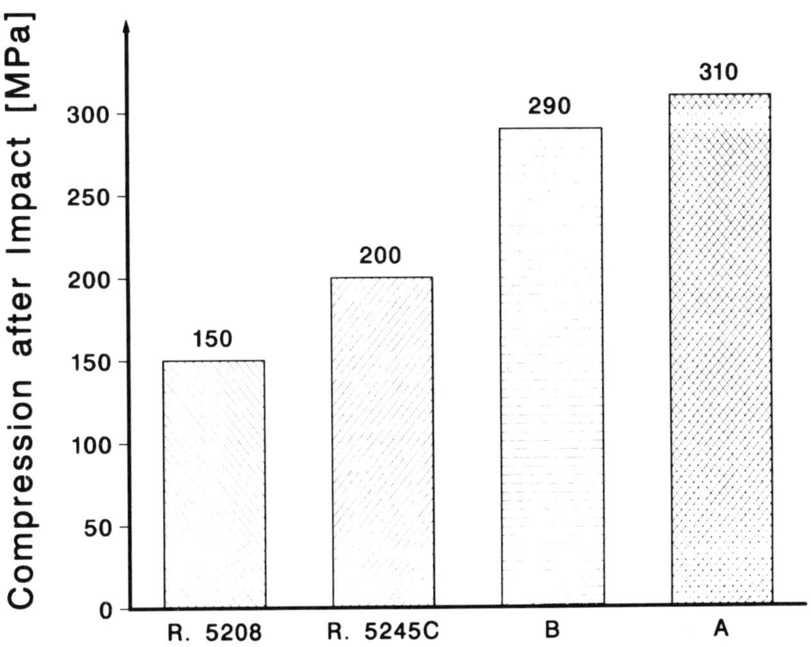

Figure 7: Comparison of compression after impact data of developmental systems A and B with Rigidite 5208 and Rigidite 5245C on IM7 fiber.

4.2.3. Hot/wet performance

While developing highly damage tolerant prepreg systems one frequently faces the problem of generating toughness at the expense of hot/wet performance. Sufficiently high softening temperatures Ts of dry and water immersed specimens of developmental system A in Table 3 indicate that a system with superior fracture toughness and damage tolerance without sacrificing hot/wet properties has been obtained. In the case of system B it was possible to achieve even higher softening temperatures with only a slight decrease in damage tolerance.

The fundamental understanding of the applied toughening concept allows to mutually adjust damage tolerance and hot/wet performance in a certain range.

For overall judgement of high temperature performance of materials hot/wet compression strength data are still to be performed.

4.2.4 Solvent resistance

Solvent resistance is frequently considered to be a major drawback of thermoplastic modified thermoset resin. With systems A and B the problem of solvent sensitivity has been minimized by optimizing the properties of the thermoplastic modifiers, which was already discussed in 4.1.1. Table 4 shows the excellent solvent resistance profile of system A. In solvents like MEK and acetone there is almost no solvent absorption even after three weeks soaking, and therefore ILSS is almost not effected. In the case of methylene chloride the solvent absorption is 2.4 % after three weeks with a 90 % retention in ILSS.

5. CONCLUS IONS

The main goal of this study was the development of a basic understanding of toughening epoxy thermosets with thermoplastics. In order to achieve this goal major contributions were made by detailed monotonic and dynamic fracture toughness and morphological investigations. Key parameters were epoxy backbone structures and functionalities, thermoplastic backbone structure, molecular weight and concentration, as well as the epoxy/curing agent ratio.

Based on these investigations and further understanding of structure-property-relationships a series of new toughened prepreg resin systems was developed. From fracture toughness, damage tolerance, softening temperature and solvent resistance properties of two experimental systems it can be concluded that these new thermoset resins meet the requirements for primary aircraft structural applications.

	Resin system / fiber	
	A / IM7	A / G40-800
G_{IC} [J/m²]	600	450
G_{IIC} [J/m²]	920	940
CAI [MPa]	310	300

Table 2: Fracture toughness and damage tolerance of system A on two different IM carbon fibers.

	Resin system / fiber	
	A / IM7	B / IM7
T_S dry [°C]	217	237
T_S wet [°C]	179	185

Table 3 : Softening temperatures under dry and wet conditions of systems A and B.

	ILSS [MPa]	Solvent absorption [%]
dry (reference)	112	---
MEK	112	0,3
Acetone	105	0,1
CH_2Cl_2	101	2,4

Table 4: Solvent resistance of resin system A/IM7 after three weeks soaking in MEK, acetone and methylene chloride.

REFERENCES

1) J. L. Hedrick, I. Yilgör, G.L. Wilkes, J. E. Mc Grath, Pol. Bull. 13, 201 (1985).
2) J. A. Cecere, J. E. Mc Grath, Polym. Prepr. 27, 299 (1986).
3) M.S. Sefton et al., 19rd Int. SAMPE Techn. Conf., 700 (1987).
4) G. R. Almen et al., 33rd Int. SAMPE Symp. 33, 272 (1988).
5) N. Odagiri et al., 33rd Int. SAMPE Symp. 33, 272 (1988).
6) E.W. Garnish et al., British Patent 1 299 177, Reinforced Composites, January 1969.
7) R.S. Raghava, 28th Nat. SAMPE Symp. 28, 367 (1983).
8) C.B. Bucknall, I.K. Partridge, Polymer 24, 639 (1983).
9) J. Diamant, R.J Moulton, 29th Nat. SAMPE Symp. 29, 422 (1984).
10) S.C. Kim, H. R. Brown, J. Mat. Sci. 22, 2589 (1987).
11) R. W. Lang et al., in "High Tech - the Way into the Nineties", Elsevier Sci. Publ. B.V., Amsterdam, 261 (1986).
12) A. Saxena, S.J. Hudak, Int. J. Fract., 14 (1978) 453.
13) V. Altstädt, R. W. Lang, in "New Generation Materials and Processes", Grafiche F.B.M., Milano, 385 (1988).
14) R.W. Lang et al., in "Looking ahead for Materials and Processes", Elsevier Sci. Publ. B.V., Amsterdam, 109 (1987).
15) SACMA RECOMMENDED STANDARDS, SRS 2-87, Supl. of Adv. Comp. Mat. Associat.
16) Standard Tests for Toughened Resin Composites, NASA RP 1092, Revised Edition (1983).
17) J.D. Barret and R.O. Foschi, Eng. Fract. Mechanics 9, 371 (1977).
18) H.G. Recker et al., 34rd Int. SAMPE Symp. 34, (1989).

Materials and Processing – Move into the 90's
edited by S. Benson, T. Cook, E. Trewin and R.M. Turner
Elsevier Science Publishers B.V., Amsterdam, 1989

MECHANICAL PROPERTIES FOR DESIGN AND ANALYSIS OF CARBON FIBRE REINFORCED PEEK COMPOSITE STRUCTURES

D R Moore, I M Robinson, N Zahlan

ICI plc
Wilton Materials Research Centre
P O Box 90
Wilton, Cleveland, TS6 8JE
UK

F J Guild

Department of Materials
Queen Mary College
Mile End Road
London, E1 4NS
UK

ABSTRACT

The engineering properties, at room temperature, for unidirectional AS4 carbon fibre reinforced PEEK composite, necessary for conducting three-dimensional structural numerical analyses have been determined. These data have been employed in predicting the in-plane engineering properties of two multi-angle laminates of the same material using a computer formulation of the Classical Laminate Theory. The predicted values of axial modulus are found to be overestimated by 7.5% for the $[+45/0/-45/90]_{5S}$ laminate, while other data showed good agreement with experimentally measured values.

Checks of the measured data highlights a desire for improved experimental accuracy, particularly when measuring out-of-plane properties directly. Furthermore, the short-comings of Classical Laminate Theory in accuracy and inability to predict much needed out-of-plane properties are noted.

INTRODUCTION

The process of designing an engineering structure includes consideration of the relationship between structural performance, component geometry and the properties of the material to be used. When the performance requirement is that of structural stiffness, for a linear elastic material and when the geometry is simple, then a closed form expression is usually available for linking these three features. Moreover, two elastic properties often adequately describe the material behaviour; for metals these would be Young's modulus and Poisson's ratio.

The association between performance, geometry and properties becomes complex when the material is anisotropic and visco-elastic and when the geometry is not simple, making a closed form analytical expression elusive. On these occasions it becomes necessary to use a numerical modelling approach in order to accommodate these aspects. To this end, the process is significantly aided by the increased accessibility of computing power. Continuous fibre reinforced composites and their structures are examples of these complex systems which benefit greatly from the application of modelling techniques.

Continuous fibre reinforced composites are inherently anisotropic. Therefore, a set of several mechanical properties is necessary for structural design, even if the performance requirement relates only to stiffness. A further complication arises for continuous fibre reinforced composites prepared from unidirectional pre-pregs. This relates to the numerous stacking possibilities that may be used in the preparation of multi-angle laminates optimised for particular service requirements.

It is accepted that numerical methods are necessary for design of intricate composite structures, but what mechanical properties are required to define the stiffness of a laminate? Moreover, how are these anisotropic properties obtained experimentally? A rigourous approach to this problem would necessitate the use of a general relationship between strain and stress as represented by a matrix containing 36 independent elements[1]. These 36 elements are compliance terms which fully describe the anisotropic nature of the composite. Because composite laminates comprise stacks of unidirectional plies, certain simplifications have been identified. Consequently, the 36 compliance terms reduce to no more than nine independent terms describing an orthotropic laminate.

The nine independent compliance terms could be obtained experimentally. For example, three coupon specimens cut from a unidirectional laminate at mutually perpendicular directions (one of which is aligned with the axis of the fibres) can be subjected, separately, to tensile and shear deformation. A total of twelve engineering properties could be measured from these tests from which nine independent compliance terms could be calculated[2]. If direction 1 is that for the fibres, direction 2 that transverse to the fibres and direction 3 that through the thickness of the coupon then the following engineering properties can be measured:

Tensile moduli: E_1, E_2, and E_3
Shear moduli: G_{12}, G_{23}, and G_{31}
Lateral contraction ratios: ν_{12}, ν_{13}, ν_{23}, ν_{21}, ν_{31}, and ν_{32}

These terms can then be used to calculate the nine independent compliance terms.

In practice, there is a problem! Tensile and torsion tests of coupon type specimens with the applied loading directed through the thickness of a laminate are difficult to achieve. Tests conducted on axial and transverse specimens are manageable. Consequently, only eight of the engineering properties can be measured:

2 Tensile moduli: E_1, and E_2
2 Shear moduli: G_{12}, and G_{23}
4 Lateral contraction ratios: ν_{12}, ν_{13}, ν_{23}, and ν_{21}

These eight engineering properties do not enable the calculation of all of the nine independent compliance terms. Therefore, further assumptions must be made and two options are available.

The first option makes the assumption that laminates are sufficiently thin relative to their other dimensions that through thickness effects can be neglected. This is the plate theory assumption where deformations will be of a plane stress type. For these conditions the nine independent compliance terms reduce to four. Moreover, the eight engineering properties that can be measured do enable calculation of these four compliance terms.

The second option makes the assumption that unidirectional laminates behave in a transversely isotropic manner. That is, the material contains one plane in which the material properties are equal in all directions. The assumption of 'transverse isotropy' reduces the number of independent compliance terms from nine to five. In addition, the eight measurable engineering properties do enable these five independent compliance terms to be calculated.

Traditional design methodology for composites has made much use of the plate approach (option 1). In particular, it uses the approach to determine the engineering properties of multi-angle laminates by application of the Classical Laminate Theory (CLT)[1]. Subsequently, these calculated properties are employed in the design of structures in continuous fibre reinforced materials, using available composites design tools and methodology.

Several issues now emerge:

(i) Is the plate theory assumption acceptable for the calculation of in-plane engineering properties of multi-angle laminates? In particular, this point should be addressed in a way that accommodates all modern day materials, especially thermoplastic composites which, when utilised to their full potential, can exhibit large through thickness deformations by nature of their enhanced toughness[4].

(ii) Is the assumption of 'transverse isotropy' acceptable for the description of three-dimensional mechanical behaviour of unidirectional composites?

(iii) What is an appropriate experimental plan for the determination of the engineering properties required to calculate the compliance terms?

These issues constitute the content of this paper.

MATERIALS AND EXPERIMENTAL METHODS

Continuous carbon fibre reinforced Poly Ether Ether Ketone (PEEK) is the subject of this experimental study. This PEEK semi-crystalline thermoplastic based composite contains 61% by volume Hercules AS4 carbon fibres and was prepared into laminates from prepreg following standard procedure[3]. All laminates were inspected and found to be of satisfactory quality using standard ultrasonic C-scan techniques.

The engineering properties described in the previous section were measured using tensile and torsion tests. The tests were conducted on three different laminate configurations:

(i) Unidirectional laminates, $[0]_{40}$ and $[0]_{16}$
(ii) An angle ply laminate, $[+45/-45]_{10S}$
(iii) A quasi-isotropic laminate, $[+45/0/-45/90]_{5S}$

The tensile tests were conducted on coupon specimens, 283 mm in length, 12.5 mm in width, and 5.2 mm thick, cut from laminates using a diamond impregnated slitting disc. End tabs were bonded to the specimens using a thin film adhesive (Redux 319A). The end tabs were prepared from a glass epoxy composite and included tapers of 15 degrees in order to reduce stress concentrations in the specimen clamping regions. The tensile tests were conducted, at 23°C, using an Instron 6025 test machine at a displacement rate of 1mm/min. Three extensometers were attached to each specimen in order to measure the strains in three mutually perpendicular directions. Details of the test technique are published in reference 5; whilst Figure 1 illustrates the instrumentation. We believe that using these procedures, a tensile modulus can be determined to within an accuracy of 3%, and a lateral contraction ratio to within an accuracy of 10%. Repeat experiments can then account for inter-specimen variability.

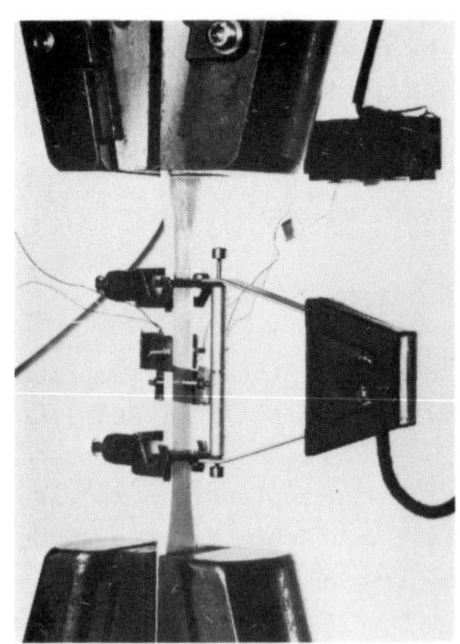

FIGURE 1
Axial Test

The shear moduli were determined from two torsion tests of rectangular section bars cut from the 16 ply unidirectional material. Two types of specimen, each about 2.2 mm thick, were machined to a length of 140 mm and a width of 10 mm with the specimen axis oriented along (0° orientation) and across (90° orientation) the fibre direction. A twisting moment was applied to each specimen mounted in an apparatus as shown in Figure 2, further details of which can be found in reference 6. The shear moduli of the material can be determined according to the formula[7]:

$$G_{ij} = \frac{TL}{K\theta}$$

Where G_{ij} is the 'ij' shear modulus, i and j being the directions perpendicular to the specimen axis,

T is the torque applied to the specimen,

L is the specimen length,

K is a specimen specific function,

and θ is the angular twist of the specimen.

The function 'K' is dependent on the specimen geometry as well as the material behaviour. In the case of the 0° orientation specimen, the transverse isotropy assumption reduces the function 'K' to its form applicable to an isotropic material, dependent only on specimen geometry. The fuction is calculated from summation of a quickly converging infinite series[7]; therefore, calculation of G_{12} is a simple matter. For the 90° orientation specimen, the orhtotropic nature of the material must be taken into account; the applicable analysis is presented in reference 8. In addition to dependence on the geometry, the infinite series is dependent on the two moduli G_{12} and G_{23}. Determination of G_{23} is achieved through iterative solution making use of the previously calculated value of G_{12}.

This procedure provides a shear modulus value at small strains (less than 0.1%) to within an accuracy of 2%.

FIGURE 2
Torsion Test Apparatus

PRESENTATION AND DISCUSSION OF EXPERIMENTAL RESULTS

The experimental results obtained from the tensile and torsion tests performed on the unidirectional and multi-angle laminates are presented in Table 1.

TABLE 1
Experimentally Determined Data For Three Laminates For AS4-CF/PEEK Composite

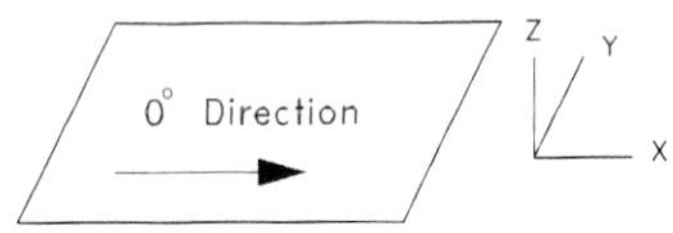

Laminate Configuration			$[0]_n$	$[+45/-45]_{10S}$	$[+45/0/-45/90]_{5S}$
Axial Modulus	E_x	GPa	135.3	18.4	48.6
Axial Modulus	E_y	GPa	9.0		
Width Contraction Ratio	ν_{xy}		0.34	0.74	0.30
	ν_{yx}		0.024		
Thickness Contraction Ratio	ν_{xz}		0.28	0.21	0.34
	ν_{yz}		0.46		
Shear Modulus, in-plane	G_{xy}	GPa	5.2		
Shear Modulus, out-of-plane	G_{yz}	GPa	1.9		

Axial moduli are determined within an accuracy of $\pm3\%$.
Lateral contraction ratios are determined within an accuracy of $\pm10\%$.
Shear moduli are determined within an accuracy of $\pm2\%$.

The tensile moduli and lateral contraction ratios were those determined within the linear regions of the stress-strain relationship, which were up to about 0.8% strain for unidirectional and multi-angle laminates, and up to 0.4% for the angle-ply laminate. Beyond these strain ranges, the relationship should not be assumed to be linear; in-deed, observation of the stress-strain curve for the angle-ply laminate indicates a particularly non-linear relationship.

The design of the experimental procedure provides more data than are necessary for the calculation of five independent compliance terms which describe a transversely isotropic material. This excess allows verification of the stated assumptions and the test methods. Additionally, it allows selection of an appropriate set of engineering properties which describe the material behaviour.

The assumption of transverse isotropy implies that the lateral contraction ratios ν_{12} and ν_{13} should be equal. The values measured, and quoted in Table 1, for these two properties appear to be different. This could be an indication that the material can not be assumed to behave mechanically in a transversely isotropic manner. Alternatively, it could be due to experimental error, particularly since measurement of very small distortions is involved. Results of further investigation of this issue will be published shortly.

Furthermore, the condition of symmetry of the compliance matrix suggests a reciprocal relation of the form[1]:

$$\frac{\nu_{21}}{E_2} = \frac{\nu_{12}}{E_1}$$

From this equation and the experimentally determined data, a value for ν_{21} of 0.022 is calculated; this compares well with the measured value of 0.024.

It is possible to calculate the value for in-plane shear modulus, G_{12}, using data obtained from a uniaxial test conducted on a specimen of $[+45/-45]_{nS}$ laminate by employing the equation[9]:

$$G_{12} = \frac{E_{xy}}{2(1 + \nu_{xy})}$$

where E_{xy} and ν_{xy} relate to the test conducted on the angle-ply specimen. Application of this formula to the present data suggests a value for in-plane shear modulus, G_{12}, of 5.3 GPa, well within the error bounds for the value measured directly and reported in Table 1 (5.2 GPa).

Guided by the above discussion and levels of confidence quoted for the individual measurements, it is possible to select a complete set of data suggested for use when modelling structures utilising AS4-CF/PEEK composites; these data are listed in Table 2.

TABLE 2
Selected Engineering Propertie Set For
AS4-CF/PEEK Composite

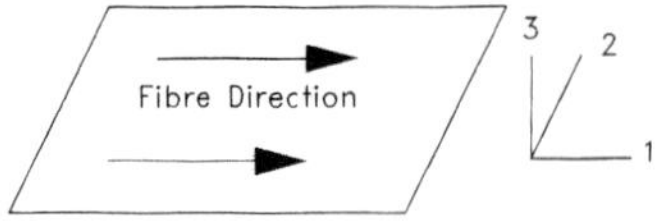

Property	Units	
E_1	GPa	135.3
E_2	GPa	9.0
E_3	GPa	9.0
G_{12}	GPa	5.2
G_{13}	GPa	5.2
G_{23}	GPa	1.9
ν_{12}		0.34
ν_{13}		0.34
ν_{23}		0.46

COMPARISON WITH THEORETICAL PREDICTIONS FOR MULTI-ANGLE LAMINATES

The benefit of data measured for a unidirectional composite material is greatly enhanced if they can be used to predict the characteristics of laminates of multi-angle form. Use of the plate behaviour assumptions in design of composite structures led to the development and propagation of the classical laminate theory[1] and employment of this in predicting the in-plane properties of multi-angle laminates. Classical laminate theory is limited by an inherent inability to consider out-of-plane properties. It is desirable to overcome this limitation by developing procedures which will satisfy the needs of three-dimensional modelling[10,11]. For the purpose of the present paper, in-plane properties predicted using the classical laminate theory will be compared to experimental measurements.

COMLAN, an implementation of classical laminate theory developed in-house using the LOTUS 123 spreadsheet has been used to predict the in-plane properties of $[+45/-45]_{nS}$ and $[+45/0/-45/90]_{nS}$ laminates of the present AS4-CF/PEEK composite. The engineering properties appearing in Table 2 were employed as input data. Predicted values of in-plane axial moduli and lateral contraction ratios, as listed in Table 3, compare well with values determined experimentally; the only noticeable discrepancy being a 7.5% overestimate of axial modulus of the $[+45/0/-45/90]_{nS}$ laminate.

TABLE 3
Comparing Experimental Data From Table 1 And CLT Based Estimates

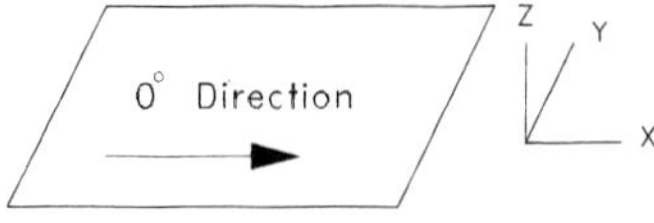

Laminate Configuration		$[+45/-45]_{10S}$		$[+45/0/-45/90]_{5S}$	
		CLT Based Estimate	Experiment	CLT Based Estimate	Experiment
Ex	GPa	18.3	18.4	52.3	48.6
ν_{xy}		0.76	0.74	0.31	0.30

CONCLUSIONS

The present work outlines a procedure for determining the engineering properties required to conduct three-dimensional analysis for design of structures manufactured from unidirectional continuous fibre reinforced composites. The procedure assumes that the material will behave in a transversely isotropic manner, the plane of isotropy being that normal to the fibre direction. Using two tensile and two torsion tests, eight engineering properties are generated which are sufficient to characterise the material stiffness. The present data do not allow comment on the validity of the assumption of transverse isotropy for characterisation of the mechanical behaviour of the present material.

Availability of the complete set of data permits the modelling, for design and analysis, of composite components while accounting for three-dimensional behaviour. Such analysis would rely on numerical modelling techniques and is necessary for certain applications.

In many instances, it is acceptable to neglect through thickness effects, following the assumptions of plate theory. Where this is structurally permissible, Classical Laminate Theory can be utilised in conjunction with some of the determined engineering properties to obtain a reasonable estimate of the in-plane properties for multi-angle laminates, although an overestimate by 7.5% was determined for one lamiante configuration.

The engineering properties describing the mechanical behaviour of prepreg based, unidirectional AS4 continuous carbon fibre reinforced PEEK are presented in Table 2.

ACKNOWLEDGEMENT

The authors gratefully acknowledge the efforts of Mr Nick Burgoyne in completing this work.

REFERENCES

1) R M Jones, Mechanics of Composite Materials (Scripta Book Company, Washington, D.C., 1975).

2) I M Ward, Mechanical Properties of Solid Polymers, 2nd edition (J. Wiley, 1985).

3) ICI Fiberite, Aromatic Polymer Composite, APC-2 product data sheets, 1986.

4) D C Leach, D C Curtis and D R Tamblin, Delamination Behaviour of Carbon Fibre/PEEK Composites, in: Toughened Composites ASTM STP 937, ed. N J Johnson, (ASTM, Philadelphia).

5) I T Bernie, D R Moore and S Turner, Developments in the Generation and Manipulation of Mechanical Properties Data, Plastics & Rubber, Processing & Applications 3, vol 4, (1983) pp 365-371.

6) M J Bonnin, C M R Dunn and S Turner, A Comparison of Torsion and Flexural Deformation in Plastics, Plastics & Polymers, December (1969) pp 517-522.

7) S Timoshenko and J N Goodier, Theory of elasticity (McGraw-Hill, 1951).

8) A E H Love, A Treatise On The Mathematical Theory Of Elasticity (Dover, 1944).

9) L A Carlsson and R B Pipes, Experimental Characterisation of Advanced Composite Materials (Prentice-Hall, 1987).

10) P L N Murthy and C C Chamis, ICAN: Integrated Composites Analyzer, Journal of Composites Technology & Research, Vol 8, 1, Spring (1986) pp 8-17.

11) N Zahlan and F J Guild, Computer Aided Design Of Thermoplastic Structures, to be presented at ICCM-VII (August 1989).

Materials and Processing – Move into the 90's
edited by S. Benson, T. Cook, E. Trewin and R.M. Turner
Elsevier Science Publishers B.V., Amsterdam, 1989

HYBRID LAMINATES FOR DAMAGE MANAGEMENT IN FIBRE REINFORCED PLASTICS

M A LEAITY, P A SMITH AND M G BADER

Department of Materials Science and Engineering, University of Surrey, Guildford, Surrey, GU2 5XH, UK.

1. INTRODUCTION

Continuous-fibre reinforced plastics composites based on glass, aramid and carbon fibres in thermosetting matrices, such as the epoxy-resins, have excellent specific stiffness and strength and are being used in a large number of applications where they have superseded high performance metallic materials. They do, unfortunately, suffer from certain intrinsic limitations such as low interlaminar shear strength and, in particular, low tensile strength in the direction transverse to the fibre axis. In multiaxial laminates failure, in the form of matrix cracks in the transverse plies, may occur at strains as low as 0.2% (2000 u strain) whilst in the longitudinal direction integrity is preserved to well over 1% strain. Conservative design practices lead to the imposition of limit strains which do not allow transverse failure. This results in the materials being utilised well below their apparent potential structural efficiency.

It is well known[1] that transverse cracking may be delayed by using thin plies to exploit the constraint effect and by using more ductile resin matrices[2]. It is also probable that different levels of interface strength will influence the phenomenon. However, in practice, it is not usually possible to utilise a more compliant matrix because the primary properties: longitudinal tensile strength, compressive strength and environmental resistance may be adversely affected.

In this work we propose to tailor each ply of a laminate to optimise its properties for the design stress state. Thus a longitudinal ply which is required to carry a tensile stress will use a combination of fibre and resin with the interface optimised for maximum tensile strength. A different combination might be needed for maximising compressive strength. The initial experiments are aimed to develop a $(0_m, 90_n)_s$ laminate with optimised tensile strength and resistance to transverse ply matrix cracking. To this end the use of a more compliant resin for the 90° plies, whilst keeping a conventional stiff resin in the 0° layers, has been assessed.

2. MATERIAL PREPARATION

2.1 Matrix and reinforcement

The matrix is a blend of a proprietary diglycidal-ester of bisphenol-A (DGEBA) epoxy cured with a boron trifluoride complex and a single component polyurethane. These are miscible in all proportions and cure to give a single-phase solid. The urethane addition increases the flexibility of the cured resin. In the first phase of the work, resins with urethane contents of 0, 5, 10 and 20% have been used. The matrix is reinforced with an E-glass fibre of modulus 70 GPa.

2.2 Prepreg Manufacture

Prepreg is manufactured using a microprocessor controlled, drum-type filament winder which was specially constructed for this project, Figure 1. The roving is passed through a bath containing a solution of the resin in dichloromethane and wound onto the drum to form a uniaxial hoop winding. After winding is complete the cylindrical array is cut across and removed, on its backing paper, from the drum to form a sheet approximately 1000 x 300 mm. The solvent is allowed to evaporate for 24h, when the prepreg is typically of 400 g/m^2 with a resin content of 35%.

2.3 Cure processing of the prepreg

Cure processing is carried out in a press-clave. A layer of release film and bleeder cloth is placed on both surfaces of the stacked plies of prepreg. This laminate pack is enclosed in a nylon-film bag which is placed on the base of the press-clave. The clave is heated in a hot press and pressure is applied to the outside of the bag from a pressure port.

A de-bulking stage, at a temperature of 60°C and a pressure of 0.7 MPa for 1 hour, eliminates air entrapped between plies during the lay-up of the laminate. It is then cured at a temperature of 190°C and a pressure of 0.7 MPa for 2 hours. The cured laminates were translucent and generally observed to be well impregnated and without excessive porosity.

3. EXPERIMENTAL

3.1 Laminates

Cross-ply laminates of $(0, 90)_s$ and $(0, 90_2)_s$ construction have been prepared as outlined above with a range of matrix compositions. Table 1 lists the lay-ups and the urethane contents, specified as percentages by weight of the matrix. The "uniform" matrix lay-up consists of one matrix throughout the laminate. The hybrid matrix lay-up consists of unmodified epoxy resin in the outer 0° plies and urethane modified resin in the inner 90° plies. The average thickness of the $(0, 90)_s$ laminates was found to be 0.77 mm and that of the $(0, 90_2)_s$ laminates to be 1.09 mm. The fibre volume fraction of the laminates was found to be 0.72 (+0.01) by a "burn-off" test. The transverse ply thickness was measured by microscopy of polished sections and found to be 0.38 mm for the $(0, 90)_s$ and 0.73mm for the $(0, 90_2)_s$. Tensile test coupons of 20 x 200 mm (140 mm gauge length) were cut from these laminates.

3.2 Tensile Testing

Strain gauged coupons are loaded in an Instron 1175 testing machine and a load versus strain plot is recorded. Tests are executed at crosshead speeds ranging from 0.5 to 2 mm/min.

Low strain damage in transverse plies during loading is detected in two ways; crack counting and acoustic emission measurements. Crack density in the transverse ply is determined as a function of applied strain by loading coupons in 0.1% strain increments and counting the cracks after each successive increment. Acoustic emission counts are measured in separate tests by placing transducers on coupons which are then loaded continuously to failure. Threshold damage strains are determined from the knee in the continuous stress-strain curves.

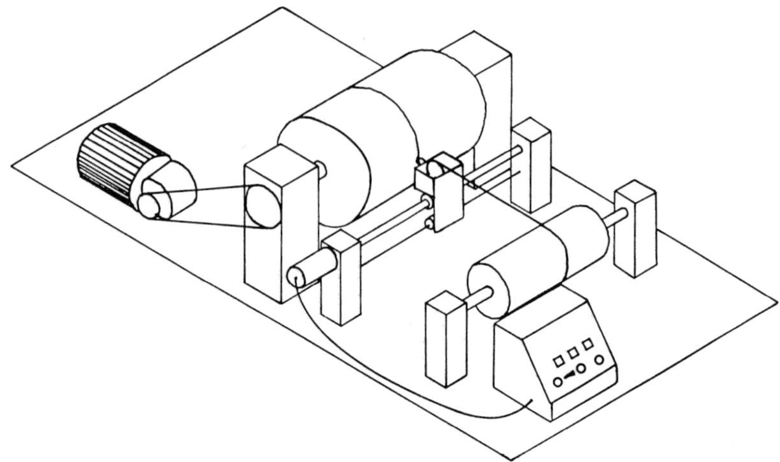

Fig 1. Sketch showing the general arrangement of the laboratory-scale drum pre-pregging machine.

Table 1 Matrix composition and lay-ups

Urethane content *(%)	Uniform matrix	Hybrid matrix
0	$(0, 90)_s$; $(0, 90_2)_s$	
5	$(0, 90)_s$	$(0, 90)_s$
10	$(0, 90)_s$	$(0, 90)_s$
20	$(0, 90)_s$; $(0, 90_2)_s$	$(0, 90)_s$; $(0, 90_2)_s$

* percent of matrix by weight

4. RESULTS

4.1 Damage onset and growth in the transverse ply

$(0, 90)_s$ laminates

Plots of crack density in the transverse ply against applied strain for the $(0, 90)_s$ range of laminates are shown in Fig 2. Two specimens were tested from each of the lay-ups. The mean curves are shown superimposed in Fig 3. The dependence of threshold damage strain on the urethane content in the matrix of the $(0, 90)_s$ laminates is shown in Fig 4. The threshold damage strain increases with an increase in urethane content as would be expected. Threshold strains in the hybrid matrix laminates compare well with the uniform matrix laminates suggesting that the urethane modified matrix in the transverse ply has not mixed significantly with the unmodified epoxy of the outer plies during processing of the hybrid matrix laminate.

$(0, 90_2)_s$ laminates

Plots of crack density in the transverse ply against strain for the $(0, 90_2)_s$ range of laminates are shown in Fig 5. As before, two tests were carried out on each lay-up and for these laminates, the repeatability is very good. Figure 6 shows the superimposed plots for these tests. For any applied strain, the epoxy matrix laminate exhibits the highest crack density and the hybrid matrix laminate containing 20% urethane has the lowest. Threshold damage strains for these laminates, in Figure 4, exhibit the same trend as the $(0, 90)_s$ laminates.

In Figure 7 the stress-strain curves are shown with corresponding measurements of acoustic activity from transverse ply cracks in the $(0, 90_2)_s$ laminates.

4.2 Initial moduli and ultimate strength

Figure 8 shows the variation of the initial tensile moduli with urethane content in the $(0, 90)_s$ and $(0, 90_2)_s$ laminates. The modulus drops slightly as urethane content increases in both lay-ups. The inferred transverse modulus of a unidirectional layer against urethane content for both lay-ups is shown in Figure 9 and, as expected, this drops with increasing urethane content. Figure 10 shows the inferred stress in the 0° plies of the laminates at failure, as a function of urethane content. There is little change with urethane content for both lay-ups but in both cases there was a tendency for failure to occur at the grips rather than within the gauge length.

5. DISCUSSION

5.1 Crack density in the transverse ply

$(0, 90)_s$ laminates

Comparison of crack density in the transverse ply as a function of applied strain for each laminate is a means of characterising damage. Figure 3 shows the superimposed curves from each of the $(0, 90)_s$ laminates. The curves originate from best fits to the data in Figure 2. Repeatability of data is good except for the unmodified epoxy matrix laminate and the uniform matrix with 20% urethane. The variability in the data is attributed to uncertainty of crack counts. This was a result of a diffuse cracking pattern in the transverse ply. (It is probable that the transverse ply thickness of 0.35 mm is less than the critical flaw size required for rapid propagation to form a transverse ply crack spanning the whole width and breadth of the ply)[3,4]. Nevertheless the average

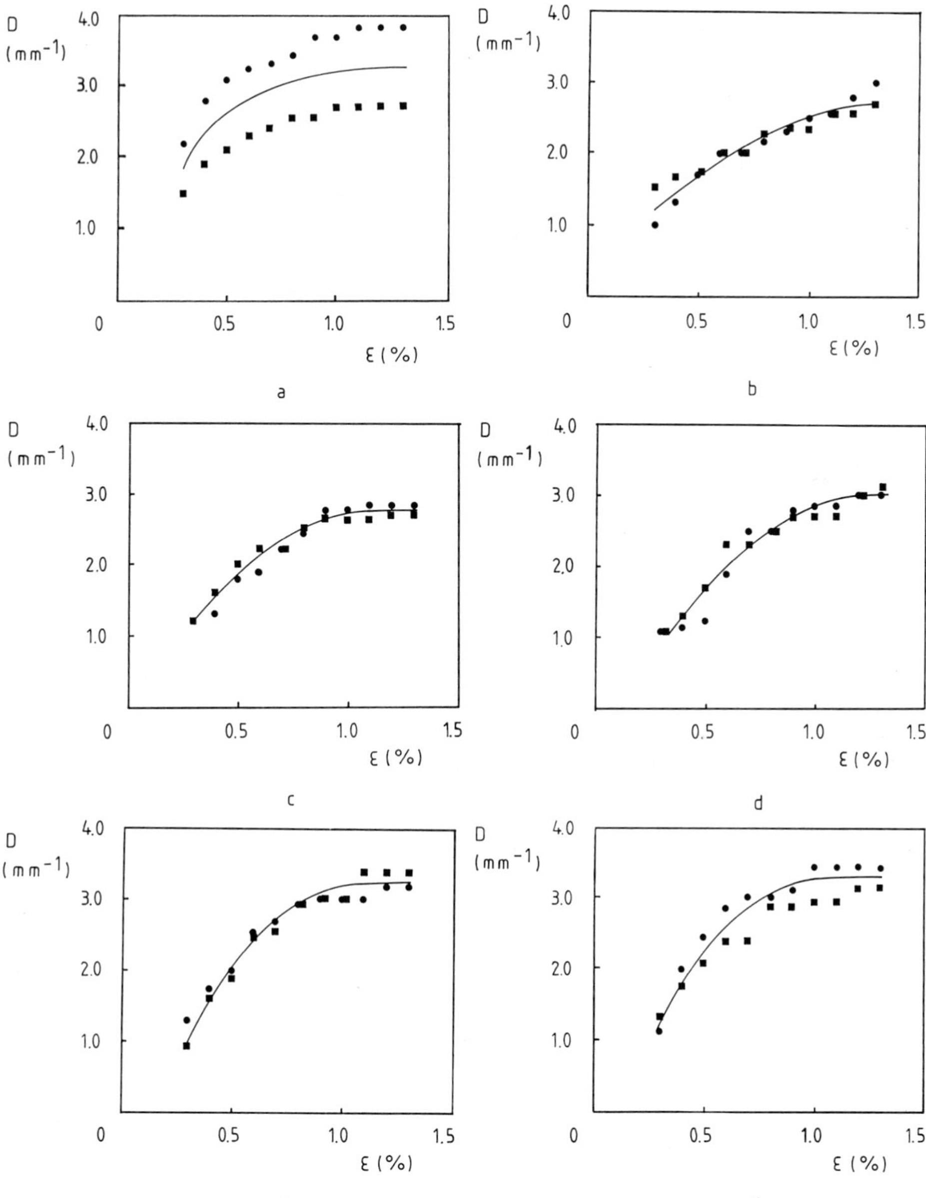

Fig 2. Transverse ply crack-density versus applied strain for $(0, 90)_s$ laminates with:

a. Unmodified epoxy matrix
b. Hybrid matrix - 5% urethane
c. Hybrid matrix - 10% urethane
d. Hybrid matrix - 20% urethane
e. Uniform matrix - 5% urethane
f. Uniform matrix - 10% urethane
g. Uniform matrix - 20% urethane

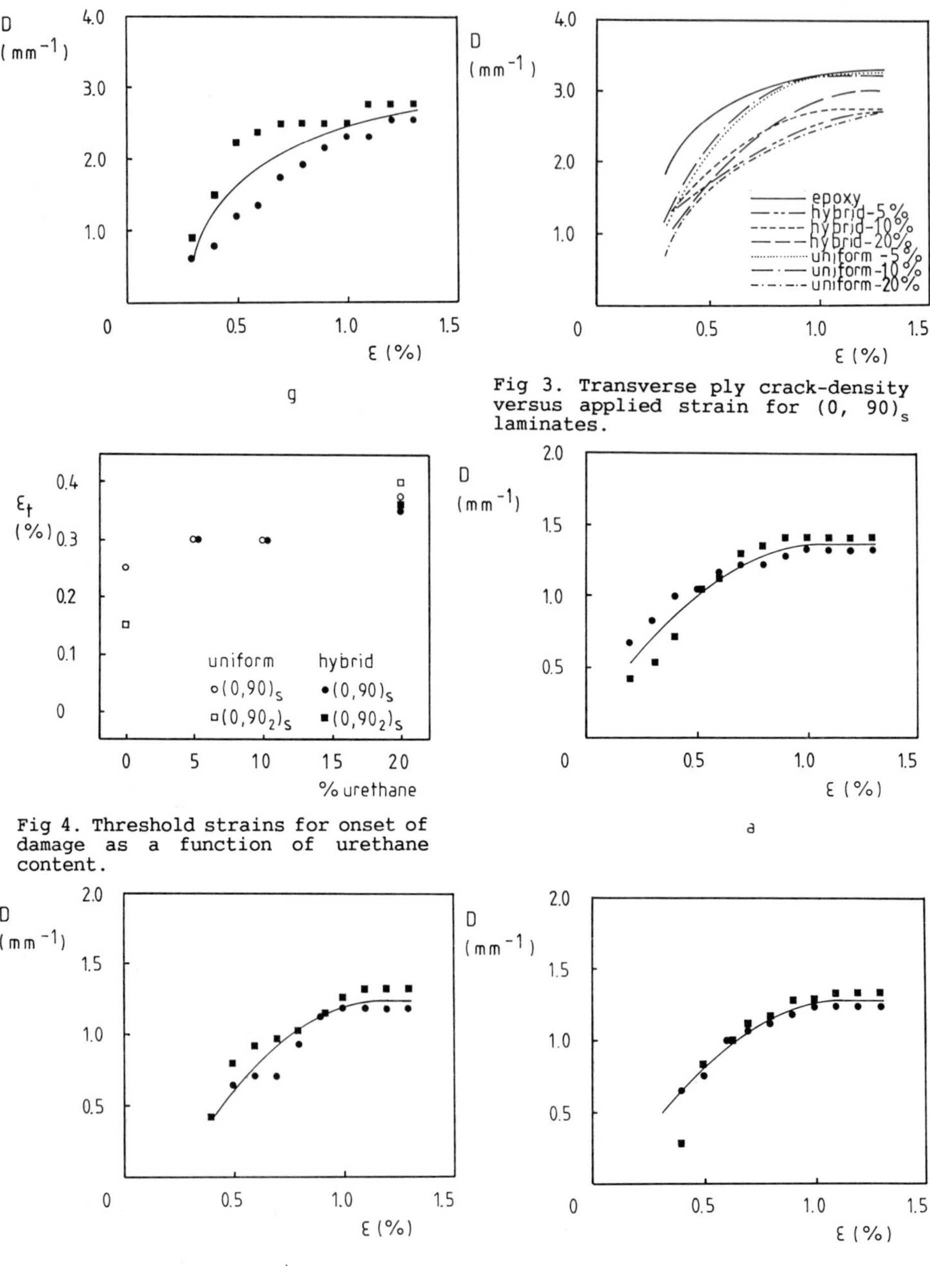

Fig 3. Transverse ply crack-density versus applied strain for $(0, 90)_s$ laminates.

Fig 4. Threshold strains for onset of damage as a function of urethane content.

Fig. 5. Transverse ply crack-density versus applied strain for $(0, 90_2)_s$ laminates with :

a. Unmodified epoxy matrix
b. Hybrid matrix - 20% urethane
c. Uniform matrix - 20% urethane

crack density in the epoxy matrix is still distinctly higher than in the modified uniform and hybrid matrix laminates, up to applied strains of 0.75%, indicating that it has incurred more damage than the modified laminates at this strain.

$(0, 90_2)_s$ laminates

In these laminates with a transverse ply thickness of 0.73 mm, crack spacings in the transverse ply are larger and hence, crack counts are more accurate. In Figure 5 the repeatability of the data is good, indicating that the comparative curves in Figure 6 are reliable. Again, the unmodified epoxy matrix laminate has incurred the most damage for a given applied strain, while the hybrid matrix containing 20% urethane has incurred the least damage. This confirms the trends observed in the $(0, 90)_s$ laminates that the modified matrix confers enhanced resistance to transverse cracking.

5.2 Threshold damage strain and acoustic emission

In Figure 4, the threshold damage strain for the $(0, 90)_s$ laminates is shown to increase with increasing urethane content, this indicates that the onset of damage in the modified laminates is delayed to a higher applied strain. The threshold cracking strains for the $(0, 90_2)_s$ laminates shown in Figure 4 were derived from the knee in the stress-strain curves in Figure 7. The epoxy matrix laminate exhibits a distinct knee in the stress-strain curve at 0.15% strain corresponding to the start of considerable acoustic activity. In comparison, the uniform and hybrid matrix laminates with a 20% urethane content show a gentle knee at approximately 0.40% strain and exhibit remarkably less acoustic activity. These data confirm the trend of the $(0, 90)_s$ laminates of an increase in threshold damage strain with increasing urethane content. The large reduction in acoustic activity indicates that the cracking in the modified laminates does not occur in as brittle a fashion as it does in the epoxy. This would indicate that the greater toughness of the urethane-containing plies has resulted in an increase of critical flaw size so that they behave essentially as "constrained" plies even in the $(0, 90_2)_s$ laminate.

5.3 Initial modulus and ultimate tensile strength

The initial moduli of the $(0, 90)_s$ and the $(0, 90_2)_s$ laminates were slightly lower in the laminates of higher urethane content, Figure 8. The drop is more noticeable in the $(0, 90_2)_s$ laminates because of the relatively greater contribution of the transverse ply to the total laminate stiffness. There was little observed difference between the hybrids and their uniform counterparts since, on the whole, the longitudinal stiffness is dominated by the 0° plies.

The inferred transverse moduli plotted in Figure 9 were calculated from the rule-of-mixtures prediction : $E_{(lam)} = V_1E_1 + V_2E_2$ where $E_{(lam)}$ is the laminate stiffness, V_1 and V_2 the volume fractions of the longitudinal and transverse plies respectively and E_1 and E_2 their moduli. E_1 was determined from tensile tests on unidirectional material and was found to agree well with the rule-of-mixtures prediction: $E_1 = 70.V_f + (1 - V_f)\ 3.3$ (V_f is the fibre volume fraction) and was independent of the urethane content. $E_{(lam)}$ was determined from the slope of the stress-strain curves in tensile tests. E_2 was then calculated from the known E(lam), E_1, $V_1$and V_2 for the range of urethane contents. This inferred transverse modulus shows the expected drop with increasing urethane content.

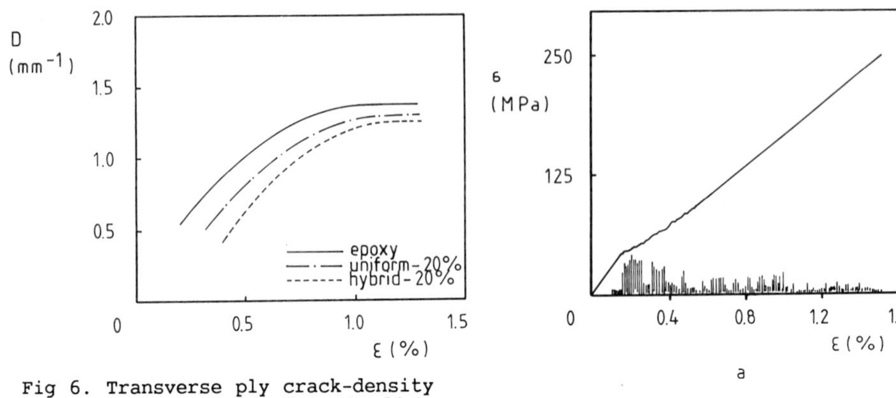

Fig 6. Transverse ply crack-density versus applied strain for $(0, 90_2)_s$ laminates.

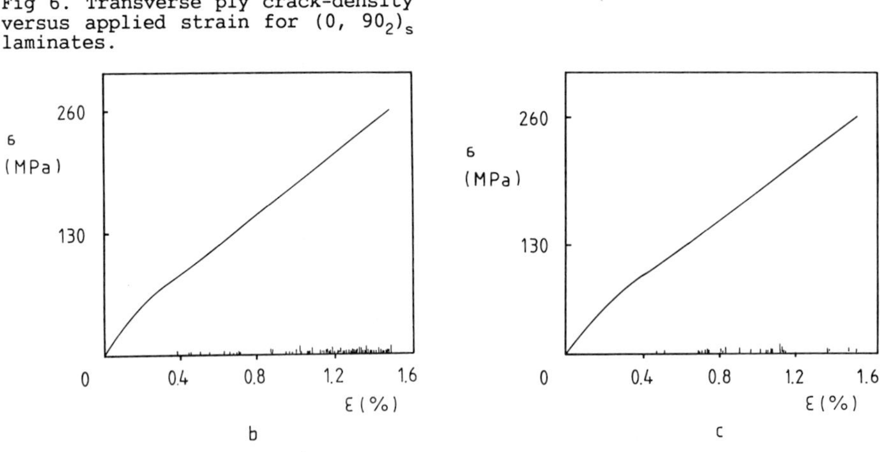

Fig 7. Stress/strain curve and acoustic emission trace for $(0, 90_2)_s$ laminates:

a. Unmodified epoxy matrix
b. Hybrid matrix - 20% urethane
c. Uniform matrix - 20% urethane

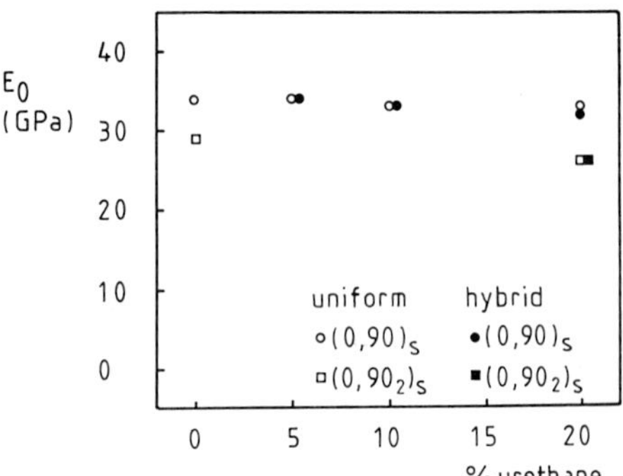

Fig 8. Longitudinal Young's modulus for cross-ply laminates as a function of urethane content.

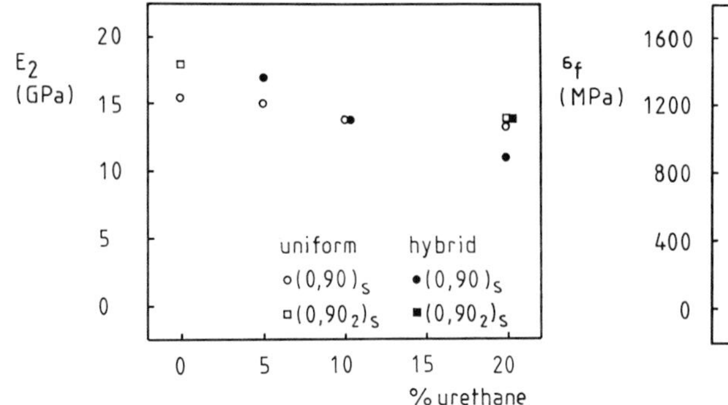

Fig 9. Inferred transverse modulus of a unidirectional lamina as a function of urethane content.

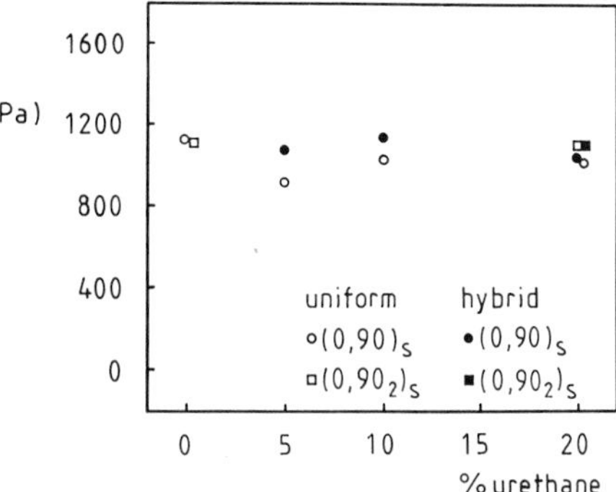

Fig 10. Inferred stress at failure in the unidirectional layer of cross-plied laminates as a function of urethane content.

Figure 10 shows that there is no decrease in the inferred failure stress in the 0° plies with an increase in urethane content. This suggests that the degree of toughening of the resin afforded by the range of urethane added does not significantly affect load transfer from failed fibres to surviving fibres during the process of failure in the 0° plies.

5.4 Concluding remarks

Both the uniform and hybrid matrix laminates containing a proportion of urethane exhibited superior damage resistance to transverse ply cracking over the unmodified uniform epoxy matrix. The onset of damage has been observed to occur at a higher strain in these laminates. The distinction between the hybrid matrix and the uniform matrix laminates containing the same proportion of urethane was not great, as might be expected under simple tension loading. In the next phase of the work the behaviour of these laminates under compression and other states of stress will be studied in order to assess the overall potential of the hybrid matrix concept.

ACKNOWLEDGEMENTS

This work was carried out with the support of the Procurement Executive of the Ministry of Defence. We would like to acknowledge the considerable help of Mr J Walker with the experimental work and also Dr P T Curtis of the Royal Aerospace Establishment at Farnborough for helpful discussions.

REFERENCES

1) A Parvizi, K W Garrett and J E Bailey, Constrained cracking in glass fibre-reinforced epoxy cross-ply laminates, J Mater Sci, 13 pp 195-201 (1978).

2) K W Garrett and J E Bailey, The effect of resin failure strain on the tensile properties of glass fibre-reinforced polyester cross-ply laminates, J Mater Sci 12, pp 2189-2194 (1977).

3) D L Flaggs and M H Kural, Transverse lamina strength in graphite-epoxy laminates, J Comp Mater, 16 pp 103-116 (1982).

4) S L Ogin and P A Smith, Fast fracture and fatigue growth of transverse ply cracks in composite laminates, Scripta Met 19, pp 779-784 (1985).

Materials and Processing – Move into the 90's
edited by S. Benson, T. Cook, E. Trewin and R.M. Turner
Elsevier Science Publishers B.V., Amsterdam, 1989

SIGNATURE MANAGEMENT AND STRUCTURAL MATERIALS

P.S. Bradshaw, B.Sc.,

Plessey Materials, Plessey-UK Ltd., Woodburcote Way, Burcote Road, Towcester, Northants. NN12 7JS. Tel: (0327) 52828

1. INTRODUCTION

This paper describes two inter-related subjects, i.e. Signature Management and Structural Materials. The Signature Management issue is of prime importance and affects all aspects of efficient defensive policy. The meaning of Signature will be described and the advantages and gains to be reaped from the active control of emissions will be evaluated. Structural Materials or more precisely Structural Microwave Absorbent Materials are now available to designers and platform builders and a general description of the types of products emerging will be presented. The success of these inherently absorbing structures and components is largely dependent on the acceptance of Signature Management as a prime operational requirement, as the materials technology is already well founded.

2. SIGNATURE

The intrinsic characteristics of a platform that are exploitable by a detection system to identify its presence, combine to produce a well defined signature. Military platforms producing or emitting such characteristic signatures include - aircraft, RPV's, missiles, ships, submarines, armoured vehicles, radar sites and fixed ground assets. It is clear that all military departments must have an interest in signature management. The signature of any platform is made up of many individual generic components as follows:- Magnetic, acoustic, visual, infrared, microwave and even seismic elements.

Detailed analysis of a signature will reveal to the observer much coveted information such as the platforms' identification, motion, readiness for engagement, intentions etc. Consequently the importance of the platform signature cannot be under-estimated.

Camouflage is one of the basic defensive elements of battle. Simple measures can be employed to reduce, enhance or control a signature in order to conceal a platform or to present a false target to confuse the observer. Properly employed, camouflage, concealment and deception techniques can spell the difference between victory and defeat.

Modern forces recognise the importance of camouflage and go to considerable lengths to conceal their movements and intentions from the enemy. However, today's surveillance devices and advanced guided weaponry take advantage of the many elements constituting the platform signature. Attention must be given to all of the major elements of the signature, for example, there is little to be gained by using visual camouflage and smoke screens to protect a tank which is an inherent thermal and radar beacon and no amount of visual camouflage will protect a ship from an active radar homing missile. Today's design and materials technology offers not only the capability to suppress the main elements of a multi-spectral signature but also, with careful use of decoy systems, to modify or enhance emissions to further confuse the observer or to mis-direct a guided weapon.

The elements of signature most utilised by long range surveillance systems and pin-pointed by guided munitions are the acoustic, infrared and microwave components.

3 MICROWAVE - RCS REDUCTION

Particular expertise in the control of the microwave element of the signature is available within Plessey. The radar cross section (RCS) of a target can be defined as:-

"The area of a fictitious perfect reflector of electromagnetic waves that would reflect the same amount of energy back to the radar as would the actual target."

In its most simple form the "radar range equation" defines a radar's received power as:-

$$Pr = \frac{Pt\ G^2\ \lambda^2\ \sigma}{(4\pi)^3\ R^4}$$

Where
- Pr = received power
- Pt = transmitted power
- G = antenna gain
- λ = Wavelength of transmitted signal
- σ = RCS
- R = radar range

To summarise, the radar's received power is directly proportional to RCS and, just as importantly, the radar range capability is directly proportional to the 4th root of RCS. It must be highlighted here that the RCS of complex targets is not a simple matter for either theory or measurement. The intensity of a reflection varies with the emitted frequency and polarisation and on the target's attitude and profile. At high frequencies the RCS depends very much on the orientation of a target's smaller components. Surface wave propagation of energy, leading to substantial emissions from a target's edges, must also be considered.

Two immediate benefits can be gained by the treatment of, for example, an aircraft to reduce its RCS ie survivability and jamming improvements. Figure 1 emphasises the importance of RCS reduction to the survivability of fighter aircraft. The scenario is of 2 fighter aircraft approaching each other in a head-on encounter at Mach 0.9. Both aircraft have the same initial RCS of $5m^2$ and have the same radar capabilities. If one fighter should reduce its RCS from $5m^2$ to say $3m^2$ then, if both aircraft have a radar range of 50km, ie they can detect $5m^2$ at 50km, the treated fighter aircraft will have more than 10 seconds in which he has already detected the enemy but remains undetected himself. Furthermore if the radar range of both aircraft is 125km then the treated fighter would have more than 20 seconds advantage in which to launch an attack whilst still undetected.

Figure 1

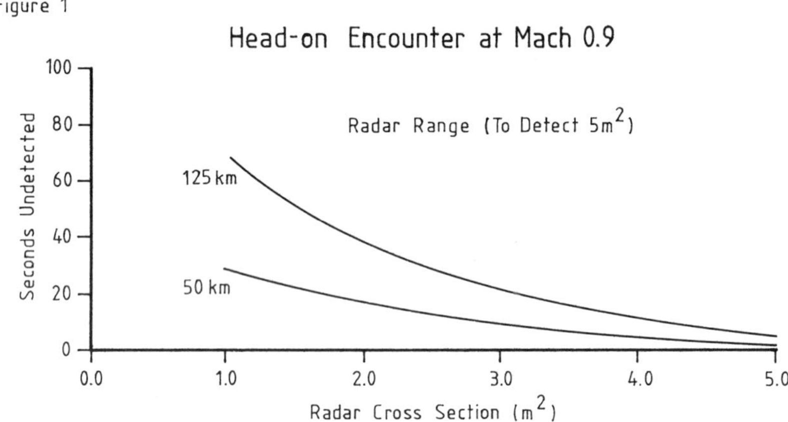

Increasingly, the reduction of radar signature is being viewed as an integral element of electronic countermeasure suites. The effect of RCS reduction can be seen from examination of the jamming equation. When a radar is searchlighting a target which has a jammer, a mode of operation referred to as burnthrough, the radar range is expressed as:-

$$R^2 = \frac{P\ to\ Gt}{4\pi} \; \frac{\sigma}{(E/N)} \; \frac{B_J}{P_J G_J}$$

Where R = radar range; P = radar average power

t = observation time; Gt = transmitting antenna gain

G_J = Jammer antenna gain

(E/N) = signal energy to noise ratio; σ = RCS

P_J = Jammer power; B_J = Jammer bandwidth

To summarise

$$P_J \propto \sigma \qquad \text{and } R \propto \sqrt{\sigma}$$

For the same burnthrough conditions the radiated jammer power required can be reduced in direct proportion to the RCS ie if the RCS of the host vehicle can be reduced by 10dB then the effective power requirement of an active ECM suite can be reduced by 90%. Similarly, when chaff is employed to protect a target, a reduction in RCS of the target of 10dB allows the quantity of chaff required, to provide the same level of protection, to be reduced by 90%. Conversely, for the same jammer power, the burnthrough range is directly proportional to the square root of the RCS.

For example, a typical strike aircraft with active ECM noise jamming, operating against a SAM threat; a reduction in the aircraft RCS of 10dB effectively reduces the SAM radar burnthrough range by over 60% and enables the aircraft to approach and avoid detection until well inside the missile minimum reaction zone.

4 STRUCTURAL MICROWAVE ABSORBENT MATERIALS

Concerted efforts in this highly desired field have led to the design, development and production of a range of materials suited to many load bearing or ballistic applications. Plessey is well known and respected for the production of a whole range of parasitic microwave absorbent materials. This product range is made up of various lightweight foams, elastomeric sheet and custom mouldings, rubberised fibre matting, camouflage netting and panels all of which are retro-fitted into the desired position. These parasitic materials are generally utilised for 3 major reasons:-

* antennae performance improvements - removing unwanted signal reflections and interference from adjacent emitters or structures.

* radar power testing - including test hoods and anechoic chamber applications.

* radar signature reduction of military vehicles and assets by fitting materials onto existing platforms.

Due to the increasing awareness of the value of military asset and vehicle signature control, radar absorbent properties are now being considered as a prime operational requirement in many equipment updates and new programmes. Technical advances at Plessey now make it possible to design structures, components and fixtures with inherent absorbent characteristics without sacrificing mechanical or environmental performance. The aim being to "design-in" the structural and microwave absorbent performance rather than to treat a platform at a later stage. Structural RAM is an integrated material designed into a vehicle platform or asset whereas parasitic RAM is bonded or fixed onto the target in a retro-fitting exercise. In general terms, structural RAM is a dual function material i.e.

load bearing <u>plus</u> microwave absorbent properties
or ballistic <u>plus</u> microwave absorbent properties

Various composite structures, without microwave absorbent properties, are already being extensively utilised in the design and manufacture of modern military vehicles, ships and aircraft. Many types of aircraft contain non-metallic components such as ribs, spars, beams, wing skins, fuselage sections, nose cones etc as well as many ancillary items such as access panels and doors and of course rotor blades for helicopters. Uses of composites on ground vehicles include lightweight armour and hatch doors and polymeric materials find extensive use on specialist ships, for example minesweepers.

Plessey manufacture 4 generic types of Structural Microwave Absorbent Materials ie

Syntactic foam, lightweight honeycomb, single skin fibre reinforced composite and various sandwich constructions

Extensive use is made of the 3 main reinforcing fibres - glass, aramid (Kevlar) and carbon. Figure 2 describes how these various materials compare with steel and aluminium with regard to specific modulus ie stiffness divided by density.

Figure 2

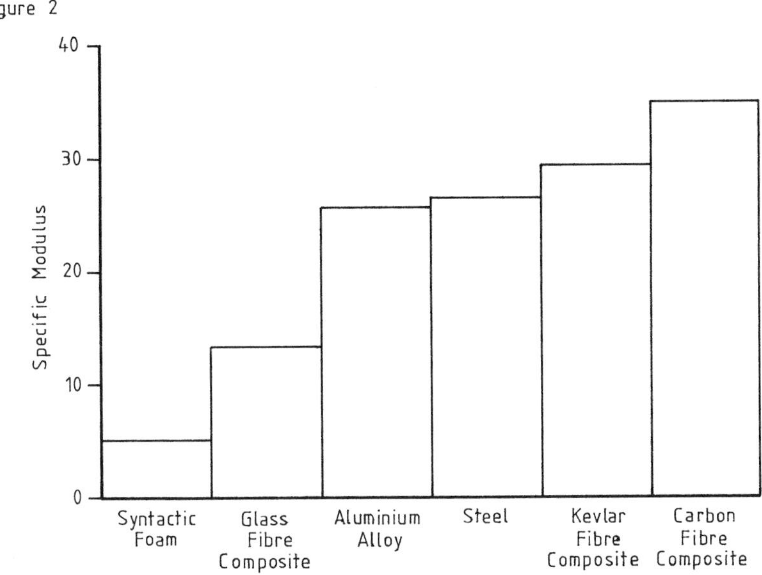

It can be seen that, weight for weight, aramid and carbon based composites compare very favourably with metals. The same can be said when examining specific strength. If you look at actual performance, a carbon fibre reinforced laminate has a modulus of 57GPa (8265 KSI) and a tensile strength of 525MPa (76 KSI) compared to aluminium alloy (L65) producing 72 GPa (10440 KSI) and 460 MPa (67KSI) respectively. Titanium provides a modulus of 110 GPa (15950KSI) and a tensile modulus of 930 MPa (135KS1).

4.1 SYNTACTIC FOAM

This material is a rigid, lightweight structural foam based on a resin system which is highly filled with hollow, ceramic microspheres producing a very fine "closed-cell" structure.

Based on Plessey's base syntactic foam material, the microwave absorbent variant, LA4, is a member of the Lightweight Absorber product range. Microwave absorbent properties are gained by the inclusion of a dielectrically loaded element within the foam structure. This produces a broadband absorber over the frequency range of 6-18GHz (figure 3), custom products can be designed with performance in the 100GHz range. In flat sheet, this material produces a minimum attenuation of 15dB (96.9% absorption) and better than 20dB (99.0% absorption) in some parts of its range.

Figure 3

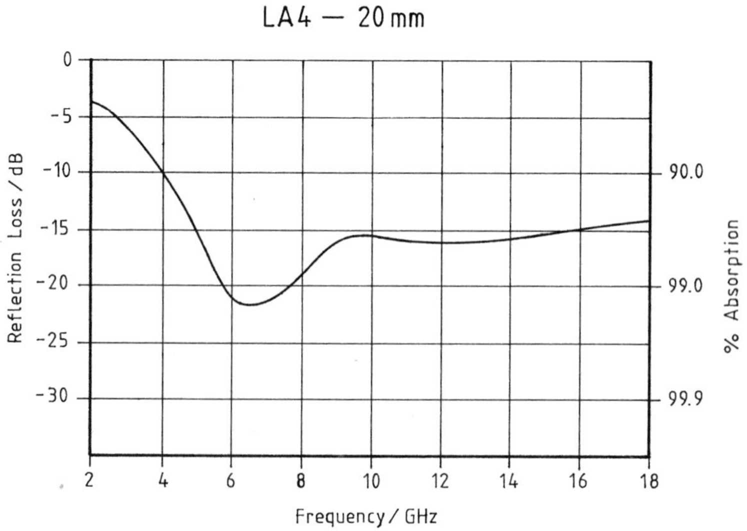

This material was specifically designed to withstand hydrostatic pressures up to 1000 psi with very low water absorbency. The material is easily moulded into complex shapes, it can be machined and will accept standard fasteners.

4.2 LIGHTWEIGHT HONEYCOMB

This material is a broadband absorber with a typical performance of 20dB attenuation over the 6-18GHz (figure 4) range. Based on commercially available honeycomb this product is highly resistant to fatigue, all forms of corrosion or chemical attack and is non-flammable. Honeycomb densities are very low (0.023 - 0.123 g/cm^3), yet this product is extremely tough and resilient and has excellent resistance to impact and handling damage.

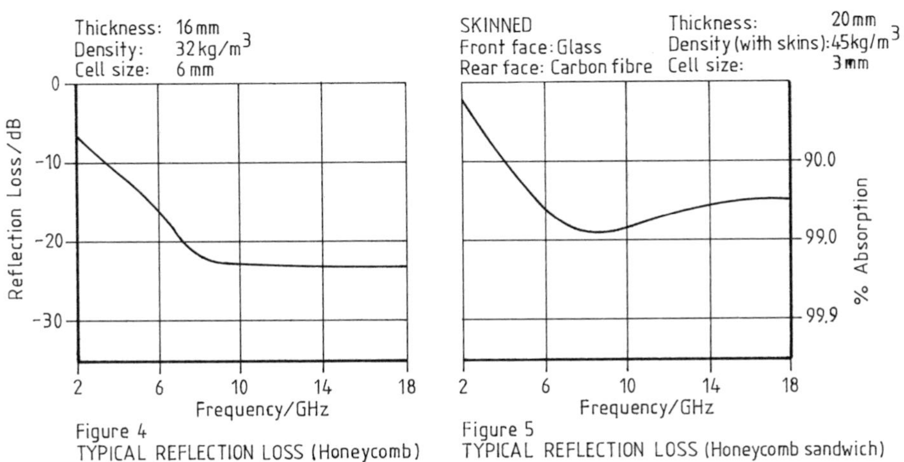

Figure 4
TYPICAL REFLECTION LOSS (Honeycomb)

Figure 5
TYPICAL REFLECTION LOSS (Honeycomb sandwich)

4.3. SANDWICH CONSTRUCTIONS

Structural, sandwich constructions with Microwave absorbent properties all have a generic composition consisting of a microwave transparent front face, a reflecting rear face and an absorbing centre.

Probably the best example of this is the honeycomb sandwich construction. This has a radar transparent front face eg aramid or glass fabric reinforced resin, absorbent honeycomb centre and a carbon fibre fabric reflective rear face. Optimisation of electrical performance of such a structure is a complex matter involving the transparent front skin thickness, the honeycomb type and thickness and the dielectric loading of the honeycomb. A 20mm thick sandwich of this type, with a glass fibre reinforced front skin and a carbon fibre rear face, having a density of $0.045 g/cm^3$ and a cell size of 3mm will provide greater than 15dB attenuation (96.9% absorption) from 6-18GHz (Figure 5).

4.4. SINGLE SKIN FIBRE - REINFORCED COMPOSITES

This range of products is based on glass fibres or aramid fibres, in the form of fabrics, encapsulated in a resin system. The thickness of laminate is built up by using several layers of the desired fabric producing a rigid single skin product. The absorbent characteristics are obtained by the incorporation of lossy fillers within the laminated construction. Due to the nature of this product standard composite manufacturing techniques are used to produce complex double curvature components and structures. "K-RAM" is an example of this type of product and can be designed to resonate at two or three bands within the 2-40GHz frequency range. Since the absorbing layer will be thinner at higher frequencies, additional layers of fabric, which have no loading, can be laminated to the rear surface to maintain the required strength without affecting the electrical performance. The product is usually backed with a layer of carbon fibre fabric which acts as an integral reflector. Electrical designs aim for 20dB attenuation at the resonant frequencies.

5. SIGNATURE MANAGEMENT

The countermeasure to todays multispectral threat is Signature Management. The Plessey Company offers a complete consultancy and implementation package for all elements of military platform signature management.

The assembled capability offers an integrated service covering:-

Theoretical Prediction
Modelling
Comprehensive Full-Scale Measurement
Systems Assessment
Materials Data-Base and
Implementation

Materials and Processing – Move into the 90's
edited by S. Benson, T. Cook, E. Trewin and R.M. Turner
Elsevier Science Publishers B.V., Amsterdam, 1989

DEVELOPMENT OF A LARGE DEPLOYABLE CARBON FIBER COMPOSITE ANTENNA STRUCTURE FOR FUTURE ADVANCED COMMUNICATIONS SATELLITES

F. Grimaldi and G. Tempesta

Selenia Spazio, Via Pile 60, 67100 L'Aquila, Italy

R. Stonier

Consultant

ABSTRACT

A very advanced antenna subsystem has been developed for future communications satellite missions. The antenna consists of a multiple beam high frequency Ka-Band (20/30 GHz) feed array with a 4-meter (13.1 ft) diameter carbon fiber composite reflector structure. The reflector has two foldable tip segments incorporating precision hinge mechanisms for deployment, so that the antenna will fit inside the Ariane launch vehicle. This antenna reflector is one of the largest carbon fiber composite spacecraft reflector structures ever developed. The manufacturing of full-scale hardware for qualification testing has recently been completed at the Selenia Spazio Composite Manufacturing Plant in L'Aquila, Italy. The engineering effort involved in the materials, processes and manufacturing technology for this advanced composite structure are described in this paper.

1. INTRODUCTION

Selenia Spazio has developed a new 4-meter (13.1-ft) diameter carbon fiber composite antenna reflector on an ESA (European Space Agency) development program. This structure (referred to as the ASTP 20/30 GHz Antenna) is the world's largest solid dish satellite antenna reflector of this type. It is part of a very advanced Ka-Band (20/30 GHz) antenna subsystem now being developed and qualified for future communications satellites. The reflector ia a rib stiffened structure with two foldable tip segments incorporating precision hinge mechanisms for deployment, so that the antenna will fit inside the Ariane launch vehicle. A new hinge mechanism for the reflector tips was developed and qualified for Selenia Spazio by Contraves Italiana(1). A new P-75S carbon fiber epoxy prepreg material was also qualified for this structure, and a semi-cocuring process was developed for manufacturing. The full-scale Engineering Model ASTP reflector structure, shown in Figure 1, has recently been completed at the new Selenia Spazio Advanced Composite Manufacturing Facility in L'Aquila, Italy. Structural qualification testing of this antenna is currently in progress.

Figure 1 Engineering Model 4-meter Diameter ASTP 20/30 GHz Antenna Reflector

2. CONFIGURATION

The offset parabolic reflector consists of a large center section (manufactured in two pieces and joined together by adhesive bonding) and two foldable tips that deploy in-orbit (Figure 2).

The design philosophy used for the reflector dish was derived from successful hardware programs at Selenia Spazio such as INSAT-I[(2)] and ARABSAT.

The same type of design was also used for the ITALSAT 20/30 GHz antenna reflectors[3]. The design consists of a thin honeycomb composite sandwich shell with thin skins and stiffening ribs. This type of construction resulted in a lighter weight design than a more conventional thick sandwich shell or ring stiffened shell configurations. The design requirements are listed in Table 1.

TABLE 1

DESIGN REQUIREMENTS FOR ASTP 20/30 GHz ANTENNA REFLECTOR

Antenna Electrical Frequency	20/30 GHz
Reflector Diameter	4.0-meter (13.1-ft)
Projected Aperture Dimension	3.7-meter (12.1-ft)
Maximum Mass	40 Kg (88 lb)
Max Deviation from Parabola in-orbit (best-fit RMS)	0.3 mm (11.8 mil)
Operating Temperature	-180°C to 100°C (-292°F to 212°F)

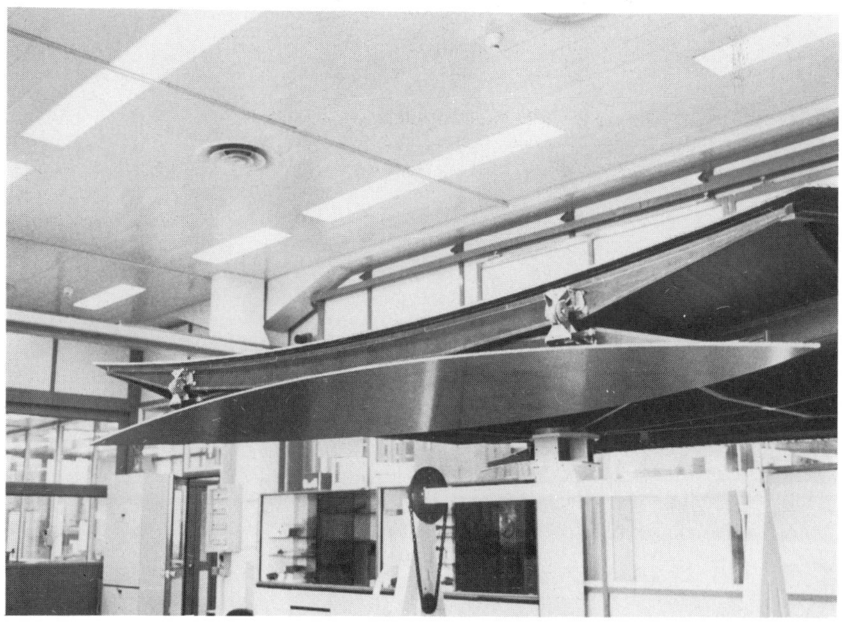

Figure 2 ASTP 20/30 GHz Antenna Reflector with Folded Tips

2.1 Reflector dish design

The reflector dish design consists of four separate reflector shell pieces manufactured on four different layup molds. The large center section consists of two pieces joined together by adhesive bonding during the final assembly of the reflector.

The two tips are separate pieces and are integrated with the deployment hinges during final assembly. Two hinges (Figure 3) are used for each tip. The interfaces with the hinge mechanisms consist of titanium brackets which are attached to the ribs by means of titanium inserts.

The reflector dish structure can be subdivided into two main categories, the reflector shell and the stiffening ribs.

Figure 3 Deployment Hinge Mechanism for Foldable Tips of ASTP 20/30 GHz Antenna Reflector

Reflector Shell

The separate pieces of the reflector shell are manufactured as follows: Each piece has two carbon fiber reinforced plastic (CFRP) skins (Pitch-75S epoxy prepreg, 4 layers with a stacking sequence of 0°/90°/90°/0°) and aluminium honeycomb core. The sandwich is manufactured using a structural film adhesive (FM-300M) and a semi-cocuring process. The 0° and 90° layers are perpendicular and parallel respectively to the symmetry axis of the reflector dish. The cured skin has a nominal thickness of 0.24 mm (0.010 in) and the honeycomb core is 6.35 mm (0.25 in) thick. The open honeycomb core along the edge of the sandwich perimeter is enclosed by means of a space qualified polyimide tape. Provision for venting is achieved by means of perforated honeycomb and needle holes in the perimeter enclosure tape.

Stiffening Ribs

The stiffening ribs which can be seen in Figure 4, consist of a sandwich

Figure 4 Stiffening Rib Detail of 4-meter Diameter Antenna Reflector

structure with two CFRP skins (Pitch-75S epoxy prepreg, 6 layers with a (0 $\pm$ 60)s stacking sequence) and aluminium honeycomb core. The main function of the ribs is to increase the stiffness of the reflector dish, to structurally join the two halves of the central body of the reflector, and to provide for the interface for the deployment arm, APM (antenna positioning mechanism) and four pyrotechnic hold-downs.

Particular care was placed on developing the proper bonding procedures and assembly tooling so that the necessary stiffness and the required surface contour RMS of the reflector dish could be achieved.

The upper edges of the ribs have precured closeout channels (CAPS). The bonding between shell and ribs is obtained by means of precured angle strips (CLIPS) on both sides of the ribs. Both the caps and clips consist of a woven fabric carbon fiber epoxy composite.

The reflector interface with the APM consists of a Ti6AL4V titanium fitting attached to the ribs by means of inserts and titanium screws. The inserts are potted into the ribs by means of a room temperature curing epoxy syntactic foam material.

2.2 Structural analysis

The antenna structure was designed to achieve the required natural frequency and survive launch loading conditions. A structural/thermal mathematical model, shown in Figure 5, was developed. Natural resonance frequencies were calculated and the first six eigen frequencies are shown in Table 2. A detailed stress analysis was performed around highly loaded areas such as the four pyrotechnic hold-down fittings.

TABLE 2

NATURAL RESONANCE FREQUENCIES
CALCULATED for ASTP 20/30 GHz ANTENNA REFLECTOR

Natural Resonance Frequencies Stowed Configuration			
f1	=	47	Hz
f2	=	64	Hz
f3	=	67	Hz
f4	=	78	Hz
f5	=	86.5	Hz
f6	=	96.5	Hz

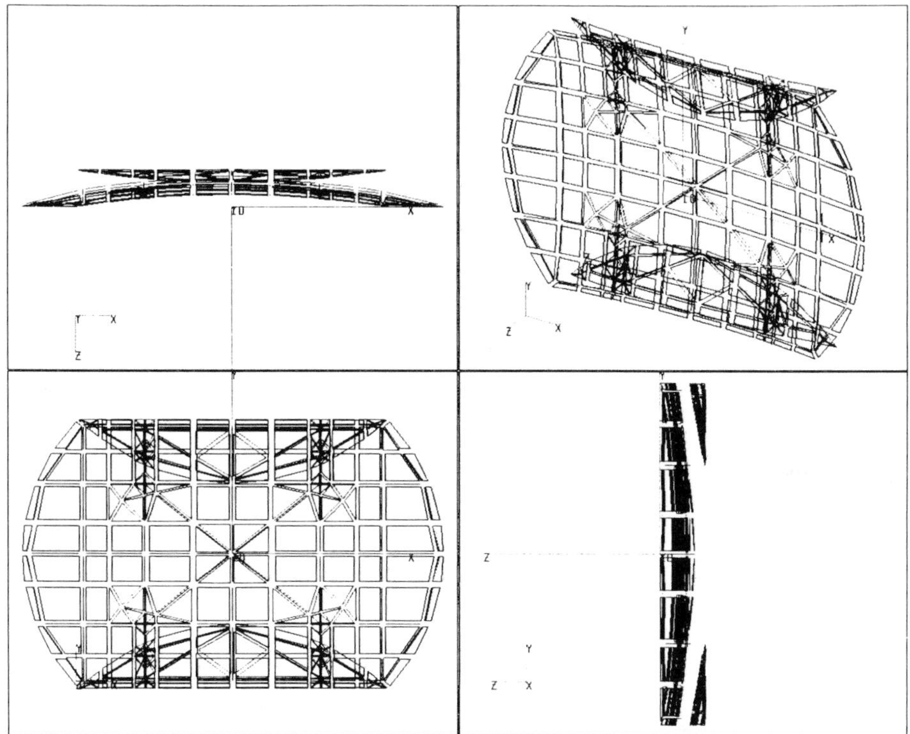

Figure 5 Structural/Thermal Mathematical Model of 4-meter Diameter Antenna Reflector with Foldable Tips

3. MATERIALS

Materials development and qualification testing was performed for the materials used on this program. Physical and mechanical properties were obtained for the P-75S/934 thin ply 0.06 mm (2.5 mil) per ply carbon fiber prepreg tape supplied by Fiberite Corp. Table 3 shows typical unidirectional and pseudoisotropic mechanical property data for this material.

Coefficient of thermal expansion (CTE) testing was performed by ESA/ESTEC for the sandwich construction of the reflector shell (center section and tips) and for the ribs. The average CTE values over the temperature range of -120 to +120°C are shown in Table 4.

A T-300/934 woven fabric broadgoods material (Fiberite Corp. designation HMF-134/34) was used for manufacturing clips for rib to shell and rib to rib joints and for upper rib close-out caps. A lighter weight fabric based

TABLE 3

TYPICAL MECHANICAL PROPERTIES OF
UNIDIRECTIONAL AND PSEUDOISOTROPIC
P-75S/934 EPOXY COMPOSITES

	UNIDIRECTIONAL PROPERTIES	PSEUDOISOTROPIC PROPERTIES
Tensile strength, MPa (ksi)	800 (116)	310 (45)
Tensile modulus, GPa (msi)	303 (44)	103 (15)
Compression strength, MPa (ksi)	379 (55)	172 (25)
Compression modulus, GPa (msi)	242 (35)	83 (12)
Flexure strength, MPa (ksi)	655 (95)	276 (40)
Flexure modulus, GPa (msi)	262 (38)	103 (15)
Short beam shear strength, MPa (ksi)	64 (9.3)	48 (7.0)

Room Temperature Data
Data based on Fiberite Corp. HY-E 2034D

TABLE 4

COEFFICIENT OF THERMAL EXPANSION
OF SANDWICH COMPONENTS OF
ASTP ANTENNA REFLECTOR

REFLECTOR COMPONENT	P75S/934 SKIN CONFIGURATION	ALUMINUM HONEYCOMB CORE TYPE	SANDWICH CONSTRUCTION	CTE, cm/cm/°C x 10^{-6} -120 to + 120°C
Central Shell & Tips	0°/90°/90°/0°	3/16-5052-.0007-2.0 pcf 6.35 mm (0.25 in) thick	Semi Cocured Skins	1.04
Ribs	(0 ± 60)s	1/4-5052-.0007-1.6 pcf 12.7 mm (0.50 in) thick	Precured skins	0.96

P75S/934 Thin Ply 0.06 mm (2.5 mil/ply) prepreg tape supplied by Fiberite Corp.
FM-300M Film Adhesive used for honeycomb core bonding supplied by American Cyanamid.

on a 1000 filament (1K) T-300 yarn (Fiberite Corp. designation HMF-341/34) was also evaluated for potential future weight saving. The typical physical and mechanical properties of these two graphite fabric materials are shown in Table 5.

Honeycomb core to carbon fiber composite skin bonding was achieved using FM-300M modified epoxy film adhesive supplied by American Cyanamid. All assembly bonding was performed at room temperature using Hysol EA-9321 structural epoxy paste adhesive, which has good elevated temperature properties. Typical shear strength values for this adhesive are shown in Figure 6.

TABLE 5

TYPICAL PHYSICAL AND MECHANICAL PROPERTIES OF THIN HIGH STRENGTH WOVEN GRAPHITE/EPOXY FABRIC COMPOSITE LAMINATES

MECHANICAL PROPERTY	HMF 134/34	HMF-341/34
Tensile strength, MPa (ksi)	550 (80)	655 (95)
Tensile modulus, GPa (msi)	65 (9.4)	66 (9.6)
Compression strength, MPa (ksi)	425 (62)	620 (90)
Compression modulus, GPa (msi)	55 (8.0)	63 (9.1)
Flexure strength, MPa (ksi)	620 (90)	869 (126)
Flexure modulus, GPa (msi)	60 (8.7)	61.3 (8.9)
Short beam shear strength, MPa (ksi)	55 (8.0)	86.2 (12.5)
Nominal cured ply thickness, mm (in)	0.18 (.007)	0.13 (.005)
Number of filaments/yarn	3000	1000
Weave style	Plain weave	Plain weave
Weave construction, yarns/in.	12.5 x 12.0	24 x 24
Fabric areal mass, g/m^2	190 $\pm$ 8	128 $\pm$ 5

Fabrics contain T-300 carbon fiber.
Prepreg based on Fiberite Corp. 934 matrix resin system.

FIGURE 6

TYPICAL SHEAR STRENGTH OF EA-9321 STRUCTURAL PASTE ADHESIVE

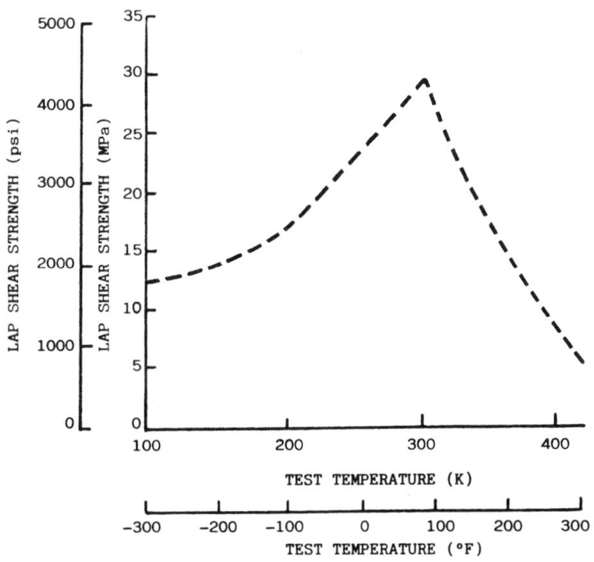

4. TOOLING

A low CTE cast iron material (GS-370-17) was used for manufacturing the layup molds for curing the carbon fiber composite parts making up the reflector shell (CTE = 11 x 10^{-6} cm/cm/°C & 5.9 x 10^{-6} in/in/°F). The mold construction consists of a thin metallic surface with integrally cast ribs. There are four different layup molds, two for fabricating the center section sandwich pieces (plus an extension mold that attaches to the central edge of each of these molds, because this edge has been cut net), and a separate mold for each of the two foldable tips. After manufacturing the two center sections of the shell, these two main molds (which are each one-half of the overall surface of the reflector) are joined together to form the assembly tool for the entire structure, shown in Figure 7.

All the layup molds were machined at the same time using an N.C. vertical turn lathe. The RMS surface contour error from the desired theoretical shape was measured using a DEA (Digital Electronic Automation) inspection

Figure 7 Assembly Tool for the 4-meter Diameter Antenna Reflector Structure

machine, and the following RMS values were obtained:

Two Main Molds :	0.048 mm, (1.9 mil)	0.037 mm RMS (1.5 mil)
Two Tip Molds :	0.021 mm (0.8 mil)	0.037 mm RMS (1.5 mil)
Main Molds Joined Together :	0.065 mm RMS (2.6 mil)	

These values are considered to be very good for conventional machining of this large a diameter (11-meter/36-ft. diameter surface of rotation due to the offset of the parabolic curve).

A tubular steel cradle structure was used to support each mold during curing and allow autoclave airflow up under the mold. A different steel support structure (Figure 8) was used for supporting the two mold pieces joined together for reflector assembly.

Figure 8 Assembly Tool with Tubular Steel Support Structure for ASTP 20/30 GHz Antenna Reflector

A number of other molds and tools were used for fabrication and trimming of clips, caps and structural brackets and for trimming the sandwich parts. Assembly tooling was used for rib location and rib to shell clip bonding. All bonding was performed at room temperature and the rib bonding tool design provided for applying pressure to all bond-lines. Additional tooling was used for locating inserts and brackets prior to installation and for integration of the hinge mechanisms.

5. PROCESSING AND MANUFACTURING

The use of a new composite prepreg material containing ultra high modulus Pitch-75S carbon fiber along with a new structural film adhesive (FM-300M), necessitated the development of completely new curing/fabrication processes. A semi-cocuring autoclave process was developed for manufacturing the various sandwich parts of this reflector structure. The front skin is first precured on the mold and then the honeycomb core bonding and back skin cocuring is performed in a second manufacturing step.

Rib assembly and bonding is performed at room temperature. All clip, cap and structural fitting bonding, and all insert potting is also performed at room temperature. Selenia Spazio utilized these proprietary processes for the laminating, sandwich cocuring and bonding operations performed in manufacturing this Engineering Model ASTP antenna reflector structure.

6. TESTING

Testing of the completed Engineering Model antenna reflector is currently being performed. The reflector structure, with all the titanium interface fittings, had a mass of 42 Kg (92.6 lb). A detailed mass summary is shown in Table 6. The as-manufactured surface contour RMS is now being determined before environmental testing (Figure 9).

Mechanical impedance non-destructive inspection (NDI) equipment has also been used to inspect the sandwich structures for any unbonded or delamination conditions. The surface contour check and NDI inspection will be repeated after structural qualifcation environmental testing.

Structural Qualification testing will consist of sine vibration, acoustic noise and thermal vacuum cycling. One reflector tip will also be deployed at the coldest temperature (-120°C/-184°F) of the thermal vacuum test.

TABLE 6

FINAL MASS SUMMARY
4-METER DIAMETER ASTP 20/30 GHz ANTENNA REFLECTOR

Reflector dish	19.3 Kg	(42.6 lb)
Ribs	9.85	(21.7)
Clips and caps (including adhesive)	6.10	(13.5)
Tips hinges	1.35	(3.0)
Inserts and screws	1.2	(2.6)
Titanium brackets and interfaces	3.1	(6.8)
Deployment arm fittings	1.1	(2.4)
	42.0 Kg	(92.6 lb)

Figure 9 Surface Contour Inspection of 4-meter Diameter Antenna Reflector

7. SUMMARY

A 4-meter (13.1-ft) diameter lightweight carbon fiber composite satellite antenna reflector has been designed, developed and manufactured on a European Space Agency (ESA) development program. This structure which is now undergoing extensive qualification testing will have a deployment arm attaching to the center of the dish. This precision structure, designed to operate at Ka-Band (20/30 GHz) electrical frequency has two foldable tips so that it can fit into the Ariane launch vehicle. New materials for this advanced structure have been qualified and manufacturing and assembly processes have been developed. This structure, the largest solid dish advanced composite satellite reflector of its type, has been developed for future communications satellite missions.

ACKNOWLEDGEMENTS

The authors would like to acknowledge the European Space Agency (ESA) and thank ESTEC engineers G. Crone, R. Garcia and G. Reibaldi for their involvement and technical assistance on this program. The authors thank Selenia Spazio engineers L. Scialino, G. Noni, A. Meschini, C. Bruno and F. Morganti for their design and analysis efforts; C. Manzi for supervising the tooling design; L. Donadoni and F. Salustri for supervising the manufacturing; S. Pesciarelli for mechanical and thermal testing; Program Manager, R. Braschi and Project Engineer, F. Bruschetta.

REFERENCES

1) D. Stella, F. Morganti and G. Nielsen, Contraves' Antenna Tip Mechanism for Selenia Spazio 20/30 GHz Antenna, 2nd ESA Workshop on Mechanical Technology for Antennas, ESTEC, Noordwijk, The Netherlands, ESA SP-261 (1986) 185.

2) A. Pace, I. La Rosa and R. Stonier, Development of Lightweight Dimensionally Stable Carbon Fiber Composite Antenna Reflectors for the INSAT-I Satellite, 103 of ref. 1.

3) F. Grimaldi, G. Tempesta, F. Pastorelli, S. Pesciarelli and R. Stonier, Development of Dimensionally Stable Lightweight Composite Satellite Antenna Structures, 34th International SAMPE Symposium and Exhibition, Reno, Nevada, 'in print'.

Materials and Processing – Move into the 90's
edited by S. Benson, T. Cook, E. Trewin and R.M. Turner
Elsevier Science Publishers B.V., Amsterdam, 1989

DESIGN, FABRICATION AND EXPERIMENTAL TEST OF HI-TEMPERATURE CFRP STIFFENED STRUCTURES.

Didier VIGNERON

Hispano Suiza, Route du Pont VIII, B.P 91, Gonfreville l'Orcher,
76700 Harfleur, France

ABSTRACT

This paper describes the design and fabrication development done for the rotating cowlings that were installed and tested on the GE36 UDF TM demonstrator engine. The rotating cowlings are the structure between the blades that define the external line of the nacelle.

The maximum weight saving required on the engine imposes the use of composites materials. The panels have to widthstand maximum service temperature up to 250°C, with a given limited deformation due to centrifugal and pressure loads.

The only materials available on the market able to meet these requirements are the condensation reaction polyimide resin systems, PMR15. Although much work has been done for the past 10 years on these materials, little experience is known about the vacuum bag/autoclave techniques processing of carbone fabrics polyimide prepregs. Moreover, associate materials such as honeycomb, or adhesive systems are difficult to use in civil aircraft applications, due to their lack of toughness and commercial availability.

The chosen concept for the rotating cowls was to design monolithic stiffened panel "cocure" without any adhesive. The external skin of each rotating panel is made of five plies of carbon-polyimide fabric prepreg (0,28 mm/ply). The thickness is increased up to 23 plies in the fastener areas on the both sides. The required rigidity of the skin is reached by adding three internal hat section stiffeners between the thickned areas.

Preliminary design allowables have been determined in order to calculate mechanical behaviour of the panels. A finite element model, using NASTRAN analysis, has enabled us to identify the highly loaded areas and choose margins of safety (static & dynamic).

The complexity of the structure have led us to develop a specific way of curing ; the standard cure cycle of the PMR15 systems consists of a first stage of imidization of the monomers present in the resin followed by a polymerisation stage. The one-stage process developped for the rotating cowls have consisted in imidizing separately the external skin and the different stiffeners. Then, the fully imidized components were positionned in a complex tool, "bonded" through a thermoplastic phase and finally cured under pressure. Ultrasonic inspection of the parts showed a low void content even in the thick areas and the imidization joint. Moreover micrography of the first article part demonstrated that no micro cracking occur in the laminates.

Static and vibration test performed on two panels showed that the maximum measured deflection of the panel is well correlated with the predicted model. There is no propagation of defect under a representative dynamic loading after 10^7 cycles.

Finally, a set of 13 panels has been manufactured and inspected. Each panel was weighed and balanced with $\pm$ 1 gram before delivery. They have been installed on the demonstrator engine without any problem of fitting and dynamically balanced within $\pm$ 0,1 gram. The engine test flight performed on the MD80 aircraft has shown a satisfactory behaviour of the rotating cowlings after more than 100 flight hours.

1. INTRODUCTION

The high performance composite materials are largely used in aircraft structures including primary parts, but still remain very limited on engines and surrounding areas. The first applications on civils engines consisted of a few small non structural parts, such as fairings, and continued by increasing the dimensions and performances of the parts and including carbon fiber reinforced composites. Nowadays, all the external area of the engine, i.e thrust reverser and overall nacelle, is mostly made of composite materials. Therefore, HISPANO SUIZA subsidiary of SNECMA Group has used for several years the composite materials in thrust-reversers on civil fan engines. An example is given on the figures 1

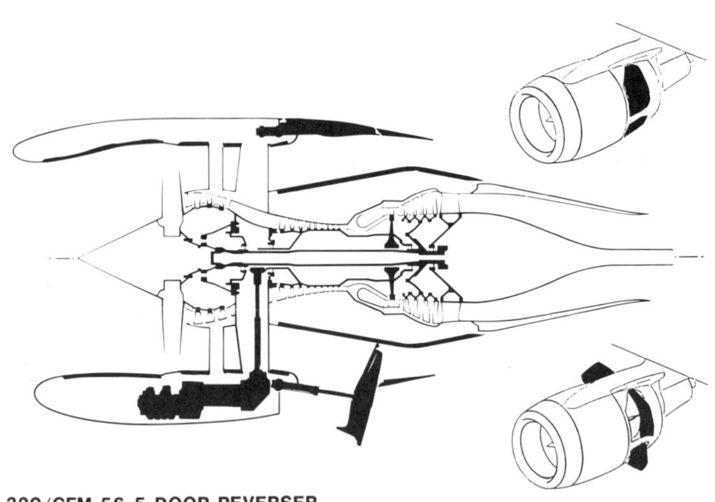

FIGURE 1 : THRUST REVERSER INSTALLATION ON AN ENGINE

These composite materials are generally carbon epoxy fabric prepregs able to widthstand service temperatures up to 150°C.

The following step consisted in increasing the service temperature capability of the composite materials up to 250°C - 300°C in order to apply them :

. on the areas closer to the engine center line for the actual engines, such as shown in figure 1,

. or, on the overall nacelle for engines currently under development, such as unducted fan engine (THR/UDF).

Many different composite systems (polyimides, bismaleimids, thermoplastics..) are proposed to cover this temperture range and seem attractive for the future. At the present time, the most interesting and available industrial product is the PMR15 resin. This polymerizable monomer reactants resin is polyimide with a molecular weight of 1500, developped by NASA, early 1970. Since, it is a significant improvement compared to the condensation polyimide, this resin is very difficult to process and present in-service problems such as laminate microcracking. Different work (1,2) mainly in the United States, have shown that PMR15 has the potential for meeting the structural requirements for military engine applications.

The development work in this program consists in designing and manufacturing the access panels of the second SNECMA/GENERAL ELECTRIC THR/UDF demonstrator engine.

THR/UDF ENGINE DESCRIPTION

The THR is a new generation engine developped jointly by GENERAL ELECTRIC and SNECMA for civil aircraft application. This ultra high bypass engine, an unducted fan engine which drives two counter-rotating bladed fans, is able to power the different new aircraft developped on the market (MD91, MD92, 757 ...). It is proposed that engine allows three-fold improvment in the ratio of fuel to payload. Overall, direct operating costs will go down by as much as 10%. Beyond this, transport equipped with THR/UDF engines will be able to operate in and out of the world's most sound sensitive airports, with a very low cabin noise level excepted.

ACCESS PANELS DESCRIPTION

In this program, currently under developement, HISPANO SUIZA has been selected in 1986 to design and manufacture, for SNECMA, the 1st stage access panels, (10 blades) on the second demonstrator engine. These panels are rotating cowlings which form the external lines of the nacelle between the blades - as shown in figure 2. The panels must withstand the following requirements :

. service temperature of 204°C (with a maximum 260°C in engine soak back condition)

. loading conditions due essentially to centrifugal forces :

- cruise conditions (1300 RPM)
- limit conditions (1500 RPM)

. the overall external line of the nacelle must be flush in cruise conditions

. the panels have to widthstand dynamic loading : mechanical (up to 25 000 cycles service life) and acoustical fatigue

. they have to be equilibrated within ± 1 gram before delivery and must be replaceable.

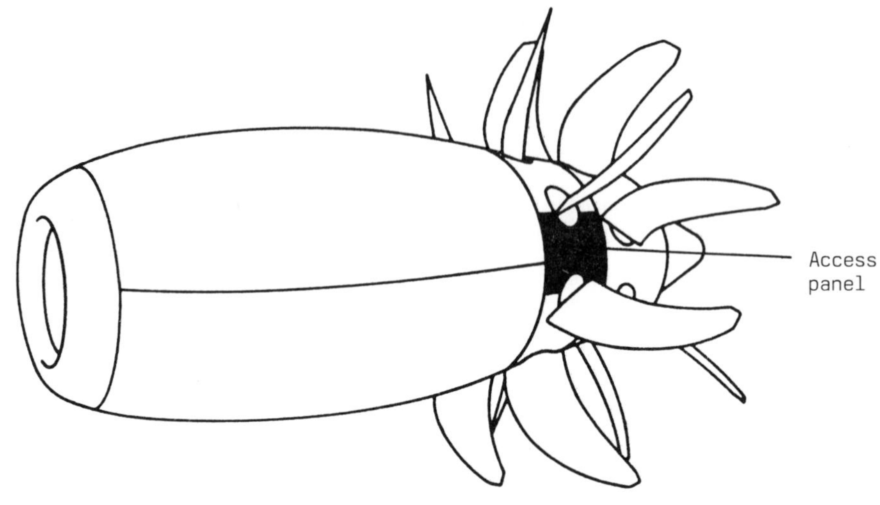

FIGURE 2 : THR / UDF ENGINE INSTALLATION

Moreover, the weight target was set at 500 grams per individual panel. Each panel, which dimensions are approximatly (430 X 360 mm), is fitted by two fastener-lines on the foward and aft frames structures, as shown in figure 3. On each side, the blade foot location inducted a circular cut-out in the panel. In that area, due to the sealing requirement between the blades and the panel, the seal surrounding the blade foot has to fit in a circular ring on the edge on the panel. The ring must be sufficiently high to contain the vertical relative displacement between the blades and the nacelle.

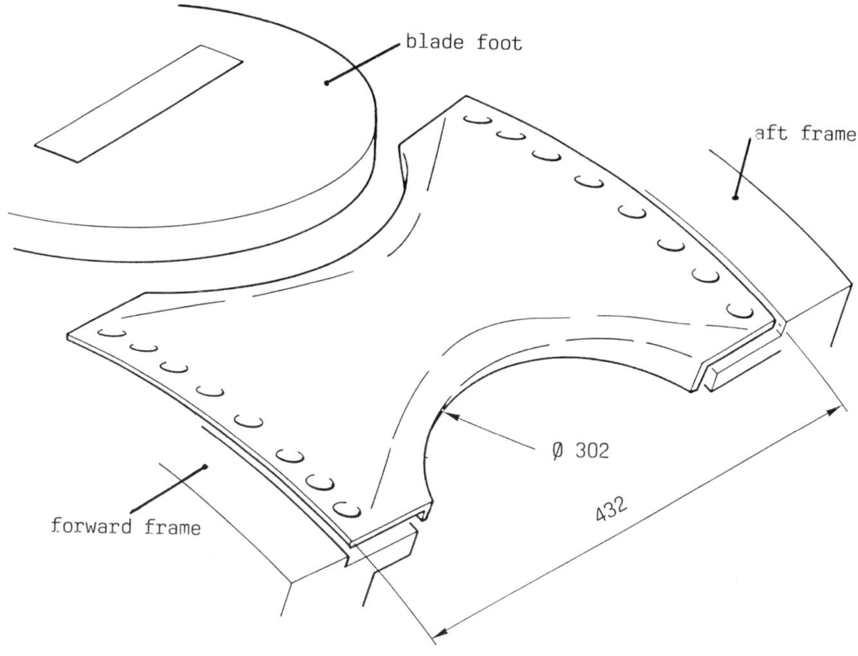

FIGURE 3 : ACCESS PANEL INSTALLATION

DESIGN PHILOSOPHY

The severe requirements due to the flush external nacelle line conditions under loading, imposed to limit the maximum displacements of the panel. We have fixed an arbitrary maximum value of 5 mm for the external surface displacement. The best way to achieve it is to reinforce the panel by two possible means :

. design a sandwich construction

. or design a stiffened monolithic construction

After a preliminary investigation of both possibilities, the first was abandonned because of the lack of confidence in the few materials commercially available. Evaluation of the core materials included metallic honeycomb (titanium, stainless steel, nickel base alloy ...) which presents severe problems of surface preparation before bonding and affect the panel weight requirement. The non-metallic core honeycomb such as glass polyimid was difficult to form and very sensitive to fatigue environment. Concerning the adhesive, the almost only available system the FM35 cyanamid polyimide adhesive film, seems very brittle. From another side, the bonding process required to bond :

. precured skins on the core that leads to difficulty to produce and control an accurate bond joint quality due to the geometry of the panel.

. Or cocured/semi cocured skins on the honeycomb that is not suitable regarding the structural requirements of the panel.

Anyway, each type of sandwich construction show an impact on the weight panel that it is sufficient to eliminate the sandwich design solution.

PRELIMINARY DESIGN

The stiffened preliminary design consists of two skins cured separately and bonded together in a second-stage process. The external skin include the fasteners areas and the blade foot edge sections. The internal skin is a omega shape section laying along the greater dimension of the panel and extending under the fastener area.

Unfortunatly, evaluation tests performed on flat specimens have shown that the mechanical performance of the adhesive film system is too poor considering shear and peeling caracteristics. Moreover, the lack of confidence was increased by detection of high porosity in bond lines joints subjected to severe arbitrary conditions. The conclusion at that time of the development was that no bonding should be used for that application.

FINAL DESIGN

The final design shown in figure 4 is different from the preliminary design essentially from a process point of view. There is no longer, any adhesive between the internal and external skins. The PMR15 resin system used was cured in a one-shot process. Preliminary physical and mechanical evaluation on samples manufacture in an autoclave process give the basic data, used for design analysis. The selected reinforcement 5H satin carbon fabric was choosen because of the easy use and drape in such contours.

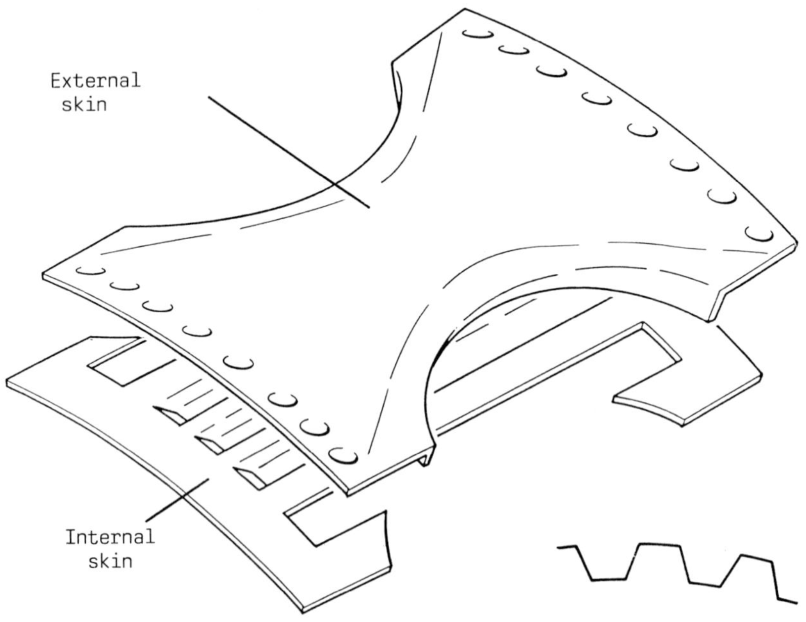

FIGURE 4 : FINAL DESIGN

Considering the paint scheme, the panel is primed with an epoxy-polyamide primer and only the outer skin is painted with a white polyurethane finish. The definitive geometry of the reinforced panel was achieved through a finite element model (FEM) analysis using NASTRAN code. The results presented in figure 5 show the maximal loaded areas and corresponding displacements of the external line of the panel in cruise conditions. The most difficult area to calculate was the transition between the end of the stiffeners and the fasteners areas. Note that the relative displacements of the forward and aft frames are included in the analysis. By removing the maximum displacements calculated from the cruise flush external nacelle line, we can determine the theoretical line of the non-loaded panel that will become the reference line for the manufacture tooling.

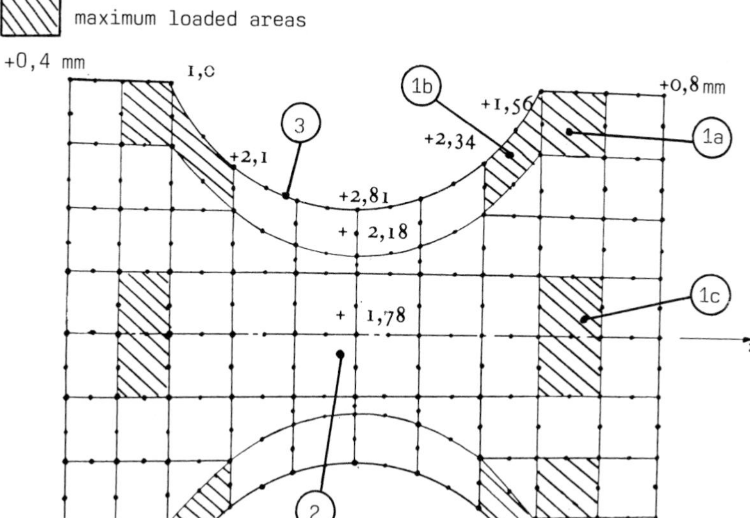

FIGURE 5 : PANEL EXTERNAL DISPLACEMENTS IN MM
(cruise conditions)

The FEM analysis uses allowables derived from bibliography correlated with internal work. The TSAI-WU quadratic failure criteria used, taking into account the thermal effects, give the safety of margin of the main areas in the panel, as shown in figure 6 The satisfactory behaviour of the fasteners avoid any problem during flight.

Areas see figure 5	Failure mode	Margin of safety
1A	Tension + flexion	1,67
1B	Tension + flexion	0,14
1C	Tension + flexion	0,63
2	Tension + flexion	0,68
3	Torsion	0,10
4	Fasteners bearing	0,81

FIGURE 6 : STRESS DISTRIBUTION IN THE PANEL (ULTIMATE CONDITIONS)

PROCESS DEVELOPMENT

We have noted that the panel has to be "cocured" for reaching the mechanical requirements. How do we achieve that difficult task ? That was the problem, knowing the high volatile content in the prepreg and the difficulty to process.

STANDARD CYCLE

Firstly the standard cycle for curing a flat standard laminate had to be determined. This had been done using physical and chemical analysis such as dynamic mechanical analysis (DMA-VANHOGRAPH). At that time, sophisticated mean of investigation such as HPLC were not yet available at HISPANO SUIZA. Figure 7 indicates the standard cure cycle used for flat, thin laminates. Different samples processed in that way have shown good reproducibility of mechanical performance associated with a low porosity content. This cure cycle was adapted for thick laminates.

No microcracking was observed on the different laminates, after curing. Note that there is no post-cure needed for the relatively low service temperature requirements.

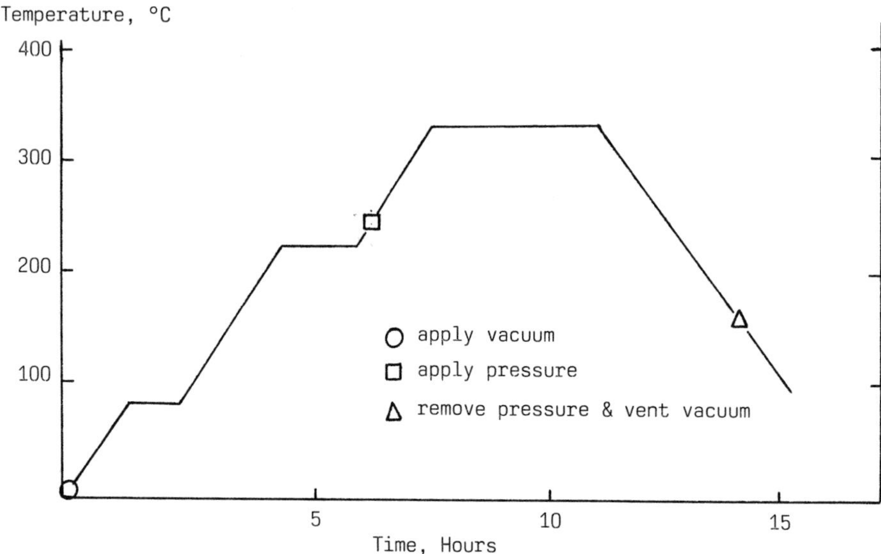

FIGURE 7 : PMR15 AUTOCLAVE CURE CYCLE

OPTIMISED CYCLE

The target was to be able to cocure together the internal and external skins that are both complex shapes locally thick. Preliminary representative trials revealed a high porosity content in the interface cocured line, but also in the overall thickness of the laminate. The basic idea is to separate the standard cure cycle into two different stages :

. The first stage acts as a debulking process and allows the high content of volatiles and water present in the prepreg lay-up to evaporate in the vacuum line of the autoclave. The other advantage is to form the lay-up in the final form and thickness required.

. The second stage is the curing cycle.

For the first stage "debulking" cycle three different temperature levels, for the imidization reaction were investigated. Both chemical analysis on debulked prepregs, supported by destructive and non-destructive tests on laminates have been performed. The results indicated an incomplete imidization at the two low temperature levels. The mechanical performances, flexure and interlaminar shear, are affected by that inadequate debulking at low temperature as shown in the figure 8. The 17 % reduction compared to the high temperature debulked specimen noted for interlaminar shearstrength tested at 260°C is confirmed by the high level of porosity of the laminate detected through ultrasonic inspection and specimen dissection.

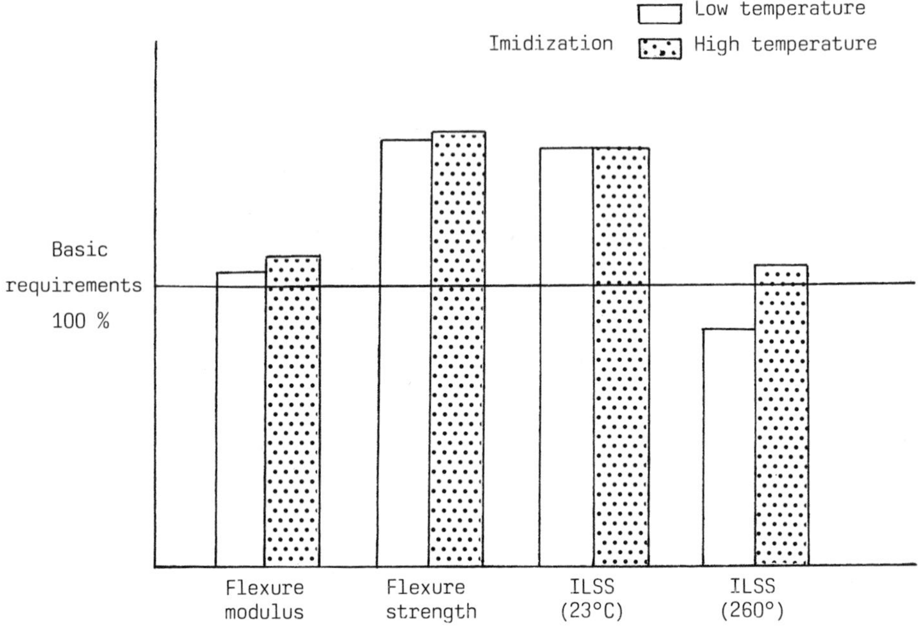

FIGURE 8 : MECHANICAL PERFORMANCE FOR TWO IMIDIZATION LEVELS

PANELS MANUFACTURING

The manufacturing process is the following :

. Lay-up separately internal and external skin on individual machined tools.

. Debulk in a vacuum bag/autoclave technique. The materials are this stage thermoforming, bled and imidized.

. Remove the hat stiffener section from his tool and transfer it on the external skin; auxiliary silicone tools are needed to fill the areas above the hat section and the complete volume under the stiffeners between the circular edge flanges. The silicone rubber acts as a liquid when the pressure is applied.

. Cure the part in the autoclave. The tool is designed in such a manner that there is no stress applied on the part during the cooling phase of the cure cycle avoiding any internal stress due to wear between the tool and the manufactured part. So, the piece can be removed very easily from the tool.

After curing, each panel is clamped on a tool for machining and trimming at the final geometry. The panel is then drilled and the dimpled inox washers are bonded using a supported film adhesive. When a complete set of panels have to be manufactured for one engine, each panel is weighed after washer bonding, and balanced within ± 1 gram. This is done by removing some amount of materials on a four place-pattern in the internal side of the panel, located at the proximity of the fasteners lines, close to the flanges. Note that the balance operation is made before painting, because of the dirty nature of this machining work.

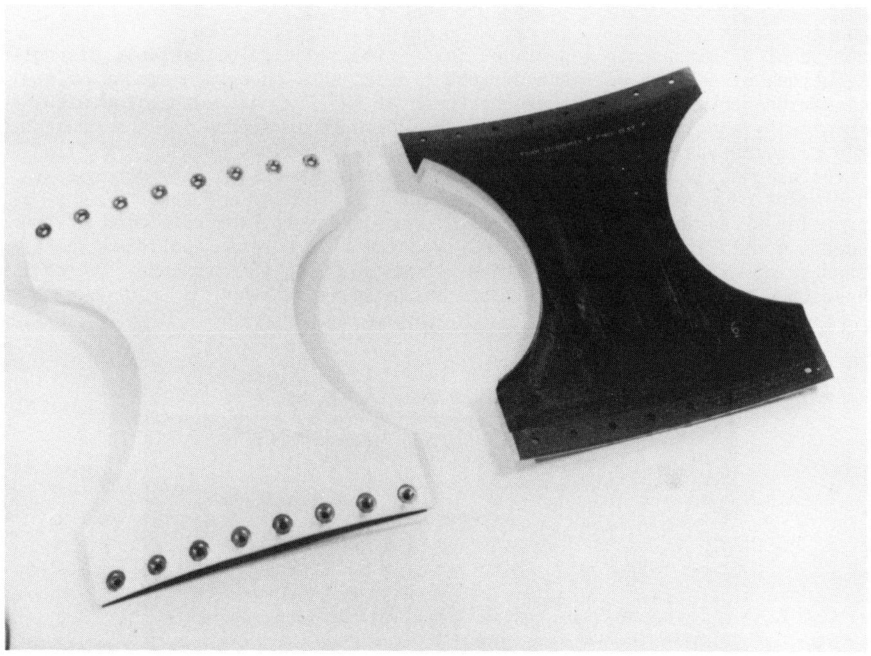

FIGURE 9

Naturally, throughout the overall process, accurate inspection is performed. Non-destructive control include visual examination during lay-up, ultrasonic inspection and final geometrical conformity. Preliminary study on representative samples including manufacturing defects, have given the acceptance criteria to use during ultrasonic "A" scan inspection. Destructive control include the mechanical characterisation of standard monolithic specimens, i.e flexure and ILSS tested at room temperature and 260°C. In addition, the mechanical behaviour of the imidization bond line was measured for each panel through a specific single lap shear specimen, representative of the process.

All the complete process has been validated before production manufacturing by a "first article" units procedure. Three parts were used to take into account the variability processing parameters (these parts were taken in a development set of panels).

In the first development panel, some defects were detected in the thick areas of the panel by ultrasonic and visual inspection (edge delamination). Micrographic examinations of samples cut from different areas show a certain amount of small delamination arranged through the thickness of the laminate associated with microporosity and local lack of resin. As indicated in figure 10 the crack tip of the delamination is filled by transversal PMR15 fillets between the "lip" edge representative of a viscoplastic behaviour of the resin. The proposed mechanism of degradation is the accumulation of volatiles and water on the prepreg lay-up during the heating cycle, under pressure. The remaining solvants induced a modification of the chemical reaction of the resin and formation of viscous bridges between the porosities. After removing the pressure, internal stresses due to gaz inclusions initiate micro delaminations.

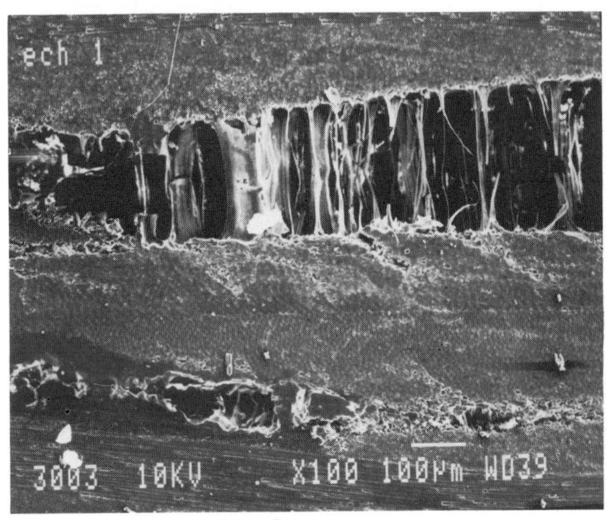

FIGURE 10 : TYPICAL DEFECT ON THICK LAMINATE

DUE TO INADEQUATE CURE CYCLE

Through the development phase, the tooling system volatiles drainage was improved in order to remove excess solvants and water. The "first" article panel shows no delamination or microcracking. Porosity range measures on representative areas show a good quality of the panel, confirmed by DSC analysis.

TESTS OF THE ACCESS PANELS

Different panels were subjected to :

. static test

. mechanical vibrations

. flight test

STATIC TEST

The aim of that test was to simulate the centrifugal forces applied to the panel and correlate experimental measured and calculated displacements. The test jig is shown on figure 11 et 12.

FIGURE 11 : STATIC TEST EQUIPMENT

The panel chosen in the development phase test was instrumented with several deformation gages and displacement transducers and tested at room temperature, non aged.

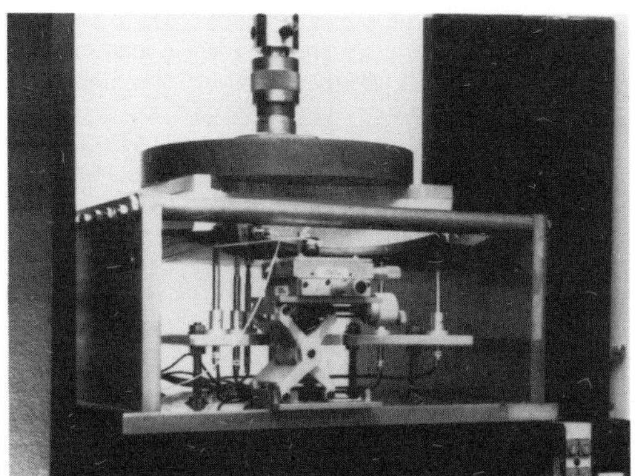

FIGURE 12 : DETAIL OF THE TEST JIG

The results, shown in figure 13, indicate that the ultimate failure of the panel occurs at 2,5 times the ultime loading conditions, close to the fasteners area.

Flight conditions	Real centrifugal panel loads (daN)	Simulated representative loads (daN)	Calculated displacement (mm)	Measured displacement (mm)
Cruise	950	465	1,81	1,80
Limite	1250	600	2,30	2,50
Ultimate	2260	1120	4,00	4,50
Panel failure	4700	2400	-	10,50

FIGURE 13 : CALCULATED AND MEASURED DISPLACEMENTS
COMPARISON FOR DIFFERENT FLIGHT CONDITIONS

VIBRATION TEST

Another development panel, installed on the same jig as is used for static test, was submitted to vibratory sollicitations in order to determine the resonance frequency in flexure and torsion mode. This panel show a delamination (manufacturing defect of 20x30 mm) on the imidization joint at 50 mm inside from the upper corner. The basic requirement was that the panel should be designed in dynamic conditions to a 15 % minimum speed margin at cruise conditions (1390 RPM i.e 370 Hz). The results, indicated a good correlation between the calculated values and the measured values on the non aged, panel tested at room temperature. For the first torsion mode, the resonance frequency is 378 Hz, so the margin is only 3% compared to the requirements.

During that resonance identification, we have measured the amplification coeficient Q for all the main resonance frequence. Moreover, a micro accelerometer was used to identify the modal lines on the surface panel. Then, the highly loaded areas were determined and 10 gages were bonded. The maximum stress measured for the panel was 6 MPa in first torsion mode. A fatigue torsion test was performed at 378 Hz, using a dynamic excitator, located in the accurated position on the modal line for that particular case. The chosen amplitude of displacement (1 mm) corresponds to the measured maximum stress of 6 MPa in the panel. After 10^7 cycles, there is no loss of caracteristics observed. Ultrasonic inspection, do not revealed any propagation of the existing delamination in that severe conditions.

FIGURE 14 : VIBRATION TEST EQUIPMENT

FLIGHT TEST

Finally, a set of 13 panels (10 + 3 spares) have been fabricated and inspected. Each panel was machined and balanced to give a equilibrated weight of 685 $\pm$ 1 gram before painting.

They were delivered in April 1987 to GENERAL ELECTRIC through SNECMA, for installation in the second demonstrator engine. After satisfactory ground test, as shown in figure 15, they were mounted on a MD80 Aircraft, in order to evaluate overall UHB (ultra high bypass) engine behaviour. After more than 100 flight hours, visual inspection showed that there was no degradation of the access panels. A complete expertise of the panel will be done at the issue of the flight campaign.

FIGURE 15 : ENGINE INSTALLATION DURING THE GROUND TEST

CONCLUSION

The ability to design and manufacture small, complex shape, carbon-PMR15 stiffened structures has been demonstrated. Particularly, a specific cure cycle consisting of two-stage process, imidization then polymerisation, have been used for cocuring differents lay-up. The laminates, show a good level of porosity and mechanical performance, with no microcracking. Further work on the PMR15 system include feasibility of great dimension parts, specially for nacelles and reversers applications where the design is mainly honeycomb sandwich because of the acoustic attenuation required. Variability in the procurement and in processing, material toxicity and in-situ processing parameters are also investigated.

REFERENCES

1/ Quiet clean short-haul experimental engine (CQSEE) - NASA CR135279, July 1988

2/ Graphite/Polyimide composites - NASA Conference Publication 2079, February 28-March 1, 1979

Materials and Processing – Move into the 90's
edited by S. Benson, T. Cook, E. Trewin and R.M. Turner
Elsevier Science Publishers B.V., Amsterdam, 1989

THE USE OF SILANE ADHESION PROMOTORS IN POLYMER INDUSTRIES

P WALKER

1. INTRODUCTION

Silane adhesion promoters or coupling agents as they are sometimes called, are monomeric species of the general formula $R\text{-}Si(OR')_3$, where R represents an organofunctional group eg vinyl or amino, and (OR') is a hydrolysable ester group. Developed initially for use in glass filled polyester composites to improve the water resistance, they are now proving to be of great interest in other polymer based industries notably adhesives[1] and surface coatings[2]. Silanes represent a relatively "low tech" method of improving the initial, wet, and recovered adhesion of adhesives and surface coatings[3] and the strength of glass filled thermoplastic composites[4]. Although there are many literature references to the value of silanes in improving adhesion, numerical data is sparse, particularly on adhesives[1], and the work carried out in the author's laboratories[5-8] constitutes the major body of data published on surface coatings. This data is consolidated and amplified in the present paper, which contains additional new information on adhesives and composites.

The aim of this present paper is to review the practical factors which are, or may be of importance in the use of silanes as adhesion promotors and to present data demonstrating the beneficial effects of using silane adhesion promotors in the surface coatings, adhesives and composite industries.

2. SILANES USED

The silanes used in a series of investigations are shown below

- Methacryloxy

Propyltrimethoxysilane-MAMS

β-(3, 4, Epoxycyclohexyl)

Ethyltrimethoxysilane-ECMS

-Glycidoxy Propyltrimethoxysilane-GPMS

-Mercaptopropyl Trimethoxysilane-MPS

-Amino Propyltrimethoxysilane-APES

N-Beta (Aminoethyl)-gamma Aminopropyl

Trimethoxysilane-AAMS

3. REACTIONS OF INTEREST IN COUPLING

The reactions of interest may be summarised as:

1) Hydrolysis of the ester group

$$R - Si\,X_3 + 3\,H_2O \xrightarrow[\text{catalyst}]{\text{pH}} R - Si\,(OH)_3 + 3\,HX$$

2) Hydrogen Bonding at the surface

```
                   |/           OH      H      |/
                   |/           |              |/
R - Si(OH)3 + HO-  |/ ------> R - Si - O    O - |/
                   |/           |              |/
                   |/           OH      H      |/
```

3) Reaction with the surface

```
             |/                          |/
      OH -   |/                   OH  -  |/
             |/                          |/
             |/                          |/
R - Si(OH)3 + OH -  |/ ---->R -  Si(OH)2  - O  -  |/  + H2O
             |/                          |/
```

4) Polymerisation

$$2nR - Si\,(OH)_3 \longrightarrow HO - \left[\begin{array}{c} R \\ | \\ Si \\ | \\ OH \end{array} - O - \begin{array}{c} R \\ | \\ Si \\ | \\ OH \end{array} - O \right]_n - + 2n\,H_2O$$

5) Reaction with the polymer

```
- C - C + R-NH2            |   |
   \ /              HO - C - C - NHR
    O                      |   |
     (primary amino
      group on silane)
```

Equations 1, 4 and 5 may occur independently of the surface and may present practical limitations on their use under particular conditions. Equation 5 is typical of the many possible reactions between silanes and organic polymers, the nature of which will be determined by the functional groups present in both. A more detailed treatment of possible reactions with epoxide and urethane resin constituents is given by Walker[5].

4. METHODS OF USE

Silanes may be used as:

1. pretreatment primers
2. pretreatments for fillers
3. formulated primers
4. additives to the organic resin

If used as a pretreatment primer in which the silane is applied from a solvent solution then several factors may be important if improved adhesion is to be achieved. These include:

the solvents used
the concentration of silane
the pH of the solution
the age of the solution
the thickness of silane film deposited.

In general the concentration should be 2% and the solution freshly made, clearly the thickness of the deposited silane film will depend on both the concentration and the manner of application. An excessively thick film may result in poor adhesion. The pH of the silane solution may be particularly important in the light of work reported by Boerio and Williams[9] in which amino functional silanes deposited from solutions of pH 8.0 to 12.5 resulted in a considerable difference in retained bond strength under water soaked conditions, and there was evidence of a structural change within the silane film. The pretreatment technique has the major advantage that a silane selected for a specific organic/substrate combination can be used to obtain optimum properties. It has the disadvantage that a further step in processing is introduced which is beyond the control of the coating or adhesive manufacturer.

When used as a surface treatment for fillers in which the silane is applied/pre-reacted with the filler prior to incorporating the filler into a matrix, the same factors are important but particle size of the filler will become important as will the presence of water.

The use of formulated primers in which the silane or silanes are blended with a film former, a solvent and perhaps a hydrolysis catalyse has been described by Plueddemann[10] and in the Dow Corning trade literature[11]. Such primers are required to be wet by the adhesive or coating and to strongly bond to it, to wet the substrate and strongly bond to it and to have an intermediate modulus to resin and substrate, "a particularly interesting primer for polysulphides to glass and aluminium substrates contains a "cocktail" of PMS, AAMS and MPS". The use of formulated primers has the same advantages as the use of silanes as pretreatments.

The final way in which silanes may be used is as additives incorporated into the organic matrix either immediately prior to use or as a part of the formulation at the manufacturing stage. The single pack concept is almost universally desirable, but several critical parameters need to be recognised if success is to achieved. Problems inherent in the use of silanes as additives include potential interactions between the solvent(s) and silanes, eg amino-functional silanes will react with oxygenated solvents; interactions between the polymeric binder or curing agent and silanes, and reaction with any water present. Depletion of silane by fillers may also be a problem and screening tests are essential if long term stability is to be achieved.

5. SILANES IN THE SURFACE COATINGS INDUSTRY

Why do we need silanes in the surface coatings industry? The short answer is because paints lose adhesion on exposure to high humidity or liquid water[12]. This loss of adhesion may be both rapid and dramatic. Silanes may be used to improve the initial, wet and recovered adhesion to a variety of metals including zinc, copper and cadmium, substrates usually regarded as being difficult.

The effect of various silanes on the initial bond strength of a two pack aliphatic isocyanate cured paint to aluminium and mild steel is shown in Table 1 where the silane was used as a $2^{w}/o$ solution in an alcohol/water solvent. Similar data for a two pack polyamide cured epoxide paint is shown in Table 2. Both tables clearly illustrate the marked improvements which can be obtained on both substrates, whether degreased or gritblasted. It should be noted that where a zero area of detachment is recorded the actual bond strength is greater than the value recorded.

The effect of silanes on the wet and recovered adhesion of the polyurethane and epoxide paints after 1500 hours water immersion is shown in Tables 3 and 4 respectively. In all cases but one the use of a silane improved both the wet and recovered bond strength by a considerable margin.

The effect of using silanes as additives to the polyurethane paint is shown in Table 5 and it can be seen that the increase in bond strength due to the silane is even more marked than when the silanes are used as pretreatments.

TABLE 1

Effect of Silane Type On Bond Strength - Polyurethane Paint

Torque Shear Method - Silane on Surface

Silane/Surface Preparation	Substrate: Aluminium MPa	psi	Area of Detachment	Substrate: Mild Steel MPa	psi	Area of Detachment
None/degreased	15.8	2300	100	22.6	3290	20-100
MAMS/degreased	41.5	6030	0	38.7	5630	5-30
ECMS/degreased	21.0	3050	100	37.7	5480	100
GPMS/degreased	24.4	3540	60-100	47.1	6850	0
MPS/degreased	34.4	4990	0-80	36.4	5300	0-10
APES/degreased	41.8	6080	0	48.0	6980	0
AAMS/degreased	40.3	5850	0	48.0	6980	0
None/grit-blasted	45.8	6650	0-20	40.0	5820	5-10
MAMS/grit-blasted	48.9	7100	0	45.8	6650	0-20
ECMS/grit-blasted	46.6	6770	0-10	46.6	6770	0-10
GPMS/grit-blasted	48.2	7000	0	48.2	7000	0
MPS/grit-blasted	46.7	6780	0-5	45.9	6680	0-5
APES/grit-blasted	48.4	7040	0	49.6	7200	0
AAMS/grit-blasted	49.6	7200	0	48.0	6980	0

TABLE 2

Effect of Silane Type On Bond Strength-Epoxide Paint

Torque Shear Method - Silane on Surface

Silane/Surface Preparation	Substrate: Aluminium MPa	psi	Area of Detachment	Substrate: Mild Steel MPa	psi	Area of Detachment
None/degreased	27.9	4050	30-90	36.2	5270	100
MAMS/degreased	42.2	6130	0	39.7	5780	10-30
ECMS/degreased	44.2	6420	0	36.2	5270	0-20
GPMS/degreased	43.9	6370	0-10	35.7	5190	0-20
MPS/degreased	42.2	6130	0	42.8	6230	0
APES/degreased	46.5	6770	0	45.0	6550	0
AAMS/degreased	41.5	6030	0	44.2	6430	0
None/grit-blasted	40.6	5910	10-30	40.0	5820	0
MAMS/grit-blasted	43.5	6320	10	42.4	6170	0-30
ECMS/grit-blasted	47.1	6840	0	46.8	6810	0
GPMS/grit-blasted	42.4	6160	0	47.3	6880	0-5
MPS/grit-blasted	43.9	6380	0	49.6	7220	0
APES/grit-blasted	47.6	6920	0	49.6	7220	0
AAMS/grit-blasted	45.8	6650	0	47.3	6880	0

TABLE 3

Effect of Water Immersion on Bond Strength

Polyurethane Paint - Direct Pull Off, 1500 Hours
Silane on Surface

	Substrate					
	Aluminium			Mild Steel		
Silane/Surface Preparation	Initial MPa/Area of Detachment	Wet MPa/Area of Detachment	Recovered MPa/Area of Detachment	Initial MPa/Area of Detachment	Wet MPa/Area of Detachment	Recovered MPa/Area of Detachment
None/degreased	12.6/100	3.8/100	9.9/100	16.7/100	5.7/100	6.8/100
MAMS/degreased	32.3/30	10.1/30	21.8/30	-	-	-
MPS/degreased	-	-	-	25.2/30	5.4/100	12.1/100
AAMS/degreased	26.3/40	11.1/30	22.8/30	38.2/0	7.4/90	12.9/90
None/degreased	28.6/10	8.5/100	13.6/80	25.7/40	11.8/90	20.8/60
MPS/grit-blasted	-	-	-	32.2/5	23.7/10	25.3/0
MAMS/grit-blasted	33.7/0	13.0/20	22.4/40	-	-	-
AAMS/grit-blasted	34.0/0	14.9/20	22.0/30	36.7/0	22.8/30	29.2/0

TABLE 4

Effect of Water Immersion on Bond Strength - Epoxide Paint

Direct Pull-Off, 1500 Hours - Silane on Surface

	Substrate					
	Aluminium			Mild Steel		
Silane/Surface Preparation	Initial MPa/Area of Detachment	Wet MPa/Area of Detachment	Recovered MPa/Area of Detachment	Initial MPa/Area of Detachment	Wet MPa/Area of Detachment	Recovered MPa/Area of Detachment
None/degreased	21.4/90	5.7/100	11.2/100	19.9/100	7.2/100	11.0/100
MAMS/degreased	30.2/0	12.0/30	19.7/50	-	-	-
ECMS/degreased	-	-	-	27.4/20	17.3/100	21.8/90
AAMS/degreased	31.2/0	11.5/30	20.5/40	32.0/0	28.1/100	29.2/10
None/grit-blasted	28.5/30	8.5/100	13.7/80	25.9/40	9.2/100	21.0/100
MAMS/grit-blasted	31.8/10	13.3/40	21.6/50	-	-	-
ECMS/grit-blasted	-	-	-	27.7/0	16.3/30	31.7/60
AAMS/grit-blasted	32.5/0	13.0/40	25.0/30	33.6/0	25.3/50	27.9/40

TABLE 5

Effect of Water Immersion on Bond Strength - Polyurethane Paint

Torque Shear 1500 Hours, Silane In Paint

Silane/Surface Preparation	Substrate: Aluminium Initial MPa/Area of Detachment	Aluminium Wet MPa/Area of Detachment	Aluminium Recovered MPa/Area of Detachment	Mild Steel Initial MPa/Area of Detachment	Mild Steel Wet MPa/Area of Detachment	Mild Steel Recovered MPa/Area of Detachment
None/degreased	29.1/90	9.3/100	6.2/100	23.9/100	12.1/100	11.3/100
0.4% MPS /degreased	33.8/90	12.6/100	31.7/100	32.5/60	19.3/100	33.9/100
0.2% AAMS /degreased	37.3/30	20.6/100	31.1/100	44.0/0	17.4/70	25.5/100
None/grit-blasted	33.1/100	21.2/10	36.6/40	35.0/15	21.3/30	29.3/40
0.4% MPS /grit-blasted	-	-	-	45.1/0	30.9/5	41.0/0
0.1% AAMS /grit-blasted	44.3/0	39.7/0	48.3/0	-	-	-
0.2% AAMS /grit-blasted	45.7/0	39.2/0	46.9/0	47.0/0	31.3/0	32.7/50

6. SILANES IN THE ADHESIVES INDUSTRY

As in surface coatings the need is to improve initial bond strength and resistance to aqueous environments. The data in Table 6 shows the effect of silanes used as pretreatment primers for structural polyurethane and epoxide adhesives on a variety of substrates. It can be seen that of the three silanes tested, GPMS did not increase the bond strength of the polyurethane adhesive on any of the substrates but did improve the bond strength of the epoxide. In general, improvements were obtained. A comparison of the effect of silanes as primers and additives is shown in Table 7 from which it can be seen that they performed better as primers. It is felt that this difference probably arises from the high intrinsic viscosity of the solventless adhesives not allowing the silane to migrate to the substrate surface and become effective. This difference in behaviour is in direct contrast to the surface coating experience where the additive approach gave superior results.

The marked improvement in water resistance of bonded joints on stainless steel is shown in Table 8 where both GPMS and MPS resulted in a marked improvement in bond strength after exposure. The values recorded are for wet adhesion ie the specimens were not allowed to dry out before test. With both silanes and adhesives the residual bond strength after two years was at least a factor of three greater than the non-silane control.

TABLE 6

Effect of Silanes as Pretreatments on the Bond Strength of Structural Adhesives - Degreased, Butt Tensile

Silane/Metal	Adhesive Polyurethane MPa	Polyurethane C of V%	Epoxide MPa	Epoxide C of V%
None/Aluminium	19.0	19.1	19.0	20.3
None/Stainless Steel	37.8	12.6	22.1	15.8
None/Mild Steel	17.5	31.1	13.6	25.5
MPS/Aluminium	29.8	27.2	35.4	18.4
MPS/Stainless Steel	42.5	21.5	36.6	18.5
MPS/Stainless Steel	24.8	24.8	15.8	10.1
GPMS/Aluminium	15.7	7.5	35.6	19.8
GPMS/Stainless Steel	32.9	33.0	18.2	18.6
GPMS/Mild Steel	14.1	13.2	16.9	14.3
AAMS/Aluminium	24.8	17.4	35.0	18.1
AAMS/Stainless Steel	40.0	25.6	26.9	15.9
AAMS/Mild Steel	19.7	12.8	18.1	22.3

TABLE 7

Comparison Of Silanes As Pretreatments And Additives Grit-blasted, 2% As Pretreatment And Additive, Butt Tensile

Adhesive/Substrate	Silane None	MPS Primer	MPS Additive	AAMS Primer	AAMS Additive
Polyurethane/Stainless Steel					
Bond Strength MPa	36.3	43.7	36.5	47.8	43.8
Coefficient of Variation %	8.9	7.7	13.6	4.7	6.5
Epoxide/Mild Steel					
Bond Strength MPa	38.3	48.2	42.0	52.3	41.8
Coefficient of Variation %	7.4	7.6	13.4	2.5	7.9

TABLE 8

Effect of Silanes As Pretreatments On Humidity Resistance Grit-blasted, Stainless Steel 100% RH, Butt Tensile

Time	Polyurethane None MPa	Polyurethane GPMS MPa	Polyurethane MPS MPa	Epoxide None MPa	Epoxide GPMS MPa	Epoxide MPS MPa
Control	43.4	45.5	57.6	50.8	51.7	57.6
1 week	44.2	44.1	47.6	43.6	44.7	51.9
2 weeks	33.7	40.9	44.7	41.0	37.5	52.8
1 month	36.1	42.2	44.6	22.9	36.9	41.4
3 months	35.7	38.8	38.8	22.5	27.0	35.7
6 months	33.3	35.2	28.5	21.6	23.2	26.4
9 months	22.1	31.5	24.7	7.0	18.4	21.4
12 months	20.2	27.4	24.8	5.7	18.1	16.9
24 months	6.6	22.2	20.4	No reading	17.9	17.3

7. SILANES IN THE COMPOSITES INDUSTRY

The use of silanes as pretreatments for glass fibres in composites is well known but they have wider applications particularly in glass microballoon syntactic foams where they can improve the compressive yield stress and strain to initial yield. This is shown in Tables 9 and 10 which show the effect of different silanes in a glass microballoon filled anhydride cured epoxide foam and the effect of increasing the silane content using APES. The effect on the compressive modulus should be particularly noted as the net effect of increasing strain to failure is to decrease the modulus. A much more valuable effect where syntactic foams are to be used as cushion materials to absorb impact energy is to improve the whole range of stress/strain behaviour as shown in Figure 1.

TABLE 9

Effect of Silanes in Syntactic Foams
(MNA/828/Silica Microballoons, cured at 100°C for 4 hours)

Silane	Density g/cc	Yield Stress MPa	Yield Strain %	Proportional Limit MPa/%	Compressive Modulus MPa x 10^2
None	0.354	6.3	1.44	6.3/1.44	44
ECMS	0.357	8.9	2.85	8.3/2.50	31
GPMS	0.360	8.0	1.96	7.2/1.76	41
MPS	0.362	12.1	3.97	11.7/3.18	30
APES	0.352	14.8	4.05	14.8/4.05	35
AAMS	0.347	12.3	2.70	12.3/2.70	45

TABLE 10

Effect of Silanes in Syntactic Foams - APES
(MNA/828/Silica Microballoons, cured at 100°C for 4 hours)

Level %	Density g/cc	Yield Stress MPa	Yield Strain %	Proportional Limit MPa/%	Compressive Modulus MPa x 10^2
0	0.360	6.3	1.44	6.3/1.44	44
1	0.332	8.5	1.80	6.3/1.80	48
2	0.360	13.1	2.50	12.8/2.02	52
3	0.355	12.3	2.45	11.7/2.01	50
4	0.358	14.5	2.31	14.5/2.31	63
5	0.352	14.8	4.05	14.8/4.05	36

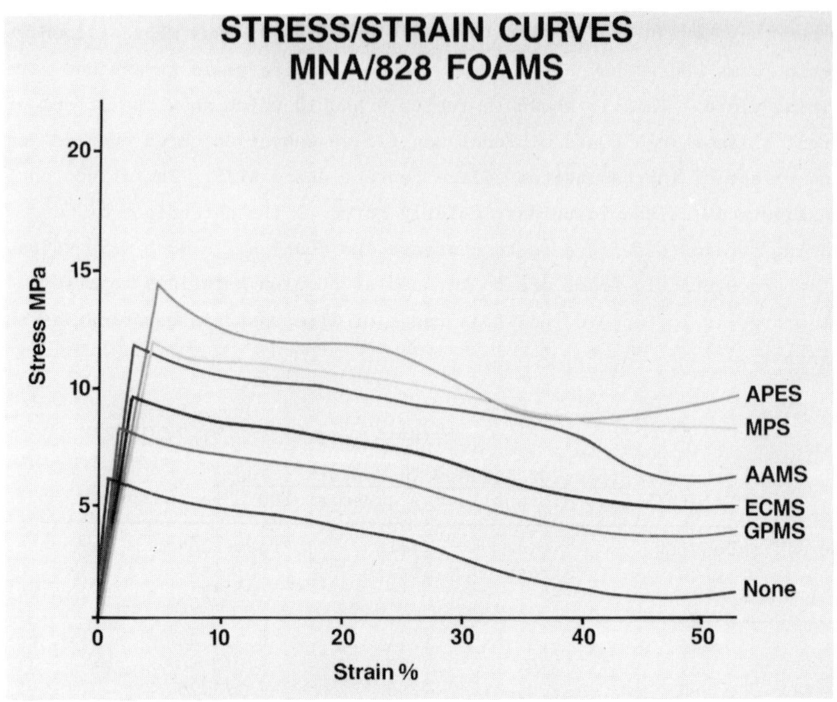

Figure 1

CONCLUSIONS

It can be seen from the present paper that organofunctional silanes can be used to improve the properties of coatings, adhesives and composites when used as pretreatments or additives and represent a relatively low tech means of product enhancement.

REFERENCES

1) C Kerr and P Walker, Some aspects of silane technology for surface coatings and adhesives, in Adhesion 11, ed K W Allen (Elsevier Applied Science, London, 1987) pp 17-37.

2) P Walker, J Coatings Technology, (1980), 52, 670.

3) P Walker, Adhesion promoters, in Surface Coatings, Vol.1 eds. A D Wilson, J W Nicholson and H J Prosser (Elsevier Applied Science, London, 1987) pp 189-232.

4) Union Carbide Technical Literature Data Sheet SC-731, 3/87-2M.

5) P Walker, J Oil & Colour Chem.Assoc., (1982), 65, 415.

6) P Walker, J Oil & Colour Chem.Assoc., (1982), 65, 436.

7) P Walker, J Oil & Colour Chem.Assoc., (1983), 66, 188.

8) P Walker, J Oil & Colour Chem.Assoc., (1984), 67, 108.

9) F J Boerio and J W Williams, Applications of surface science, Vol.7, (North- Holland, Amsterdam, 1981).

10)E P Plueddeman, Prog.Org.Coat., (1983), 297.

11)Dow Corning Technical Literature.

12)P Walker, Off.Digest, (1965), 37, 1561.

Materials and Processing – Move into the 90's
edited by S. Benson, T. Cook, E. Trewin and R.M. Turner
Elsevier Science Publishers B.V., Amsterdam, 1989

DEVELOPMENT OF IMPROVED 121°C (250°F) CURE ADHESIVES FOR AEROSPACE APPLICATIONS - FM® 300-2 ADHESIVE SYSTEM

Dalip K. Kohli, Ph.D

American Cyanamid Company, 1300 Revolution Street, Havre de Grace, Maryland 21078 U.S.A.

A new state-of-the-art 121°C (250°F) cure epoxy based film adhesive for aerospace bonding applications is described. This new adhesive, designated FM® 300-2 is a 121°C (250°F) cure version of FM® 300 adhesive and is designed for bonding metallic and non-metallic substrates. This new adhesive is formulated to cure at 121°C (250°F) and match the stress-strain and other mechanical properties of the 177°C (350°F) curing FM® 300 adhesive system.

1. INTRODUCTION

Adhesive bonding is the most suitable method of joining both metallic and non-metallic structures where strength, stiffness and fatigue life must be maximized at a minimum weight. Use of tough, moisture resistant, flow controlled epoxy based adhesives to bond composites to composites or composites to metals offers a number of advantages.[1] Because of their high elongation and high ultimate shear strength, these adhesives are particularly suitable for redistributing the high shear stress concentrations of composite to metal bonds and in accommodating the low interlaminar shear strength of the composite.

For aerospace bonding applications, FM® 300 epoxy based film adhesive is widely used for bonding metal to composite structures. Some of these applications range from bonding the wing-root assemblies (Titanium to graphite epoxy) on F-18 fighter aircraft and composite sandwich structures on B-1 bomber aircraft to bonding and surfacing applications on commercial and military transport aircraft. This 177°C (350°F) curing adhesive is used both in co-cure and secondary bonding applications.

In the case of bonding metal to composite parts, the 177°C (350°F) cure temperature for secondary bonding can lead to significant thermal stresses due to the difference in coefficient of thermal expansion between the metal and the composite. These induced stresses can result in the loss of dimensional stability, disbonds, or delamination in these parts. In order to overcome these problems, the aerospace bonding community expressed a need for a 121°C (250°F) curing adhesive, with the performance of FM® 300.

This paper describes development of a new state-of-the-art 121°C (250°F) curing epoxy based adhesive for aerospace bonding applications.

2. MATERIALS AND PROCEDURES

2.1 Adhesive Cure Cycle

For all metal or secondary bonding of composites the following cure cycle was utilized:

90 minutes at 121°C (250°F), 40 psi (0.28 MPa), heat-up rate 1.7°C (3°F) per minute

2.2 Primer Details

On metal substrates, BR® 127 adhesive primer was used at 0.0025 to 0.0075 mm thickness. The primer was air dried 30 minutes at room temperature and then cured at 121°C (250°F) for 60 minutes.

2.3 Metal and Core Details

Metal Skins: 2024-T3 bare aluminum, FPL etched, unless otherwise stated.

Metal Core: 6.35 mm cell, 0.1 mm 5052 DURACORE® II, 1.28 gm/cc, 15.9 mm thick

Nomex Core: 3.18 mm cell, 1.44 gm/cc, 12.7 mm thick.

2.4 Secondary Bonding of Composites

a. Thermosetting Composites

For secondary bonding work, three epoxy/graphite or epoxy/glass prepreg systems with cure temperatures ranging from 121°C to 177°C (250°F to 350°F) were selected. The prepregs and their cure cycles are listed below.

CYCOM® 919/3K70P - cured 60 minutes at 121°C (250°F), 50 psi (0.35 MPa)

CYCOM® 985/3K70P - cured 120 minutes at 177°C (350°F), 75 psi (0.52 MPa)

CYCOM® 1827/6681 glass - cured 120 minutes at 177°C (350°F), 75 psi (0.52 MPa)

The heat-up rate was 1.7°C (3°F) per minute in all cases. The surface preparation involved use of Nylon peel ply which was co-cured with the laminate. The peel ply was removed just prior to bonding of the laminate. No additional surface preparation was used unless otherwise stated. The laminates were bonded with FM® 300-2 adhesive and tested as lap shear coupons.

b. Thermoplastic Composites

For secondary bonding of thermoplastic laminates APC-2 (PEEK) or KIII polymer, laminates were used as obtained from the manufacturers. Types of surface preparation are listed with the tabulated data.

2.5 Co-cure Studies

Table I

FM® 300-2
Baseline Comparison With FM® 300

Property	FXM 300-2K Cured 90 min. at 121°C	FM® 300K Cured 60 min. at 177°C
Lap Shear Strength (PSI)(MPa)		
24°C	5600 (38.6)	5500 (37.9)
121°C	3900 (26.9)	4000 (27.6)
150°C	2300 (15.9)	2700 (18.6)
Floating Roller Peel Pli, (KN/m)		
24°C	36 (6.4)	35 (6.2)
Honeycomb Sandwich Peel (ipp 3in)(Nm/m)		
24°C	50 (74)	45 (67)
Flatwise Tensile (psi)(MPa)		
24°C	1100 (7.6)	1000 (6.9)
150°C	400 (2.8)	460 (3.2)
Flow (%)	450-550	450-550
Tg °C (TMA)	144	148

Primer: BR® 127, 0.005 mm thick, cured 60 min at 121°C; Metal: 2024T3, FPL etched.
Adhesive Weight: 392 gm/m^2

Table II

Effect of Cure cycle on Physical Properties of FM® 300-2K (392 gms/m^2)

Test	Test Temp(°C)	Cure Cycle 90' at 121°C	60' at 150°C	60' at 177°C
Lap Shear Strength, psi (MPa)	24	6100 (42.0)	6455 (44.5)	6280 (43.2)
	107	4660 (32.1)	4680 (32.2)	4560 (31.4)
Floating Roller Peel, Pli(KN/m)	24	30 (5.3)	34 (6.0)	34 (6.0)
	107	43 (7.6)	45 (7.9)	49 (8.6)
HCSP, in lb/3in (Nm/m)	24	48 (71)	52 (77)	50 (74)
	107	50 (74)	55 (81)	60 (89)
Flow (%)	----	575	525	575
Tg (°C)	----	143	140	140

Primer: BR® 127, 0.005 mm thick, cured at 250°F for 60 minutes.

Figure 1

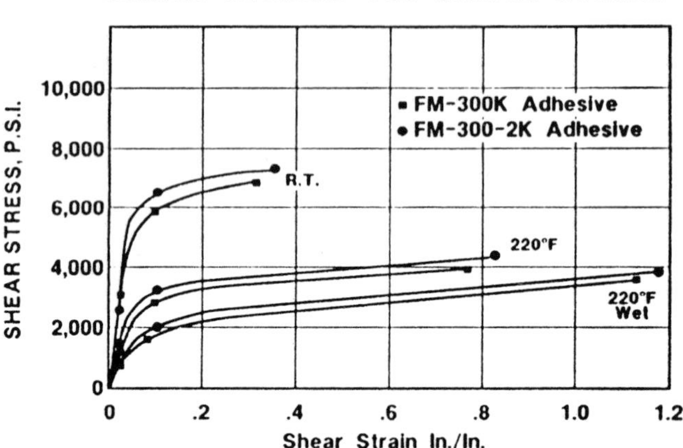

Figure 2

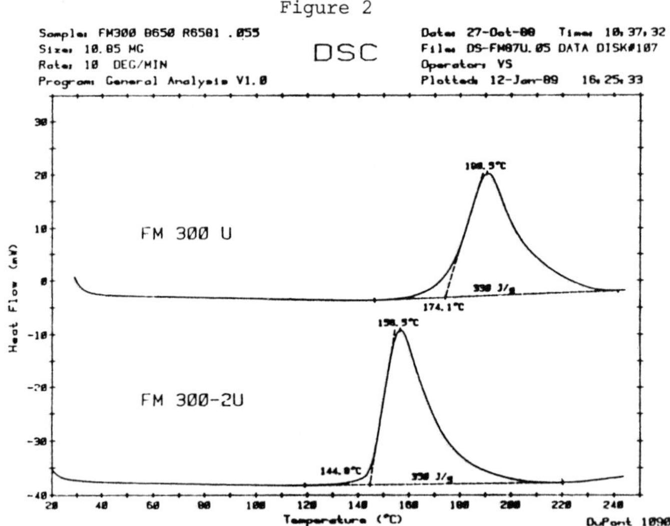

For co-cure work CYCOM® 985/AS-4, a 177°C (350°F) curing epoxy/graphite prepreg was used. The adhesive was placed between the 8th and 9th plies of a 16 ply unidirectional laminate and was co-cured at 177°C (350°F) and 75 psi (0.52 MPa). For surface ply studies, the same adhesive was used as the top layer and co- cured with the laminate.

3. RESULTS AND DISCUSSION

Utilizing an innovative materials approach, we have developed a 121°C (250°F) cure version of FM® 300 adhesive for various bonding applications where higher performance is needed but the cure temperature can not be higher than 121°C (250°F).

This adhesive system designated as FM® 300-2 is based on epoxy chemistry and is designed for bonding metallic and composite structures as well as structures fabricated from metallic, Nomex or fiberglass honeycomb. This new adhesive system has been formulated to fully cure within 90 minutes at 121°C (250°F) and has flow and handling properties similar to FM® 300 adhesive.

Comparison of physical and mechanical property data between these adhesives is shown in Table I. It is notable that the 121°C (250°F) curing adhesive has a glass transition temperature of 144°C (291°F) as compared to 148°C (298°F) for the 177°C (350°F) cured FM® 300. Because of this high glass transition temperature, this new film adhesive is capable of significant retention of its strength up to 150°C (300°F).

The toughness properties of this adhesive as indicated by peel strength are very similar to FM® 300. Both these systems have metal to metal peel strength of 35 pli (6.4 KN/m). Because of their similar toughness and flow properties, both adhesives have similar sandwich properties.

One of the key objectives in the development of this new adhesive was that it should have stress-strain properties similar to FM® 300 adhesive. The comparative stress-strain properties of these two adhesives are shown in Figure 1. Note that the 121°C (250°F) cured FM® 300-2 adhesive has the same stress and strain properties as FM® 300 adhesive not only under dry but also under hot/wet conditions up to 104°C (220°F). This means that the lower temperature curing adhesive can be used to minimize stresses caused by differential coefficients of expansion when bonding dissimilar substrates.

The DSC profiles for these adhesives are shown in Figure 2. The onset temperature for FM® 300-2 adhesive is approximately 120°C (250°F) as compared to 150°C (300°F) for the FM® 300 system. These onset temperatures indicate the different cure chemistry operating in these two systems.

All the comparative data generated thus far indicates that this new adhesive provides similar stress-strain and mechanical properties to its higher temperature curing version. The comparative composite bonding data between these two adhesives is discussed later.

3.1 Effect of Cure Cycle

The ability of adhesives to cure over a wide temperature range becomes significant when these adhesives are to be used for co-cure and secondary bonding of 121°C to 177°C (250°F to 350°F) curing epoxy prepreg systems. Since this new adhesive is designed to be used for these applications, we evaluated the effect of changing the cure temperature on its physical and mechanical properties. This data is summarized in Table II. Examination of the data in Table II shows no discernable trends. Even the glass transition temperature does not significantly change with the increase in cure temperature from 121°C (250°F) to 177°C (350°F). This implies that 121°C (250°F) cure temperature is sufficient to complete most of the crosslinking in this adhesive system. We have also evaluated the effect of thermal cycling at 177°C (350°F) on the lap shear and peel strength of this adhesive system. In this study bonded coupons were exposed to 177°C (350°F) for two to eight hours and then tested at various temperatures. Results from this study indicate that there is no drop off in the shear or peel strength after short-term thermal cycling at 177°C (350°F) (Table III).

Results from the cure cycle study indicate that this adhesive can be cured over a wide temperature range of 121°C to 177°C (250°F to 350°F) without change in its bonding performance.

3.2 Effect of Humidity

The effect of prebond humidity was evaluated by exposing the adhesive film and the metal substrates to two weeks at 80% relative humidity at 24°C (75°F). For postbond humidity studies, individually cut coupons were exposed to 71°C (160°F) and 100% relative humidity for 30 days. The data is shown in Table IV. The FM® 300-2 adhesive shows excellent retention of its strength upto 121°C (250°F) after prebond and postbond humidity exposures.

The outstanding moisture resistance of this adhesive is also demonstrated by its ability to bond to wet Nomex and maintain its strength up to 121°C (250°F) even after humidity exposures. The wet Nomex honeycomb bonding data is shown in Table IVA.

4. COMPOSITE BONDING

A major goal in the development of this new adhesive was that it should be suitable for use in co-cure applications with 121°C (250°F) and 177°C (350°F) cure prepregs as well as for bonding to the surfaces of properly prepared ther-

Table III

Effect of 177°C Cycling on Lap Shear and Bell Peel Strength of FM® 300-2K (392 gms/m²) Adhesive

Bonded Coupon Exposure	LSS Psi (MPa)		Floating Roller Peel Pli (KN/m)	
	24°C	104°C	24°C	104°C
No Additional Exposure	6370(43.9)	5250(36.1)	28 (4.9)	40 (7.0)
Exposed to 1 cycle* at 177°C	7160(49.3)	5380(37.0)	24 (4.2)	34 (6.0)
Exposed to 2 cycles* at 177°C	6840(47.1)	5466(37.6)	27 (4.6)	36 (6.3)
Exposed to 3 cycles* at 177°C	6925(47.7)	5410(37.2)	26 (4.6)	36 (6.3)

*Cycle: Heat 24°C to 177°C, hold at 177°C for 2 hours, cool to 24°C.
Adhesive Cure Cycle: 90 minutes at 121°C, 40 psi (0.27 MPa), 1.7°C/min. heat-up rate
Primer: BR® 127, 0.005 mm thick, cured 60 minutes at 121°C
Metal Substrate: 2024-T3 bare aluminum

Table IV

Effect of Postbond Humidity on Peel Strength FM® 300-2K (392 gm/m²) Adhesive/BR® 127 Adhesive Primer

Prebond Film Exposure	Bonded Coupon Exposure	Floating Roller Peel PLI (KN/m)	
		24°C	121°C
None	None	36(6.3)	36(6.3)
	30 days at 71°C/100% RH	35(6.2)	39(6.9)
15 days at 80% RH at 24°C	None	40(7.0)	37(6.5)
	30 days at 71°C/100% RH	32(5.6)	36(6.3)

Table IVA

Wet Nomex Bonding with FM® 300-2K (392 gm/m²)

Nomex Core Exposure	Bonded Coupon Exposure	Flatwise Tensile, psi (MPa)		
		75°F	180°F	250°F
3 weeks at 65% RH at 24°C	None	780(5.4) (Core)	775(5.3) (Core)	625(4.3) (Core)
	30 days at 71°C/ 100% RH	700(4.8) (Core)	570(3.9) (Core)	---
Immersed in water for 24 hours at 60°C then bonded within two hours	None	720(5.0) (Core)	745(5.1) (Core)	615(4.2) (Core)
	30 days at 71°C/ 100% RH	725(5.0) (Core)	545(3.7) (Core)	---

Failure Mode: 100% Core

moset and thermoplastic laminates.

The effect of humidity exposures on the lap shear strength of precured thermosetting composite substrates bonded with this adhesive film is shown in Table V. In the secondary bonding of 177°C (350°F) cured graphite/epoxy or glass/epoxy laminates, it shows excellent retention of its strength up to 93°C (200°F) even after 30 days at 71°C (160°F) and 100% relative humidity. Data presented for 121°C (250°F) cure laminates demonstrates good strength retention at room temperature. The service temperature for CYCOM® 919 is 71°C (160°F) and although bonded specimens were not tested at 71°C (160°F) after humidity exposures, the strength retention is expected to be satisfactory.

For the secondary bonding of thermoplastic substrates (PEEK and KIII polymer), the knit supported adhesive film was used. This data shows that the 121°C (250°F) curing adhesive provides the same performance as 177°C (350°F) cured FM® 300 adhesive over the temperature range of -55°C to 150°C (-67°F to 300°F) in bonding of PEEK laminates. The composite bonding data on Dupont's KIII thermoplastic composite substrate adhesive is shown in Table VIA. Strength retention is excellent up to 121°C (250°F) dry and 82°C (180°F) after humidity exposures.

4.1 Co-cure Studies

For co-cure and surfacing studies, adhesive film with a mat carrier was co-cured with 177°C (350°F) curing prepreg. After curing, the laminate was cut into one inch wide double notched wide area lap shear coupons. The test results are tabulated in Table VII.

Results were excellent under both dry and wet conditions over the temperature range of 24°C (75°F) to 104°C (220°F). The mode of failure in all cases was cohesive within the laminate plies. The surface ply studies were done with mat supported (147 gm/m^2 or 245 gm/m^2) adhesive films. Although 147 gm/m^2 adhesive film provided good surfacing properties, even better surface smoothness was obtained with the heavier 245 gm/m^2 adhesive film. Use of surface ply not only provides smooth ready to paint surfaces but also can reduce moisture penetration into composite skins.

4.2 Interleafing Applications

For interleafing applications in composites, we have developed a low flow adhesive. Designated as FM® 300-2 Interleaf, it is a 121°C (250°F) curing, low flow, modified epoxy based adhesive. Because of its low flow, and lower cure temperature, it resists intermixing with the prepreg resin and provides a high strain, low modulus layer between the plies of the laminate. This results in better stress distribution within the laminate and higher impact strength.[2]

The interleaf adhesive can be co-cured with most epoxy prepreg systems and

Table V

Bonding of Precured Thermosetting Composite Substrates with FM® 300-2M (245 gm/m^2) Adhesive

Precured Composite Substrate	Bonded Specimen Exposure	Lap Shear Strength Psi (MPa) 24°C	82°C	93°C
CYCOM® 985 3K70P (Graphite/Epoxy) 177°C Cure	None	3440(23.7)	4600(31.7)	4430(30.6)
	30 days at 71°C/ 100% RH	3450(23.8)	3300(22.8)	3000(20.7)
CYCOM® 1827/6781 (Glass/Epoxy) 177°C Cure	None	2320(16.0)	2400(16.6)	2350(16.2)
	30 days at 71°C/ 100% RH	2150(14.8)	2000(13.8)	1900(13.1)
CYCOM® 919/3K70P (Graphite/Epoxy) 121°C Cure	None	3525(24.3)	----	----
	30 days at 71°C/ 100% RH	4345(30.0)	----	----

Table VI
Comparative Data Between FM® 300-2 and FM® 300 Adhesives
Secondary Bonding of Thermoplastic Composite Substrate

Adhesive System	Precured Composite Substrate	Lap Shear Strength, psi (MPa) -55°C	24°C	149°C	82°C Wet[1]
FM® 300-2K (392 gm/m^2)	PEEK (APC-2)	3075 (21.2)	3655 (25.2)	1940 (13.4)	3220 (22.2)
FM® 300K (392 g/m^2)	PEEK (APC-2)	3100 (21.4)	3650 (25.2)	1820 (12.6)	3165 (21.8)

Surface Preparation: Plasma Etched

Cure Temperatures: FM® 300-2: 90 minutes at 121°C, 40 psi (0.28 MPa), heat-up 1.7°C/min.
FM® 300: 60 minutes at 177°C, 40 psi (0.28 MPa), heat-up 1.7°C/min.

[1]Individual coupons exposed 30 days at 60°C and 100% relative humidity

Table VIA

Bonding of DuPont KIII Laminate Substrate with FM® 300-2K (392 gm/m^2) Adhesive

Effect of Postbond Humidity Exposure

Bonded Specimen Exposure	Lap Shear Strength psi (MPa) 24°C	82°C	121°C
None	2420 (16.7) (70% C) (30% LF)	2570 (17.7) (60% C) (40% LF)	2340 (16.1) (85% A) (15% C)
30 Days at 71°C and 100% Relative Humidity	2600 (17.9) (LF)	2600 (17.9) (50% C) (50% LF)	1550 (10.7) (100% C)

Surface Preparation: Sandblast/Solvent Wipe

LF = Laminate Failure; C = Cohesive; A = Adhesive Failure

Table VII

Composite Bonding Data

FM® 300-2 Adhesive Co-Cured with CYCOM® 985/AS-4 Prepreg System

Adhesive	Bonded Specimen Exposure	Wide Area LSS, Psi (MPa) 24°C	104°C
FM® 300-2M ($147\ gm/m^2$)	None	2510 (17.3) (LF)	2180 (15.0) (LF)
	30 Days at 60°C/100% RH[1]	3250 (22.4) (LF)	3140 (21.7) (LF)

Cured Laminate Thickness: 3.56 mm
LF: Laminate Failure

Table VIII

Co-Cure Data - FM® 300-2 Interleaf or FM® 300 Interleaf
Adhesives Co-cured with 177°F Curing Epoxy Prepreg Systems

Adhesive	Prepreg System	Bonded Specimen Exposure	Wide Area Lap Shear Strength psi 24°C	104°C
FM® 300-2M Interleaf ($245\ gm/m^2$)	CYCOM® 985/AS-4 (graphite/epoxy controlled flow system	None	3010 (20.8) (LF)	3140 (21.7) (LF)
		30 days at 71°C/ 100% RH	2515 (17.3) (LF)	2730 (18.8) (LF)
FM® 300M Interleaf ($245\ gm/m^2$)	CYCOM® 985/AS-4 (graphite/epoxy controlled flow system	None	2720 (18.8) (LF)	2600 (17.9) (LF)
		30 days at 60°C/ 100% RH	2040 (14.1) (LF)	2760 (19.0) (LF)
FM® 300-2U Interleaf ($59\ gm/m^2$)	CYCOM® 985-1/AS-4 (graphite/epoxy) High Flow System	None	3100 (21.4) (LF)	2875 (19.8) (LF)
		30 days at 60°C/ 100% RH	3390 (23.4) (LF)	2800 (19.3) (LF)

LF = Laminate Failure

Cured Laminate Thickness = 3.56 mm

provides excellent surface ply properties. Because of its lower flow, the interleaf adhesive provides even better surfacing properties than the standard adhesive.

The effect of humidity exposure on the lap shear strength of mat supported interleaf adhesive is shown in Table VIII. Data on 177°C (350°F) cure mat supported FM® 300 Interleaf is also included for comparative purposes. These results show that the interleaf adhesive can be co-cured even with a high flow 177°C (350°F) curing epoxy prepreg system CYCOM® 985-1 and provides performance similar to FM® 300 Interleaf adhesive. We are currently evaluating the effect of interleafing on the impact strength of 177°C (350°F) cured prepreg systems.

5. SUMMARY

Development of FM® 300-2, a new state-of-the-art 121°C (250°F) curing epoxy based adhesive for aerospace bonding applications has been described. This adhesive is a 121°C (250°F) cure version of 177°C (350°F) curing FM® 300 adhesive and has similar stress-strain properties as FM® 300 under both dry and wet conditions. The new adhesive has been shown to have similar shear, toughness and flow properties as FM® 300 and has a service temperature of -55°C to 155°C (-67°F to 300°F). Due to its low moisture absorption and high glass transition temperature, this adhesive shows excellent performance when bonding to wet Nomex honeycomb and in bonding precured thermoset and thermoplastic composite substrates. It can be co-cured with most 177°C (350°F) curing epoxy prepreg systems and provides excellent bonding and surfacing properties. In addition, a low flow grade designated FM® 300-2 Interleaf has been developed for surfacing, interleafing and repair of composite structures.

REFERENCES

1) Politi, R. E., Factors Affecting the Performance of Composite Bonded Structure, 19th International SAMPE Technical Conference, Vol. 19, October 1987.

2) Hirschbuehler, K. R., An Improved Performance Interleaf System Having Extremely High Impact Resistance, SAMPE Quarterly, Vol. 17, No. 1, October 1985.

Materials and Processing – Move into the 90's
edited by S. Benson, T. Cook, E. Trewin and R.M. Turner
Elsevier Science Publishers B.V., Amsterdam, 1989

CHARACTERISATION OF ADHESIVE/ADHEREND INTERFACES IN THE ALUMINIUM-LITHIUM BONDED JOINT

J.A.BISHOPP, Bonded Structures, Ciba-Geigy Plastics, Duxford, Cambridge, England
D.JOBLING and G.E.THOMPSON, Corrosion and Protection Centre, University of Manchester Institute of Science and Technology, Manchester, England.

Transmission electron microscopy examination of ultramicrotomed sections taken through a bonded joint, allows both the effect of standard chemical pretreatments and an initial adhesive/adherend interfacial analysis to be made for structural joints prepared using T3 BA 8090C aluminium-lithium alloy substrates.

These observations, when compared with earlier work using Alclad 2024-T3 adherends, show that broadly similar films are grown on the aluminium-lithium surface; a rougher topography has, though, been noted on some specimens. In general, these films appear to be wetted and, in some cases, penetrated by the adhesive used in the same manner as are the Alclad controls. Initial measurement of bond strengths shows that the pretreated surfaces are amenable to adhesive bonding.

1. INTRODUCTION

For today's aircraft designer, one of the most important criteria when developing a new aeroplane, is weight. A reduction of five tonnes in the manufacturer's empty weight, in an aircraft initially designed to have a maximum certificated take-off weight (MTOW) of 200 tonnes, can reduce the final MTOW by nearly nine tonnes and the engine thrust required by about 4.5%[1]. This, amongst other savings, can lead to a significant reduction in fuel consumption. Little[1] lists several possible approaches to enable this weight saving to be made in an entirely new aircraft: advances in aerodynamics leading to lighter wing structures, improvement in power plants leading, directly to fuel saving, new system concepts, e.g. fly-by-wire, reducing structure weight and novel, lighter, metallic or non-metallic structural materials.

Considering the last of these options, the move from conventional aluminium alloys to carbon fibre or aramid reinforced composite materials is now an established route to weight saving; carbon fibre reinforced epoxy composites have been successfully used in the manufacture of control surfaces and engine cowls and are now used on primary structures such as vertical and horizontal stabilisers, helicopter rotor blades and the wing of the ATR 72.

Much work has also taken place in the field of metallic structures: steel sections have been replaced by titanium and the more conventional aluminium alloys by the newer, higher strength versions (for example the 7000 series alloys). However, in the early 1980's, alloys of aluminium and lithium were developed as yet further possible materials for the fabrication of aerospace structures. These alloys offered lower densities (10% reduction) and higher stiffness than the then current aluminium materials of comparable strength.

Although considerable evaluation work has taken place to determine the feasibility of using these aluminium-lithium alloys in aircraft structures, it is still too early to say whether such materials will make a significant impact in this area; product availability and cost effectiveness being two critical considerations.

Nevertheless, it is already important that such materials should be characterised and their suitability for joining, by adhesive bonding, be determined.

2. STRUCTURAL BONDING OF AEROSPACE MATERIALS

Since the early 1940's, considerable knowledge has been gained on joining standard aluminium/aluminium alloy components, for structural, aerospace applications, by means of synthetic, high strength adhesives[2,3,4]; so much so that it is now possible to use, for example, a modified epoxy, structural adhesive on an aluminium alloy adherend, such as Alclad 2024-T3, as a bench mark when characterising other substrates or adhesives.

Using this approach to evaluate the use of titanium and fibre reinforced composites as structural materials, has shown the importance of using the correct pretreatment to ensure as good an assumption as possible of the adhesive properties on the novel substrates. This has proved of particular importance with carbon fibre reinforced epoxy resin composites [5], where, even using the optimum method of pretreatment (a carefully controlled abrasion technique) some reduction in strength levels has to be accepted.

Any evaluation as to the suitability of joining aluminium-lithium substrates by adhesive bonding must, therefore, commence with the characterisation of the effects of surface pretreatment and an analysis of the adhesive/adherend interfaces produced on bonding.

3. METHOD OF CHARACTERISATION

The method used, both to characterise the effects of pretreatment as well as to analyse the interfaces, was one which had already proved invaluable in gaining similar insights into bonded joints using the more conventional alloy - Alclad 2024-T3 [6,7,8]. This was the use of transmission electron microscopy to examine ultramicrotomed slices taken through the joint, perpendicular to the plane of the bond (Figure 1).

The fact that ultramicrotomy could be used on aluminium-lithium alloys had already been established by Malis [9], who had used this technique in characterising the lithium and zirconium distribution in the 8090 alloy.

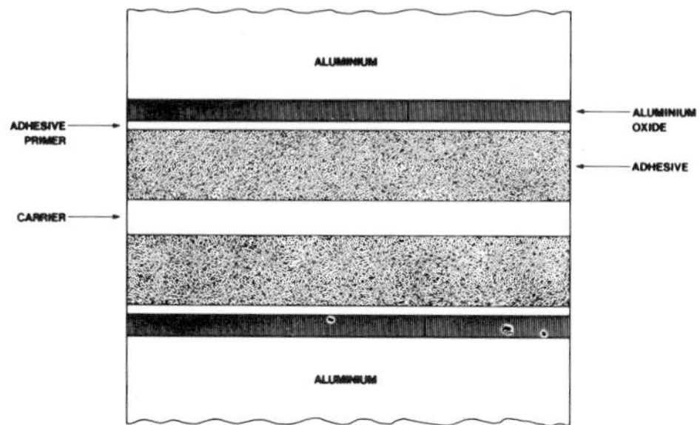

FIGURE 1: Schematic representation of an ultramicrotomed section through an adhesive joint

4. MATERIALS

Substrate : T3 BA 8090C aluminium-lithium alloy ex British Alcan; sheet thicknesses 0.55 mm and 1.6 mm. The typical composition, as determined by Colvin and Starke [10], is given in Table 1.

Comparison substrate : The substrate used for comparison purposes was Alclad 2024-T3 [11]. The composition of this alloy (core and cladding) is given in Table 1.

Adhesive : A toughened epoxy, structural film adhesive curing at 120°C.

Adhesive Support : A knitted nylon cloth (Nylon 6).

Primer : None used at this stage of the work.

5. EXPERIMENTAL

A set of standard lap-shear [12] and floating roller peel panels [13] was produced for each of the following chemical pretreatments given to the aluminium-lithium substrates:

A. Potassium dichromate/sulphuric acid pickle (CSA) in accordance with DTD915b (ii)[14].

B. Potassium dichromate/sulphuric acid pickle followed by chromic acid anodising (CAA) to DEF STAN 03-24/1 [15].

C. Potassium dichromate/sulphuric acid pickle followed by phosphoric acid anodising (PAA) to BAC 5555 [16].

The standard cure cycle for the bonded joints was 1 hour at 120°C under a bonding pressure of 0.275 MPa. Test specimens were cut out and loaded to failure. Selected areas of these samples were then mounted and prepared for ultramicrotomy in the usual manner [17]. 10 nm thick specimens were cut, using a Du Pont Sorval or Reichert Ultracut ultramicrotome, and examined by transmission electron microscopy (Philips EM 301 and EM 400 microscopes).

In the work reported here, the specimens were always taken from the peel joints, well in advance of the crack tip (i.e. in the unruptured, essentially unstressed area of the joint) as earlier work [7, 8] had shown this to be ideal both for characterising the effects of the surface pretreatment as well as enabling some degree of interfacial analysis to be achieved.

6. RESULTS

The lap-shear and peel strengths recorded are shown in Table 2, which also gives a comparison with results obtained previously [8], with the same adhesive, on Alclad 2024-T3 substrates.

The TEM micrographs of the ultramicrotomed sections through the peel specimens are shown in Figures 2a - 4a, 5 and 6; direct comparisons with Alclad 2024-T3 can be seen in Figures 2b - 4b.

7. DISCUSSION

Arrowsmith et al [18] reported no difficulty in anodising aluminium-lithium alloys, a finding supported by this work. No "burning" effect was seen on any of the anodised panels although a significant, uniform colour change - to dark grey - was noted with the chromic acid anodised specimens.

Arrowsmith examined the lap-shear performance of toughened acrylic and toughened epoxy adhesives on aluminium and aluminium-lithium substrates; performance appeared to be essentially independent of substrate, pretreatment and, to a certain extent, adhesive.

Table 2 shows this not to be the case here. The marked difference between the lap-shear strengths on Alclad and aluminium-lithium can, though, be explained by examination of the load-extension graph generated during testing. On Alclad substrates the total strain at failure is 2-3%. With the aluminium-lithium joints, however, this rises to 10-13%. Obviously, under tensile load the aluminium-lithium alloy is far more ductile; it, therefore, deforms more easily and hence far more significant peeling stresses are generated leading to an apparently lower shear value for the adhesive. This work shows the critical strength level, before onset of ductile behaviour, to be about 30 MPa; the maximum loads recorded in the Arrowsmith work were below this level and hence this effect would not have been seen.

The peel strengths show less dependency on substrate type, although the levels on aluminium-lithium adherends appear to be slightly higher. Examination of transmission electron micrographs of ultramicrotomed sections permits the film grown, during pretreatment, to be characterised and some analysis of the interfacial/interphasial structures to be made.

7.1 CSA Pickled Substrates

Figure 2 compares typical interfaces for bonded joints with (a) aluminium-lithium and (b) Alclad substrates. The similarities between the two are striking. Both exhibit finely spaced, 25-50 nm high whiskers which, from the evidence of the micrographs, appear to be extensively penetrated by the adhesive.

Transmission electron micrographs of sections through joints produced using CSA pickled adherends

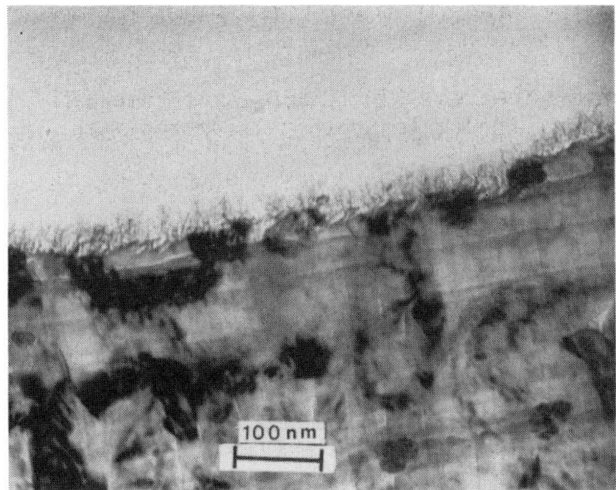

FIGURE 2a: T3 BA 8090C

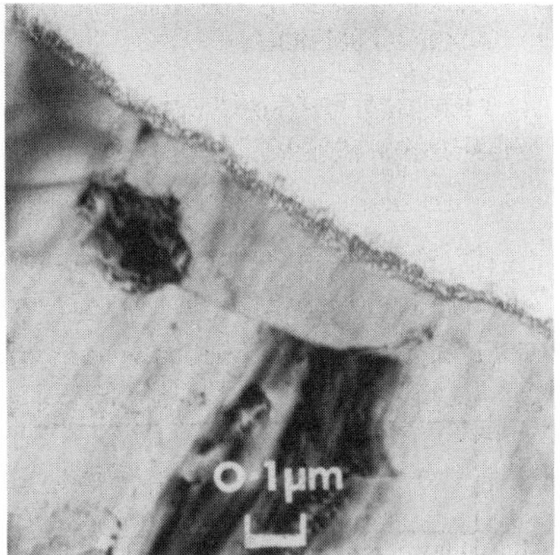

FIGURE 2b: Alclad 2024-T3

7.2 CAA Pretreated Substrates

Figure 3 again compares typical interfaces. Sections from the aluminium-lithium specimens have proved very difficult to produce which could indicate either a more brittle film is produced on this material or it is a tougher substrate causing more knife damage on microtoming. Nevertheless, from the evidence which is available in these micrographs, it is possible to conclude that both adherends generate somewhat similar film structures following chromic acid anodising.

A porous surface film is present which, in the case of Alclad 2024-T3 is about 2-4 micrometres thick but is rarely thicker than 2.5 micrometres in the case of aluminium-lithium. Further, there appears to be more "texture" to the film grown on the aluminium-lithium alloy. In both cases, however, the characteristic planar interface is evident - i.e. the grown film is well-wetted but almost certainly not significantly penetrated by the adhesive.

Transmission electron micrographs of sections through joints produced using CA anodised adherends

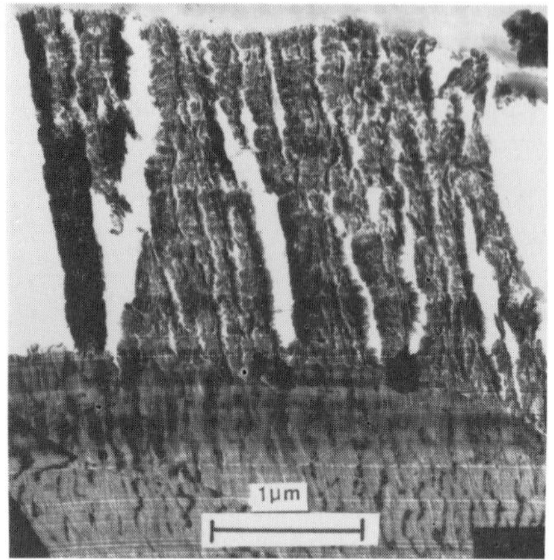

FIGURE 3a: T3 BA 8090C

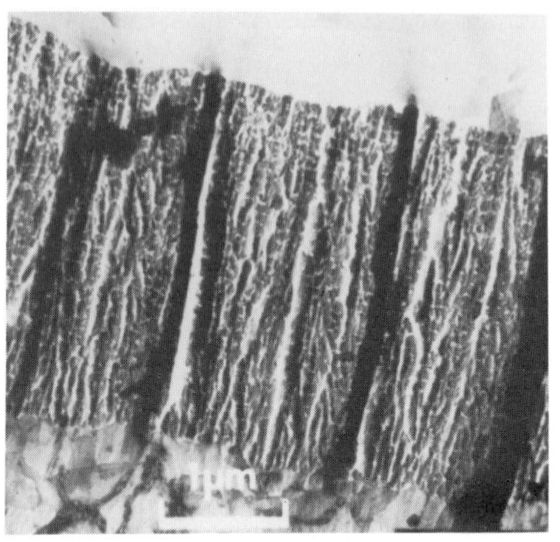

FIGURE 3b: Alclad 2024-T3

7.3 PAA Pretreated Substrates

Figure 4 shows the comparison between the two substrates. Again, similar structures are evident. A porous surface film is present in both cases, that on the Alclad is generally about 0.5 - 1.0 micrometres thick whilst on aluminium-lithium it is only about 0.35 - 0.55 micrometres.

Transmission electron micrographs of sections through joints produced using PA anodised adherends.

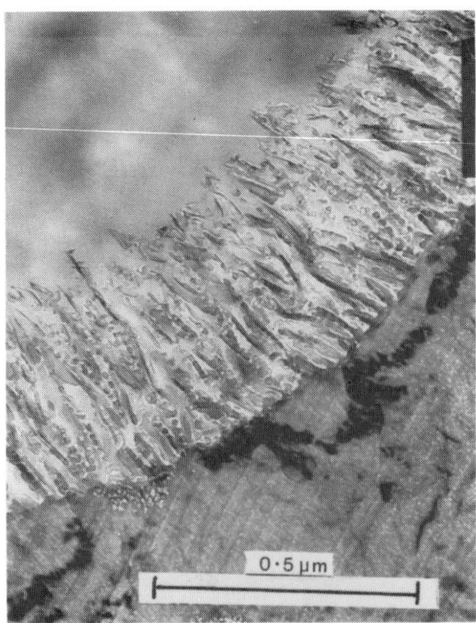

FIGURE 4a: T3 BA 8090C

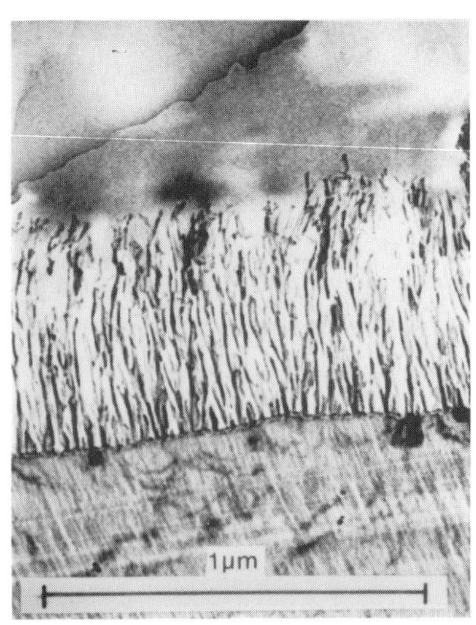

FIGURE 4b: Alclad 2024-T3

Here, for both substrates, any original surface roughness is enhanced due to film material collapse during anodising (caused by progressive thinning of the cell material adjacent to the pore wall). Careful examination of the micrographs shows that the adhesive not only wets the surface and revealed cavities well but also penetrates into the depths of the anodic film. As in the case of the CAA adherends, the anodic film grown on the aluminium-lithium appears to exhibit more "texture" than for the comparable Alclad film.

The above shows, in general, how similar the film structures, developed on aluminium-lithium surfaces are to those on Alclad. The differences in anodic film thickness can be explained by the lower current efficiencies when anodising alloys as opposed to clad materials where the cladding is 99.3% pure aluminium. However, some more significant differences have been occasionally observed - Figures 5 and 6 are typical.

Transmission electron micrographs of sections through joints produced using PA anodised aluminium-lithium adherends

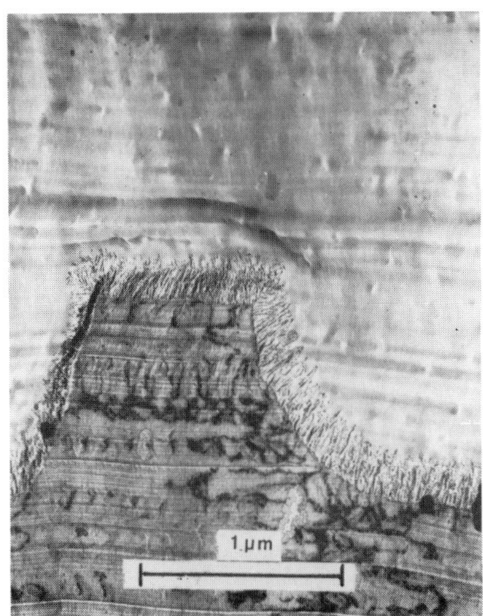

FIGURE 5: Gross surface cavity formation

FIGURE 6: Copper segregates at the interface

Figure 5 shows what appears to be a highly convoluted surface - a degree of surface roughness not experienced with the Alclad. One possible explanation is that during the heat treatment given to the aluminium-lithium alloy some of the alloying elements - particularly lithium - could relocate on the metal surface. When the metal is pretreated it is initially immersed in chromic acid which could well etch away some or all of these surface deposits; this would cause random pitting. The exposed aluminium, during the later stages of pretreatment, would still be capable of growing an anodic film and, hence, give rise to the sort of structure seen in Figure 5.

Figure 6 shows another unusual feature seen with these aluminium-lithium substrates during microscopy. This is the presence, in the region of the boundary layer and/or film/adhesive interface, of metallic segregates - Auger Electron Spectroscopy has identified them as being essentially copper.

More work needs to be carried out to determine how they got there as, at present, it is not known whether they are artifacts - i.e. they have been swept out of the bulk of the alloy by the microtome and deposited in the interfacial region - or whether the anodising process itself has caused their relocation to the surface.

8. CONCLUSIONS

Although this work is very much in its infancy with considerable investigations still to be carried out, it is already possible to draw some conclusions and make some speculations from the above.

1) No difficulties have been found in using any of the standard pickling or anodising pretreatments to prepare aluminium-lithium surfaces for bonding.

2) The film structures formed after pretreatment are, in essence, similar to those already well-characterised on Alclad 2024-T3. However, the anodic films on the aluminium-lithium do appear to be more highly "textured".

3) The thinner anodic films would be expected due to the known, lower current efficiency when anodising aluminium alloys rather than the 99.3% pure aluminium coating on the 2024-T3.

4) Unlike Alclad substrates, some surface cavity formation is noted after pretreatment; possibly due to loss of lithium- and/or copper-containing second phase particles, at or near the surface, during pretreatment.

5) The wetting/penetration of the grown film by the adhesive appears to be the same as for the Alclad adherends. This would allow the speculation that, from the arguments already advanced for bonded Alclad[8], the performance under high humidity conditions could also be similar.

6) Chemically pretreated aluminium-lithium surfaces are very amenable to being joined by adhesive bonding. However, if tensile loads of >30 MPa are applied to simple structures then severe deformation of the substrate around the bonded area can take place.

ACKNOWLEDGEMENT

The authors gratefully acknowledge the help of the following:

Short Brothers, Belfast and in particular Dr.W.Mcgarel and R.Hanna for supplying anodising facilities at very short notice.

E.K.Sim, a former student of UMIST, for preparing the ultramicrotomed Alclad 2024-T3 specimens.

Miss J.A.Underwood and P.J.Hirst for preparing the test specimens and generating the mechanical strength data.

TABLE 1: Percentage Composition of Aluminium and Aluminium-Lithium Alloys

ELEMENT	T3 BA 8090C	ALCLAD 2024 - T3	
		CORE	CLADDING
COPPER	0.86 - 1.36	3.80 - 4.90	0.10 max
LITHIUM	2.28 - 2.58		
MAGNESIUM	0.89 - 0.90	1.20 - 1.80	0.05 max
MANGANESE		0.30 - 0.90	0.05 max
ZIRCONIUM	0.13		
IRON	0.13 - 0.17	0.50 max	} 0.70 max
SILICON	0.04 - 0.06	0.50 max	
CHROMIUM		0.10 max	
ZINC		0.25 max	0.10 max
TITANIUM		0.15 max	0.03 max
OTHERS (each)		0.05 max	0.03 max
ALUMINIUM	Balance	Balance	Balance

TABLE 2: Comparison of Mechanical Strengths for Aluminium and Aluminium-Lithium Bonded Joints

PRETREATMENT	TYPICAL ADHESIVE STRENGTHS ON			
	ALCLAD 2024-T3		T3 BA 8090C	
	Lap-Shear at 22 Deg C	F-R Peel	Lap-Shear at 22 Deg C	F-R Peel
CSA PICKLE	40 MPa	230 N	35 MPa	250 N
CA ANODISE	37 MPa	200 N	32 MPa	235 N
PA ANODISE	41 MPa	270 N	35 MPa	290 N

REFERENCES

1) D.Little, Overview, in : Proceedings of the Third International Aluminium-Lithium Conference, Vol. 1, (The Institute of Metals, London, 1986) pp 15-21.

2) N.A. de Bruyne, Bonded Aircraft Structures (Bonded Structures Ltd., Cambridge, 1957).

3) R.L.Patrick, Treatise on Adhesion and Adhesives, Vol. 4, (Marcel Dekker Inc., New York, 1976)

4) A.J.Kinloch, Durability of Structural Adhesives (Applied Science Publishers, London, 1983)

5) B.M.Parker and R.M.Waghorne, Composites **13** No. 3 (1982) 280

6) J.A.Bishopp, Inten. J. Adhesion and Adhesives **4** No. 4 (1984) 153

7) J.A.Bishopp, E.K.Sim, T.V.Smith, G.E.Thompson and G.C.Wood, The Use of Electron Microscopy for the Analysis of the Adhesive-Adherend Interface in the Aluminium-Aluminium bonded joint, in : Adhesion 12, ed. K.W.Allen (Elsevier Applied Science, London, 1988) pp 248-264.

8) J.A.Bishopp et al, J.Adhesion **26** (1988) 237.

9) T.Malis, Characterisation of Lithium Distribution in Aluminium Alloys, in : Proceedings of the Third International Aluminium-Lithium Conference, Vol. 2, (The Institute of Metals, London, 1986) pp 347 - 354.

10)G.N.Colvin and E.A.Starke, SAMPE Quarterly **19** No. 4 (1988) 10.

11)Federal Specification QQ-A-250/5F, Aluminium Alloy Alclad 2024 Plate and Sheet, (1971 : amended 1974 and 1983).

12)Federal Spcification, MMM-A-132A (1981 : amended 1982 and 1984)

13)A.E.C.M.A. Standard, pr EN 2243-2 (1980 : Draft - Issue 1).

14)Aircraft Process Specification DTD 915b, (Ministry of Supply. 1956).

15)Defence Specification, DEF STAN 03-24/1, Chromic Acid Anodising of Aluminium and Aluminium Alloys, (Ministry of Defence, 1984).

16)Boeing Process Specification, BAC 5555, Phosphoric Acid Anodising of Aluminium for Structural Bonding, (Boeing, 1974).

17)R.C.Furneaux, G.E.Thompson and G.C.Wood, Corros. Sci. **18** (1978) 853.

18)D.J.Arrowsmith, A.W.Clifford, D.A.Moth and R.J.Davies, Adhesive Bonding of Aluminium-Lithium Alloys, in : Proceedings of the Third International Aluminium-Lithium Conference, Vol. 1, (The Institute of Metals, London 1986) pp 148-151.

Materials and Processing – Move into the 90's
edited by S. Benson, T. Cook, E. Trewin and R.M. Turner
Elsevier Science Publishers B.V., Amsterdam, 1989

AUTOMATIC DIGITISER

D B PAYNE

Design Technologies Limited, Bradfield Road, Finedon Road Industrial Estate, Wellingborough, Northants, NN8 4HB.

This paper describes how any irregular shape in two dimensions may be translated into a series of rectangular co-ordinates using an automatic process which gives many advantages over the equivalent manual technique.

1. INTRODUCTION

In many industries where there is a need to cut out irregular shapes from a variety of materials, vast libraries of master shapes have accumulated to be reproduced by hand or press cutting often necessitating the manufacture of steel knives. During the transition to modern cutting methods such as laser, waterjet, ultrasonic etc. which are usually controlled by CNC machines, there is a need to translate the master shapes in to a mathematical form of rectangular co-ordinates. These co-ordinates are then used directly in part programmes for the CNC machines or are increasingly used in CAD/CAM systems so that further processing of a shape or multitude of shapes may take place before the cutting process.

It was in response to this need that the concept of automatic digitising arose. Manual techniques have been available for many years but to an industry faced with a rapid tansition from an old technique to a new one, the manual technique represents an impossible bottleneck due to its inherent slowness. Automatic digitising as conceived in this paper is a remote operation such that no pressure is applied to the shape being digitised and therefore no deformation takes place. It is extremely fast compared to its manual equivalent and of a reasonably high level of accuracy commensurate with the requirement in the aforementioned industries.

Data acquired by this technique can be stored for future processing or used directly by copying machines but in any case fully describes the shape in mathematical terms.

At present there are several techniques by which an irregular shape may be translated into a series of rectangular co-ordinates. In its simplest form, a manual digitiser consists of a cross wire cursor attached to two feedback devices which record the co-ordinates of the centre point of the cursor. The operator can move the cursor to any position on the periphery of the shape that is being digitised and press a key to record that point. The method is inherently slow and the accuracy depends solely on the ability of the operator. It does have the advantage that the operator can visually assess the shape and reject those points which are insignificant and capture those that are important. If the shape has thickness i.e. a template rather than a paper pattern or drawing; then a better method is to use a tracing pen which is moved around the pattern manually collecting co-ordinates from a grid which is incorporated in the support surface. This technique is accurate and reasonably fast but cannot be used for thin or soft patterns.

Automatic digitisers do exist, however and a typical example would be a line scan camera which is passed over the shape at a fixed speed taking co-ordinates from both sides at once. This technique is simple and fast but is limited in accuracy because the number of pixels in the camera represents the total width of the shape. The accuracy therefore varies with the width of the shape. Secondly the co-ordinates are not stored sequentially and therefore a post processing exercise is required to sort them out.

This background provided the incentive to develop a machine which could significantly improve on these existing techniques and be useful over a wide range of materials.

The concept of this automatic digitiser is that the pattern to be digitised is placed onto a flat table which can be moved in two directions which are orthogonal to each other. These two directions are referred to as X and Y axes and each is controlled by a local computer. The surface of the table is illuminated so that the pattern stands out in contrast to the surface colour of the table. A fixed solid state camera is focused onto one small area of the pattern edge and from the image formed can derive control information so that the next section of the pattern edge can be presented to the camera. In order to do this the computer must analyse the image and subsequently cause the table to move along a vector which will maintain visual contact between the camera and the pattern edge. Since the image contains data relating to the pattern edge then co-ordinates of the edge can also be calculated.

The generalised requirements for an automatic digitiser can be summarised as follows:-

(a) The digitiser speed should be several times faster than that of the manual technique.

(b) Accuracy should be comparable to that obtained manually and selectable depending on the application. It may be necessary in order to achieve higher accuracy to compromise other features such as speed.

(c) Repeatability should be well within the target set for accuracy.

(d) The method of sensing the object should be remote rather than tactile so that a greater range of material can be accomodated.

(e) It should be easy to operate and not require any special operator skills or operating conditions.

(f) The digitising techniques should be applicable to any size of object without loss of accuracy or speed.

2. SYSTEM DESIGN

A prototype has been produced using a personal computer, a ballscrew driven XY table with DC drives, a solid state camera system and an electronic interface designed and built to combine all the elements into one overall system. The XY table was supplied with servomotors and encoders were fitted to the free end of the ballscrew to provide positional information.

The computer receives feedback information from the individual encoders at all times and therefore knows the position of each axis. In a manual control mode the operator can command each axis to move from its present position to a given target co-ordinate at a given feedrate. The computer then calculates the interpolation points along the specified vector and issues them as control data. The result is that the axes move at different speeds towards the defined co-ordinate but resulting in the vector speed being that defined by the operator. He can now select another pair of co-ordinates, command a movement, then repeat the exercise and effectively describe a shape comprising a series of short, straight vectors. If the shape is closed it becomes a polygon. Any shape can be described in this way and it is clear that the smaller the length of the vector, the more representative of the true shape the sequence of co-ordinates becomes.

The principle of the digitiser is the reverse of this whereby a number of points on the periphery of an irregular shape are found by a sensing device and then used as the target co-ordinates to pull the sensing device from one to the next. This sensing device could take many forms but in order to conform to the design criteria, a solid state two dimensional camera was chosen. This has the advantage of remoteness thereby applying no distorting forces to the object being digitised and has a fixed geometry so that an image formed on its sensing surface can be analysed with precision. The sensor is made up of rows and columns of photosensitive pixels, the central row and column of which can be considered as a cross wire which can be targeted onto the edge of the object being digitised. The other pixels can then be used to survey the surrounding area which can be changed in size according to the magnification introduced by the lens system. The larger the area exposed, the larger the area attributed to each pixel and the lower the accuracy of the co-ordinates produced. The image thus formed on the photosensitive array is compared to a library of predefined segments and when a match is established the computer instigates a more detailed search of the image to obtain the exact target co-ordinate.

A number of ambiguties can arise and have to be verified by special techniques. However it is interesting to note that sharp corners, re-entrant curves, notches etc. can all be dealt with effectively. One particular feature of the software is its ability to look ahead and find, for example a sharp corner. The digitiser responds by slowing down and reducing the co-ordinate spacing until the corner has been negotiated whereupon it regains its preset speed.

3. LIGHTING

The use of a solid state camera necessitates the provision of good even lighting. This can be achieved by flooding the object from above and using the light reflected from its surface to create an image. This technique is particularly useful for large areas where only the small area being digitised needs to be illuminated. It has the disadvantage or complication of creating shadows if the object being digitised has a significant thickness.

The second technique involves the use of a light box which supports the object and presents a silhouette to the camera which is unaffected by object colour. There is also no shadow effect from the component edges with this technique and consequently is to be prefered if its use is not prohibited by practical size consideration.

4. RESULTS

The test pieces were chosen mainly for their regular geometric shape which could be easily measured by other conventional techniques. In the case of the circle, no attempt was made to select the ideal material for digitising but one that was realistic. The white rectangle is a more idealised application but does show the ability of the system to cope with small intricate features.

The system works on a fixed algorithm period so that as the feedrate increases the spacing of co-ordinates also increases. At the start of this period, the camera effectively takes a snapshot of the object in view and the rest of the time is spent in analysis and control whereupon the process is repeated. Small errors are introduced due to the relative movement of

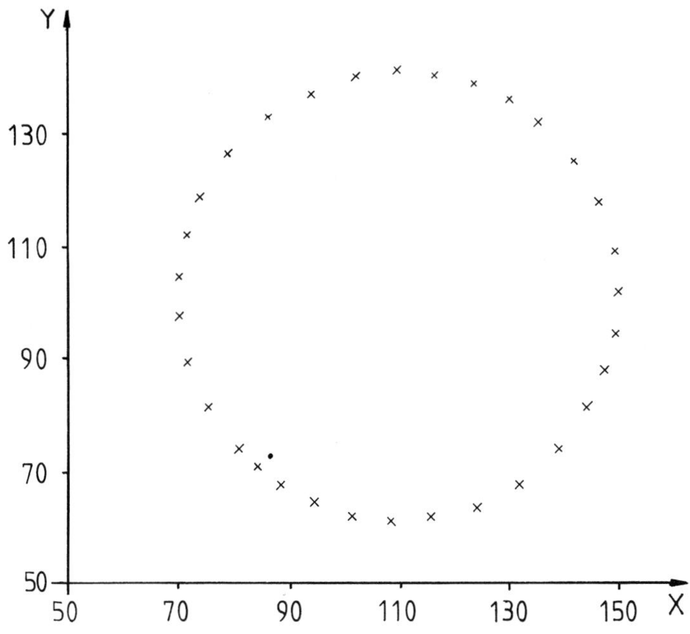
Y
130
110
90
70
50
50
70
90
110
130
150
X

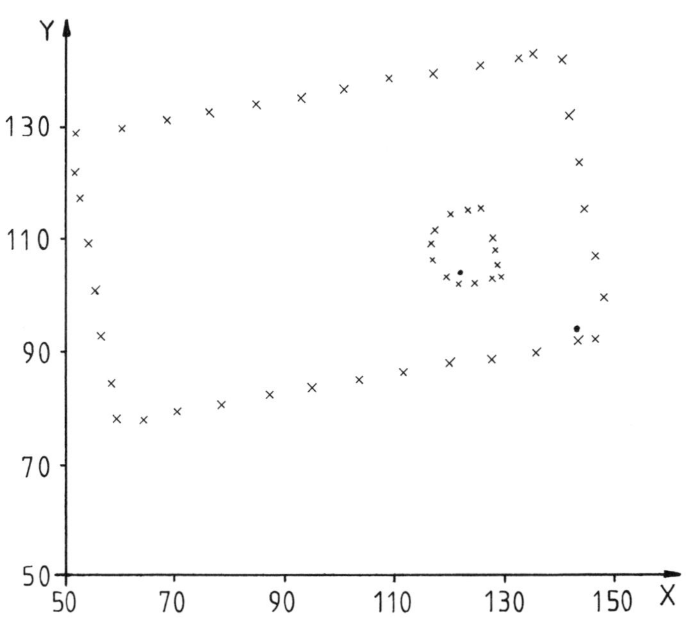
Y
130
110
90
70
50
50
70
90
110
130
150
X

camera and object which causes a blurred image but this can be compensated as the amount of distortion is known.

Other accuracy limitations are due to the optical magnification and distortion and backlash errors in the ballscrew. However, to date an accuracy of +/- 0.15mm is achievable at speeds in excess of 0.5m/min. In the future a faster microprocessor will be used which will improve the speed to 5m/min with no loss of accuracy.

The figures overleaf show the results obtained with the prototype digitiser. Only one out of every fifth co-ordinate is plotted for clarity but serves to indicate the amount of data produced. The slow down algorithm is also excluded from these results.

5. FUTURE MODIFICATIONS

The volume of data produced can be detrimental as well as beneficial. Around complex curves it is necessary to have co-ordinates close together in order to adequately define the shape whereas on straight lines or arcs then two or three co-ordinates will usually suffice. At present data is transmitted along a serial line to a remote nesting system as it is produced. In the next version we intend to store data in the digitiser as a file on a hard disk and then to post process it using data reduction algorithms. This will change the short straight line vectors into arcs and hence make the data more acceptable to most nesting systems.

The magnification produced by the optical system has a direct bearing on system accuracy as the area viewed has to be equally shared amongst the pixels in the photosensitive array. By reducing the area it is possible to achieve a very high level of accuracy albeit at the expense of speed. However by doing this it should be possible to follow lines as well as edges.

At present internal holes are digitised in exactly the same way as the periphery of the component. Whilst this is acceptable for large holes or holes of irregular shapes, it is rather cumbersome for an object containing a large number of small round holes. It is therefore intended to incorporate an automatic scanning mode which sequentially looks at the whole object and only records the centre and radius of the holes.

6. CONCLUSION

It is evident that to a large extent the design philosophy has been observed and reasonable accuracy and feedrate has been obtained.

The automatic digitiser has already proved its value in many applications. It is often sufficient to place the shape to be digitised onto the table and to align the edge visually with the centre of the camera lens. If the digitiser mode has been selected then the processor will begin immediately and will stop as soon as the shape is removed from the table. Using the digitiser in this way is extremely quick and the only disadvantage is that the co-ordinates produced are not related to the machine datum point. It is sometimes necessary to adjust the aperture setting to cater for different material surfaces. It may also be necessary to change the background from matt black to white in order to enhance the contrast. However, experience to date has shown that most materials and shapes can be accomodated, the only problems tending to be with shiny or dimpled surfaces which behave as a series of reflectors. It is evident therefore that the operator does not have to have a great deal of skill or knowledge but an eye for the situations which will provide the best operating conditions for the camera.

Patent application has been made for the techniques embodied in this paper.

Materials and Processing – Move into the 90's
edited by S. Benson, T. Cook, E. Trewin and R.M. Turner
Elsevier Science Publishers B.V., Amsterdam, 1989

TREATMENT OF CURE KINETIC DATA FOR USE IN FLOW MODELLING OF CURING MATRIX FIBRE COMPOSITES

T PORTWOOD

Koninklijke/Shell-Laboratorium, Amsterdam, Badhuisweg 3, 1031 CM Amsterdam, The Netherlands

Quantitative modelling of fibre reinforced thermoset resin composite processing can give insight into the interactions of process parameters that cannot be obtained by empirical experiments alone. As such, modelling is an important aid in the development of composite applications.

It is necessary to describe the cure reaction kinetics of the matrix during processing, because many of the properties of the system are cure dependent. If the chemistry of the curing reaction is well known, then mechanistic rate equations can be derived and the rate constants determined. Alternatively, if the cure reaction is known to belong to a class of reactions that can be characterised by a known empirical rate law, then the cure can simply be fitted to that. If, however, the chemistry of the cure is unknown or if there exists no valid empirical law to represent the reaction then neither of the above approaches can be used.

In these circumstances, the cure reaction can be characterised by the adoption of general data fitting techniques. This paper presents a method for the treatment of cure data, which is suitable for quantitative modelling studies of heat and mass flow during cure of resin matrix/fibre composites. The kinetic data are fitted to a minimal, weighted bicubic spline function by a least squares method.

In principle, data can be collected by any measurement technique that relates exothermic heat generation to degree of cure and temperature. As differential scanning calorimetry (DSC) is commonly used, a method to analyse data collected from this source is outlined. Also, as a consequence of the general nature of the method, cure data for a wide variety of matrix systems can be interpolated.

To illustrate the use of the method in processing and matrix system design, cure data for a typical epoxy system developed for pultrusion with glass fibre are analysed. The data are used in a model to predict the state of the composite along the die.

1. INTRODUCTION

Fibre reinforced thermoset resin composites are playing an increasingly important role in structural applications. The full potential of these materials in mass market applications such as the automotive industry is as yet not realised. This is due not only to limited market acceptance in general but also to the embryonic state of some of the processing techniques as applied to thermoset resin fibre composites. The quantitative modelling of composite processing is playing an important part in putting the developing technology on a sound footing. The quantitative chemorheological description of reactive systems dates back well over ten years[1], and at the present time there is modelling activity in pre-pregging, reaction injection moulding (RIM), filament winding and pultrusion[2–10].

An adequate understanding of the matrix cure reaction is central to a quantitative and predictive modelling scheme of heat and mass flow. Not only will the chemical curing reaction be exothermic, but the properties of the matrix will depend on its chemical state.

This paper describes a general numerical method for treating cure data. The technique is based on a bicubic spline interpolation that makes no assumptions about the form of the cure reaction. As such, the treatment is quite general and can be applied to a wide range of cure reactions, avoiding some of the pitfalls associated with the use of specific cure models. Data can be obtained from any suitable measurement technique, for example DSC, FTIR and dielectric methods.

As an illustration, the treatment of DSC cure data is considered and the cure for an epoxy resin based system characterised. To show the application of the technique in process modelling and matrix system design, we will use these data in a pultrusion model to calculate the state of the composite along the die.

2. GENERAL FITTING METHOD

In describing the cure of a reactive matrix system, several approaches are possible, depending on how much is known about the cure reaction.

If the chemical mechanisms of the cure reactions are known, and the rate constants determined, then one has a complete and fundamental description of the cure. Investigating the chemistry of the cure gives detailed predictions about the fates of the reacting groups involved[3,11], and variations in initial reactant concentrations can then be taken into account. The drawback of this approach is that it is specific to only one type of reaction. Any new reactions have to be mechanistically characterised and the rate constants determined. This can be laborious to perform and if there are many reactions taking place simultaneously it can be near to impossible.

Using a general empirical rate equation sacrifices chemical knowledge for relative ease and speed of determination. This makes the approach popular in modelling studies. The danger of this is that injudicious application of standard empirical equations can lead to erroneous results. For example, the Borchardt–Daniels equation is often used[12] in the modelling of epoxy resin based composites. However, if this equation is used to model the numerous epoxy systems that are auto-catalysed, then the cure behaviour will be misrepresented. The solution to this is to use equations with more adjustable parameters[13].

In situations where a detailed chemical description is impractical and a standard empirical equation cannot be found, the third approach of general numerical fitting can be considered. This is versatile as, in principle, any form of kinetic data can be interpolated. The price paid for this versatility is that the errors in the method rest on the quantity and reliability of the input data. To ensure reliable interpolation, sufficient data must be collected. The method does not permit any extrapolation, so data must be collected over the entire cure/temperature range required.

If known, then a chemical or empirical equation approach will be sufficient to describe the cure kinetics. However, the general numerical technique allows chemically complex matrix systems to be kinetically characterised where the other methods are not applicable.

3. BICUBIC SPLINE FITTING

When curve fitting to data, it is often the case that the mathematical form of the fitting function is not important to the problem. All that is required is that the function chosen should adequately represent the form of the data.

Spline fitting is one versatile approach to general curve fitting as the functions involved are straightforward to handle as well as versatile. For functions dependent on one variable, cubic splines can be used. A cubic spline fit consists of a number of cubic polynomial segments joined together at each of their ends, such as to preserve continuity in the first and second derivatives. Usually the third derivative is discontinuous. The places where the polynomials join are called knots.

Where there are two independent variables, the problem becomes one of surface fitting. Here it is possible to define a bivariate cubic polynomial over a rectangle which covers some part of the surface to be fitted. Then the entire data surface can be represented by several of these rectangular panels. By analogy with the simple cubic spline case, the panels are joined by knots parallel to the two axes of the independent variables. Around these interior knots, the first and second derivatives are again continuous, and the third is, in general, discontinuous.

The bicubic spline is defined over a rectangular area Q in the (x,y) plane. The edges of Q are parallel to the x and y axes. The panels divide Q, with the panel edges parallel to the axes. Then for each panel the bicubic spline takes the form of a bicubic polynomial

$$f(x,y) = \sum_{i=0}^{3} \sum_{j=0}^{3} a_{ij} x^i y^i \tag{1}$$

It is computationally faster and more accurate to rewrite equation (1) in the form

$$f(x,y) = \sum_{i=1}^{p} \sum_{j=1}^{q} C_{ij} M_i(x) M_j(y) \tag{2}$$

where M (x) i=1,2,..p and N (y) j=1,2,..q are normalised B splines[14-15].

The data are fitted to form a minimal, weighted least-squares solution with the prescribed knots. A program was written in PASCAL to perform this fitting. This program calls standard NAG library routines to calculate the B splines[16,17]. The results of the fitting are viewed graphically.

4. DATA COLLECTION BY DSC

4.1. Theory

For the purposes of heat and mass flow calculations, the chemical reaction can be regarded as providing a source of heat in a unit volume of sample. As an example, we can take a straightforward reaction of an epoxy resin with a certain curing agent. The rate of reaction of the epoxy, $\partial\alpha/\partial t$, is in general a function of the state of cure, α, and temperature T.

$$\frac{\partial\alpha}{\partial t} = f(\alpha,T) \tag{3}$$

The output from DSC for kinetic studies is heat flow as a function of temperature and time. Heat flow has the dimensions of energy/(time*mass) and can be related to the heat producing chemical reaction by

$$\text{Heat flow} = H_T \frac{\partial\alpha}{\partial t} \tag{4}$$

If the total heat of reaction per unit mass is H_T then the fraction reacted is given by

$$\alpha = \frac{1}{H_T} \int (\text{Heat flow})\, dt \tag{5}$$

For an isothermal DSC experiment the integral of the heat flow curve after a given time t gives the fractional extent of the reaction, as given in equation (5). Thus by integrating the temporal DSC heat flow curve, the form given in equation (3) can be obtained. The cure reaction can then be characterised with a series of isothermal experiments at different temperatures covering the entire temperature range of interest. To interpolate these data the function $f(\alpha,T)$ can be obtained by fitting the data to a least squares bicubic spline surface, as described above.

When choosing the isothermal temperatures at which to characterise the cure, the general trend of the reaction should be taken into account. For example, more data should be taken in regions where the reaction rate is changing fastest. Also, data must be given over the entire temperature and conversion data space required, because no extrapolation is permitted by the fitting method. For this degree of experimental design one needs no great *a priori* knowledge, rather only a very rough idea about the matrix system reactivity.

The extent of cure, as defined by fractional heat output, can then be obtained by a numerical integration of the fitted bi-spline surface. The fitted surface can thus be used in calculations of temperature fields in the composite, and to calculate physical properties which have a temperature/cure dependence.

4.2. Experimental

The method involves measuring the net heat flow of the matrix system at various isothermal temperatures which together span the processing data range of interest. A Dupont 912 DSC apparatus was used in these current studies for the experimental determination of the reaction cure kinetics of epoxy resin matrix systems.

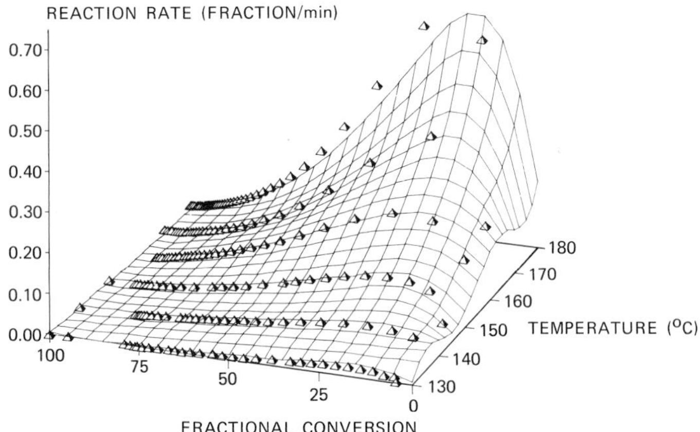

FIGURE 1
DSC Fitted Data for Epoxy-Based Matrix System

Errors in the calculated area can arise either from the position or shape of the baseline, or from the fact that there has been significant pre-reaction during the warm-up time. The baseline position can be a function of chemical state of the resins[15], but a horizontal baseline is a fair approximation for the current studies.

Pre-reaction can happen with reactive systems at high temperature. In practice, the largest errors are introduced during the very first part of the isothermal run, when the sample temperature is equilibrating with the cell temperature. For reactive systems at high temperature this becomes the most dominant error and in extreme cases an offset correction must be applied.

Figure 1 shows the result of fitting five isothermal DSC experiments for an epoxy based matrix system. The vertical axis shows the fractional reaction rate per minute of the matrix system as a function or the temperature and fractional degree of cure, defined with respect to the total heat of reaction. The points are experimentally determined and the grid is the fitted surface $f(\alpha,T)$.

5. APPLICATIONS

Once the cure function $f(\alpha,T)$ has been fitted, then it can be used in models that require a chemical reaction heat source term, and have cure and temperature dependent properties.

As an example, consider the pultrusion of an epoxy glass composite. There have been several published models of this process[2–8] using either a chemical mechanistic approach or an empirically based rate equation.

The model developed here will predict the cure and temperature profiles along the pultrusion die using the bicubic spline kinetic data method. The partial differential equation to solve is

$$v_z \frac{\partial T(x,z)}{\partial z} = Dp \frac{\partial^2 T(x,z)}{\partial x^2} + \frac{H}{Cp} \frac{\partial \alpha(x,t)}{\partial t} \tag{6}$$

Here T denotes temperature, t time , x position, v_z the pulling speed, D_p is the thermal diffusivity, C_p is the specific heat capacity of the composite, H is the heat of reaction of the composite.

Equation (7) can be solved with the appropriate boundary conditions to give predicted temperature and cure profiles through the composite along the length of the die. The speed of pultrusion can be varied in the model and for the current work we have chosen a speed of 0.5 m/min. The die section we have chosen to use is rectangular , 2 mm by 19 mm , and 80 cm long. The system used in the simulation was an epoxy resin with the addition of a reactive diluent.

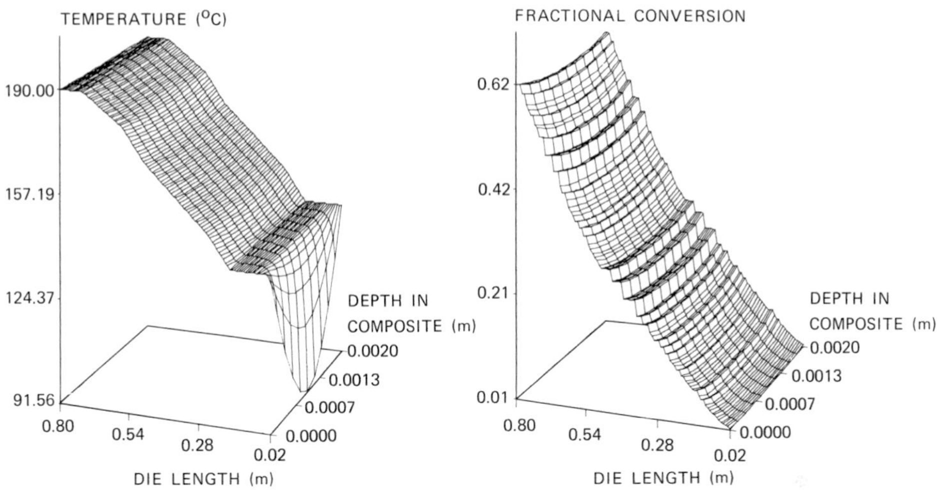

FIGURE 2
Simulation of Composite Temperature Pultrusion

FIGURE 3
Simulation of Composite Cure-Pultrusion

Figures 2 and 3 show the results of a simulation for a composite containing 60 % by volume of glass fibre and a temperature profile of the die which varies approximately linearly along the length of the die between 140 °C and 190 °C. In Figure 2, the temperature profile through a cross section of the 2 mm thickness is shown, as a function of position along the die length. It can be seen that for the chosen conditions, the temperature in the middle of the pultruded profile closely follows the value at the die walls.

Similarly, Figure 3 shows the corresponding fractional degree of cure distribution. This has been calculated using bicubic spline fitted kinetic data. Again it can be seen that the cure profile through the sample thickness has only a small curvature for a given die position.

6. DISCUSSION AND CONCLUSIONS

A general data fitting method has been applied to the analysis of cure kinetic data which can be used in heat and mass flow modelling of reactive composite systems. In this scheme, kinetic data are fitted to a bicubic minimal weighted least squares surface, with fractional degree of cure and temperature as the two independent variables.

The method treats the reaction only as a source of heat, and so can be applied in principle to any exothermic cure reaction that can be represented by the parameters of temperature and fractional degree of cure. It then follows that matrix systems with multiple independent reactions can be modelled by the same method.

It has the ability to treat data that would either to laborious to treat by a chemical approach or not be appropriately modelled by standard empirical equations. The errors in the method arise directly from the quality and quantity of the input data. Fitting errors can be made small by diligent splitting of the data into panels, more data and smaller panels being required for regions where the function is rapidly changing.

The cure of an epoxy resin based matrix system was characterised by fitting to a series of isothermal DSC experiments covering the required temperature range. The application of the technique has been shown by using the cure data in a pultrusion model to calculate the state of the composite along the die.

In future work, the technique will be used not only in predictive composite process modelling, but also as an aid in the design of new resin based matrix systems for mass production applications.

REFERENCES

1. M.B. Roller, Polym. Eng. Sci. 15 (1975) 406.
2. C.D. Han and H.B. Chin, Polym. Eng. Sci. 28 (1988) 321.
3. H.J. Buck, L.T. Blankenship and R.P. Shirtum, SAMPE J. May/June (1988) 21.
4. L.T. Manzione, Polym. Eng. Sci., 21 (1981) 1234.
5. A.C. Loos and G.S. Springer, J.Compos. Mater. 17 (1983) 135.
6. T.H. Hou, ANTEC (1986) 1300.
7. R. Dave, J.L. Kardos and M.P. Dudukovic, Polym. Compos. 8 (1987) 29.
8. C.D. Han, D.S. Lee and H.B. Chin, Polym. Eng. Sci., 26 (1986) 393.
9. C.N. Lekakou, S.M. Richardson, Polym. Eng. Sci. 26 (1986) 1264.
10. Y.A. Tajima and D.G.Crozier, Polym. Eng. Sci. 28 (1988) 491.
11. C.S. Chern and G.W.Poehlein, Polym. Eng. Sci. 27 (1987) 788.
12. H.J. Borchardt and F. Daniels, J. Am. Chem. Soc. 79 (1957) 11.
13. S. Sourour and M.R.Kamel, Thermochem. Acta. 14 (1976) 41.
14. J.G. Hayes Bull. Inst. Maths. Applics. 10 (1974) 144.
15. J.G.Hayes and J. Halliday, J. Inst. Maths. Applics. (1974) 89.
16. NAGFLIB:1830/0: Mk 9: January 1982.
17. NAGFLIB:1635/0: Mk 11: November 1983.

Materials and Processing – Move into the 90's
edited by S. Benson, T. Cook, E. Trewin and R.M. Turner
Elsevier Science Publishers B.V., Amsterdam, 1989

ADVANCED COMPOSITE PROCESSING WITH BISMALEIMIDE RESINS

H.D.STENZENBERGER, W.RÖMER, M.HERZOG, P.KÖNIG[1] and K.FEAR[2]

[1]Technochemie GmbH-Verfahrenstechnik, D6915 Dossenheim, West Germany
(A member of the Royal/Dutch SHELL Group of Companies)
[2]Shell Chemicals, UK, Chester

ABSTRACT

Low Pressure Autoclave Moulding (LPAM), Resin Transfer Moulding (RTM) and Filament Winding (FW), as processing techniques for advanced composites, are currently of great interest due to their economics and flexibility. Traditionally, RTM has been performed on resin systems which have low viscosity at room temperature (less than 1000 cps) and can be transferred easily at that temperature. Wet filament winding requires resin of a similar characteristic to achieve optimum fibre wetout during impregnation. Resin systems, such as bismaleimides (BMI), have been excluded from these techniques due to their high melting points and solid state at room temperature. This problem could be overcome by using BMI resins with low uncured glass transition temperature (COMPIMIDE 65FWR, COMPIMIDE 796/TM123 blends) and low viscosities at temperatures, still suitable for RTM and wet filament winding. The processing parameters acceptable for both RTM and wet filament winding, using COMPIMIDE-BMI resins, are outlined. Mechanical properties on carbon reinforced composite specimens, manufactured using these techniques, are presented.
Low pressure autoclave moulding (LPAM) using prepreg tape is still the most widely used processing technique for advanced composites in the aerospace industry. New Bismaleimide resins, based on COMPIMIDE materials, cure at 175°C at low pressure (4-7 bars) in reasonably short periods of time and provide processing equivalent to high temperature epoxies. The processing and properties of a model BMI system are provided and mechanical properties on carbon reinforced laminates are discussed.

INTRODUCTION

Polyimides have attracted a great deal of interest because of their unique properties at elevated temperatures and extreme environments, such as high humidity and fire conditions. However, until recently the history of polyimide resin development has been one of materials difficult to process but with outstanding performance.

Polyimides may be classified into three distinct groups:

- condensation

- thermoplastic and
- addition

Addition polyimides generate no volatiles during cure, are relatively easy to process and pass through a liquid or at least high viscosity phase during cure. They have therefore been successfully used to fabricate fibre reinforced composites. Thermosetting polyimides are characterized in that they are low molecular weight at least difunctional monomers or prepolymers or mixtures thereof which carry imide moeties in their backbone structure and are terminated by reactive groups, which undergo homo- and/or copolymerisation by thermal or catalytical means.

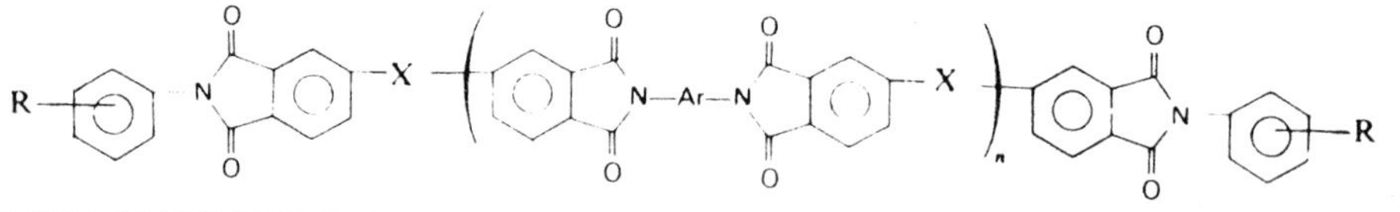

R =

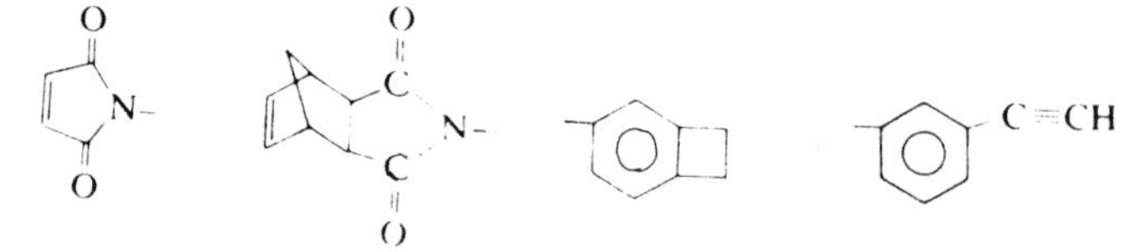

According to this definition thermosetting polyimides are classified by the chemical nature of their reactive endgroups. The most important families are

Acetylene terminated polyimides

Benzocyclobutene imides

Endomethylenetetrahydrophthalimides (PMR-resins)

Bismaleimides.

Out of these, the bismaleimides are the class of thermosetting polyimides which have found widespread application mainly as matrix resins for advanced composites.

Bismaleimides of the general formula I

I $\quad R = -CH_2-, -O-, -S-$

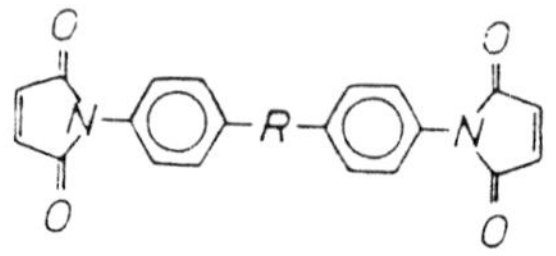

are the main building blocks for commercial resin formulations. They are synthesized from maleic anhydride and aromatic diamines in high yield. They can be cured thermally to high heat resistant materials. However, they suffer from lack of processability and therefore have to be formulated into resins which can be used in prepregging, low and high pressure cure, in filament winding, resin injection moulding (RIM) and injection moulding.

Many chemical approaches have been developed to meet these processing requirements. Modifications are also mandatory because the bismaleimide homopolymers are very brittle and need improvement.

The maleimide group can be involved in many chemical reactions therefore numerous functional monomers have been found or specifically synthesized to be used as comonomers. The target is to achieve copolymer networks with improved properties in the areas of temperature resistance (high glass transition temperature, Tg), toughness and moisture resistance. Out of the many chemical concepts for bismaleimide modification the following are most promising

Copolymerisation with vinyl-monomers (1)
Diels-Alder copolymerisation with
- styrene (2)
- divinylbenzene (3)
- bis(o-propenylphenoxy) compounds (4,5)
Copolymerisation with bis(allylphenyl) compounds (6)
Copolymerisation with allylphenols (7)
Michael Addition Copolymerisation with
- aromatic diamines (8)
- aminobenzoic hydrazide (9)
Copolymerisation with low molecular weight thermoplastics (10)
Copolymerisation with reactive elastomers and (11)
modification with thermoplastics (12)

Technochemie, now a member of the Shell group, has been active in bismaleimide resin technology for 18 years.

COMPIMIDE is our general trade mark of a family of thermosetting bismaleimide resins which show an outstanding balance between processability and temperature performance. This paper outlines their use for reinforced composites, via wet- and tow-preg filament winding (FW), resin transfer moulding (RTM) and low pressure autoclave moulding (LPAM)

2. PROCESSING OF BISMALEIMIDE (COMPIMIDE) RESINS.

2.1 Filament winding

Basically, filament winding consists of wrapping resin-impregnated continuous fibres on a mould or mandrel surface in a precise geometry pattern. The wrapped fibres are cured to form a solid composite structure following a defined time/temperature heat treatment as required by the resin cure chemistry. Usually, the mandrel is removed, but some are left in place to form a permanent liner or core for the filament wound product.

In principle there are three possibilities for impregnating the reinforcement with resin.

- Wet impregnation, i.e. passing the fibres through a resin bath which contains the resin in liquid form. Resins used for on line wet impregnation must have a low viscosity to assure thorough wet-out of the fibres. They must also have a long pot life to avoid premature gelling in the impregnation bath.
- Solvent/solution impregnation, i.e. resins that have desirable mechanical properties but are too viscous for wet impregnation can be preimpregnated onto the filaments via solution techniques. Solvent is stripped off to leave a single tow prepreg which can then be employed in a standard filament winding operation. Tow prepreg winding is advantageous because it relieves the user from handling resins and curing agents. Also, the tow supplied by the formulator is a net system which carries the desired amount of resin to allow clean processing.
- Hot melt impregnation, standard polyester and epoxy filament winding resins are liquid at room temperature. However, many high performance thermosets and thermoplastics are solids at ambient temperature. In these cases, hot melt techniques may be suitable for manufacturing a single tow prepreg.

The key operation during filament winding is the thorough impregnation of the reinforcement. With respect to our research and development objectives, this meant formulating bismaleimide into

- a processable product in terms of viscosity, pot life, gel and cure time
- a product with a use temperature of at least 200-250°C in terms of resin dominant properties being greater than 50% of room temperature properties.

2.1.1 Resin formulation (Chemisty)

Bismaleimide building blocks, like 4,4 Bismaleimide diphenylmethane (COMPIMIDE MDAB) or our base resins COMPIMIDES 353 and 796 are solids at room temperature and therefore require formulation, blending with liquid comonomers, in order to reduce the viscosity. Such reactive diluents are

- Diallylphthalate (DAP), triallylisocyanurate (TAIC), trimethylolpropanetriacrylate (TMPTA) etc.
- Epoxy resins
- Unsaturated polyesters, and/or
- Allylphenyl type comonomers

Shell/Technochemie formulated a one package (ready to use) BMI-resin COMPIMIDE 65FWR, for wet- and tow-preg filament winding. Details on the chemistry of this formulation have been published elsewhere (13)

2.1.2 COMPIMIDE 65FWR resin properties

Viscosity

The viscosity-time correlation of Compimide 65FWR is presented in Figure 1 at four different temperatures. For wet filament winding the ideal processing temperature is between 90 and 100°C. At these temperatures the pot life is approximately 3-5 hours (time to double the viscosity).

Gel Time

Typical gel times for Compimide 65FWR are 25 ± 5 mins at 170°C(specification 20 mins at 170°C).

Differential Scanning Calorimetry

A typical DSC scan is given in Figure 2. For a heating rate of 10°C/min, the cure exotherm maximum appears at ca. 245°C with a shoulder at 282°C. The heat of polymerisation is typically around 235 joules/gram.

Neat Resin Properties

The mechanical properties of neat Compimide 65FRW are given in Table 1 for different cure cycles. Cure cycles 1 and 2 differ only in the post-cure temperature. Interestingly, all properties seem to be superior for the post-cure temperature of 250°C. A shortened cure cycle of 3 hrs at 110°C plus 2 hrs at 160°C, plus a post-cure of only 3 hours at 210°C was used to simulate a more realistic situation as would be the case in the production of a filament wound part. The high temperature properties for this cycle 3 are lower as compared with the higher temperature cycles 1 and 2. However, the tensile properties are very attractive with a room temperature tensile elongation of 2.25%. This investigation also shows that Compimide 65FWR allows a wide range of cure cycles and the properties obtained do not seem to be too sensitive to changes in the cure temperature and times. A post-cure temperature of 250°C is recommended for parts requiring superior high temperature performance.

2.1.3 Filament winding parameters and composite properties

Compimide 65FWR can be employed for wet filament winding, but fibre impregnation has to be performed at elevated temperatures. The following parameters have been found to be reasonable.

Impregnation temperature	90 - 100°C
Mandrel temperature	50 - 60°C
Gel temperature on mandrel	130 - 170°C (adjustable)
Cure temperature on mandrel	170 - 190°C
Cure time	4 - 8 hours (variable)
Post-cure (after demoulding)	3 - 5 hours at 210 - 250°C depending on Tg requirements

As the resin is soluble in solvents such as methylene chloride or acetone it can be used for the fabrication of tow prepreg by use of solvent/solution dip coating techniques. Once the solvent is dried off, the tow prepreg is drapeable but not tacky.

Composite mechanical properties

Tensile strength was determined according to ASTM D 2290 from NOL-rings by use of the split disc method. The samples were prepared via filament winding. A steel mandrel with an external diameter of 146 mm and with an internal heating capability was used to fabricate a pipe (length 150-200 mm) consisting of circumferential windings from on-line impregnated carbon fiber tow. The impregnation bath and resin were heated to ca. 90 - 100°C for thorough fibre wet out and the mandrel temperature was adjusted to ca. 50 - 60°C. The pipe was cured for 2 hours at 170°C, plus 2 hours at 190°C, and post-cured for 10 hours at 210°C.

To avoid resin squeeze-out during cure, a bleed fabric (glass/PTFE) and shrink tape (polyester) were wrapped around the pipe. The pipe was post-cured after being removed from the mandrel and separated from bleeder and shrink tape. The thickness of the pipes used for evaluation was ca. 1,5 mm (contrary to ASTM D 2290 requirements) and NOL-ring specimens were cut off the pipe with a diamond cutting wheel. The NOL-rings were examined photomicroscopically (100x magnification) and were found to have a $<$0,5% void content).

The mechanical properties measured for Compimide 65FWR composites are shown in Table 2. Direct comparison is possible for the T300 and Tenax HTA-7 fibres, because they are both in the 200 GPa modulus range. However, these two fibres differ in their fibre size chemistry. T300 carries an almost fully cured epoxy size whereas Tenax HTA-7 is sized with uncured epoxy resin.

The short beam shear and flexural properties are almost identical for T300 and Tenax HTA-7 fibres both at room temperature and at elevated temperatures.

Interestingly, T300 fibres provide somewhat higher room temperature 90° flexural strength. Surprising is the great difference in the critical strain energy release rate (G_{Ic}). Tenax HTA-7 laminate provides more than double the delamination resistance of T300 laminates in the double cantilever beam test. Work is under way to explain this phenomen. We have experienced the same difference for the two fibres with other BMI resins. Scanning electron microscopic examinations of tested specimens show different fracture surface morphologies which indicate different fibre/resin interfaces.

IM-6, intermediate modulus fibre, provides the expected laminate flexural moduli; however, 90° flexural strength is low when compared with T300 and Tenax HTA-7. NOL-ring tensile strength is satisfactory for both T300 and IM-6 fibres. The 200°C strength is higher than the room temperature strength which indicates the presence of built-in cure stresses. Modification in cure cycle should help to overcome this problem.

2.2 Resin Transfer Moulding (RTM)

Resin Transfer Moulding (RTM) as a composite processing technique is currently of great interest due primarily to its economics and flexibility. The process consists of injecting (transferring) liquid resin into a mould which already contains the reinforcement preferably from the bottom of the mould, whilst air is displaced through vents on top of the mould with subsequent curing of the part while in the mould. Resin Transfer moulding is a potentially inexpensive process for the production of a variety of parts but suffers from some restrictions:

- low cost mould is required which is capable of withstanding the curing temperature of the BMI-resin.
- moulds are quite complicated and have to be opened up to permit the lay up of dry fibre and fabric to the quantity and orientation needed.
- preforms of orientated reinforcement are required to make the process economical.
- difficulties may be encountered in placing sufficient fibre in the mould in order to get a high fibre volume fraction in the finished part.

Traditionally, RTM has been performed on resin systems which have a low viscosity at room temperature and can be transferred easily at that temperature. Resin systems such as bismaleimides (BMI), have been excluded for the most part from this promising processing method due to their high melting points and solid state at room temperature.

2.2.1 Resin formulation

Since dry fibres are placed into the mould and low void content components are to be produced the resin has to have a very low viscosity. Optimum viscosities are between 200 - 300 mPa.S; if the resin viscosity is higher, i.e. 500 - 1000 mPa.S, much higher injection pressure is required. Also high pressure moulds are then needed which makes them more expensive. Developments in processing techniques and in bismaleimide resin chemistry now allow the use of this class of high performance resins in the potentially economical RTM process. (14, 15)

BMI resin formulations which may be employed in the RTM process are COMPIMIDE 65FWR, and blends of COMPIMIDE 796 with COMPIMIDE TM121 and COMPIMIDE TM123. Chemically Compimide TM121 is a bis(allylphenyl) compound which copolymerises with BMI via an ene-type chain extension reaction.

The chemistry of the BMI/compimide TM121 copolymerisation has been described in one of our recent publications (16). Compimide TM123, chemically bis(o-propenylphenoxy) benzophenone, reacts with BMI via a Diels-Alder reaction to provide tough copolymers.

Formulation I:	Compimide 796	65 parts
	Compimide TM121	35 parts

If required for further viscosity reduction 5 to 10% by weight of TAIC (triallylisocyanurate) may be employed.

Formulation II:	Compimide 796	70 parts
	Compimide TM123	30 parts
	TAIC	10 parts

The viscosities, viscosity time profile and the gel times for these two resin formulations are given in table 3. The Compimide 796/TM123/TAIC formulation shows a very low viscosity at 100 - 120°C, however the working life of this resin is relatively short (45 mins). Compimide 796/TM121 blend is a much more latent system with a much longer working life.

2.2.2 Processing conditions

Typical processing conditions for formulation I and II are

	Formulation No. I	Formulation No. II
Injection temperature	100 - 110°C	115 - 125°C
" pressure	variable (1)	variable (1)
" time (max)	2 - 3 hours	30 - 45 mins.
cure cycle in the mould	3 h 170 2 h 200°C	3 h 160 2 h 200°C
Post cure	12 h 210°C	5 - 10 h 230°C

(1) depends on component geometry, fibre type and fibre content

2.2.3 Properties of RTM - Laminates

Typical properties of flat carbon fibre laminates prepared via resin transfer moulding as described above are compiled in table 4. Formulation no. II is superior in high temperature property retention. All laminates were essentially void free. The fibre loading was 50% by volume.

Although the properties of the laminates prepared from the two different formulations are not too different, formulation I has the much longer working life and would therefore be preferred for components which require longer resin injection times. Formulation II is to be preferred for smaller parts with more stringent temperature requirements.

2.3 Low Pressure Autoclave Moulding (LPAM)

Laminate moulding is the technology of stacking prepreg material (fabric or tape) in a given way and subsequently converting the stack into a dense voidfree laminate through the application of pressure and temperature. Through this process, the resin is cured i.e. converted into a solid upon gelliation followed by vitrification. The cure process can be performed in a heated platen press at high pressure (30 - 70 bars, typical for glass fabric laminate moulding) or at low pressure in our autoclave, vacuum assisted. This latter technique is known as the Low Pressure Autoclave Moulding (LPAM) process and is used by the aerospace industry to mould large area components sometimes of complex shape. The autoclaves used generally operate at a miximum pressure of 10 bars and temperatures not exceeding 180°C.

Standard epoxy - based prepregs, both unidirectional and fabric are produced via a hot melt process. Carbon tape is achieved by first casting a resin film onto a carrier into which collimated carbon fibre strands are pressed to provide a tacky drapeable prepreg between release paper.

This prepregging and cure technique requires BMI-resin offering the following features.

- a processible product in terms of viscosity, potlife cure temperature and cure time.
- high glass transition temperature combined with a high fracture toughness of the cured composite.

With respect to bismaleimide resin technology it is again the problem of the formulator to convert crystalline BMI building blocks ar partly formulated resin (COMPIMIDE 796) into a prepreg systems. Formulation means blending BMI with reactive diluents, comonomers, additives and viscosity modifiers. Comonomers should improve cured resin toughness while reactive diluents, although coreactive with the BMI, should improve processability and, in particular, reduce viscosity. Ideal comonomers are liquid at room temperature and thus act as reactive diluents and comonomers at the same time.

2.3.1 Resin formulation (Chemistry)

Shell/Technochemie recently introduced a reactive comonomer, COMPIMIDE TM123 4,4'-bis(o-propenylphenoxy)benzophenone , which, when used in combination with COMPIMIDE 796 bismaleimide, provides tough cured copolymers. The synthesis and cure chemistry of this system has been described elsewhere (4). As with other BMI-resin formulations liquid reactive diluents like TAC (triallylcyanurate) or TAIC (triallylisocyanurate) have to be used to improve tack and drape of the prepreg.

Typical hot melt BMI-resin formulation

Compimide 796	70 parts
Compimide TM123	30 - 35 parts
TAIC (triallylisocyanurate)	10 - 15 parts

flow control (Carbosil) 2 - 3 parts

A typical procedure to blend the ingredients consists of

a) preparation of a COMPIMIDE TM123/TAIC blend and heating to 90 - 100°C

b) addition of the COMPIMIDE C796 resin (coarse particle size) while stirring at 90 - 100°C

c) addition of the flow control agent by use of a high shear mixer

The resin mixture thus obtained is suitable for the fabrication of hot melt type prepreg.

2.3.2 Composite cure cycle

Controlled bleed lay up and bagging methods are identical to those used for 175°C curing epoxies. A typical cure cycle is given in Fig. 3. Normally a 5 - 10 hour postcure at 230°C is used to fully develop the outstanding high temperature properties of the system.

2.3.3 Composite properties

A model resin system comprising 65 parts COMPIMIDE 796 and 35 parts COMPIMIDE TM123 besides other additives as decribed above was evaluated with various types of carbon fibres.

(T300/6000 Torayca, T800/6000 Torayca, Besfight HTA-7 and Grafil XAS). The fibres were taken as delivered by the vendors. T800, however, was available in two different sizes: a bismaleimide size and a standard epoxy size. The mechanical properties obtained are compiled in Table 5.

As a conclusion, the numerical results show that:

- the matrix dominant properties are influenced by the fibre size;
- cured epoxy size on T300-50B and partially cured BMI-size on T800-BOB fibres provide the best hot wet properties and the better edgewise delamination values;
- G_{Ic} fracture toughness (double cantilever beam test) is excellent for Tenax HTA-7 and Hysol Grafil XAS-12K.

From this preliminary data, it is abvious that bismaleimide modified with the COMPIMIDE TM123 comonomer can be the basis for tough BMI-prepreg formulations.

2.3.4 Composite toughness improvements

The most critical disadvantage of high Tg bismaleimide resins is their relatively low fracture toughness which leads to low composite damage tolerance in terms of low post compression after impact resistance. R & D effort is directed towards improving this deficiency without a concomitant loss in thermal stability.

One approach in achieving this target is to use ductile thermoplastics such as polyetherimide (Ultem 1000) and/or polyhydantoin (Resistherm PH 10) blended into the bismaleimide to improve toughness (17). Although the processing properties of thermoplastic modified BMI's are inferior damage tolerance of composites can be significantly improved.

The CAI data for polyhydantoin modified Bismaleimide/T800-carbon fibre composites are given in table 6 and are self-explanatory.

3. SUMMARY

Bismaleimide building blocks and resins are crystalline substances (COMPIMIDE MDAB) or amorphous solids (COMPIMIDE 796) but can be formulated into products which meet the processing parameters of wet and tow-preg filament winding (FW), resin transfer moulding (RTM) and low pressure autoclave moulding (LPAM).

The problem of adjusting the rheological properties of the BMI-resin to suit the specific composite fabrication process is achieved through comonomers such as COMPIMIDE TM123 and COMPIMIDE TM121 and reactive diluents such as triallylisocyanurate and triallylcyanurate. The cured neat resin properties show that fracture toughness of the modified systems is significantly improved versus the BMI-homopolymer. Laminate properties demonstrated that the modified bismaleimide resins perform well with a series of commercially available carbon fibres.

4. REFERENCES

1) H.D. Stenzenberger, M. Herzog, W. Römer, R. Scheiblich, N.J. Reeves and S. Pierce, 29th Int. SAMPE Symp. Proceedings Vo. 29, p. 1043 (1984).

2) C.L. Segal, H.D. Stenzenberger, M. Herzog, W. Römer, S. Pierce, M. Canning, 17th Nat. SAMPE Conference 17, p. 147 (1985).

3) S. Street, 25th Nat. SAMPE Symp. Proceedings Vol. 25, p. 366 (1980)

4) H.D. Stenzenberger, P. König, M. Herzog, W. Römer, S. Pierce, M. Canning, 32nd Int. SAMPE Symp. Proceedings 32, p. 44 (1987)

5) H.D. Stenzenberger, P. König, M. Herzog, W. Römer, M.S. Canning, S. Pierce, 19th. Int. SAMPE Techn. Conf. 19 (1986) 500

6) H.D. Stenzenberger, P. König, M. Herzog, W. Römer, S. Pierce, K. Fear, M. Canning, 19th Int. SAMPE Techn. Conf., 19, 372 (1987).

7) S.A. Zahir, A. Renner, US Patent 4 100 140 (1978)

8) M. Bergain, A. Combet, P. Grosjean, Brit. Patent Spec. 1,190,788 (1970)

9) H.D. Stenzenberger, US Patent 4,211,861 (1980)

10) G.D. Lyle, D.K. Mohanty, J.A. Cecere, S.D. Wu, J.S. Senger, D.H. Chen, S. Kilie, J. Mc Grath, 33rd Int. SAMPE Symp. 33 (1988) 1080.

11) A.J. Kinloch, S.J. Shaw, ACS, Polym-Mat. Sci. & Eng. 49 (1983) 307

12) H.D. Stenzenberger, W. Römer, M. Herzog, P. König, 39th Int. SAMPE Symp. Proceedings 33, 1546 (1988)

13) W.V. Breitigam, H.D. Stenzenberger, 33rd Int. SAMPE Symp., 33 (1988) 1229

14) F.C. Robertson, Brit. Polym. J., 20 (1988) 417

15) L.M. Dane, R. Brouwer, 33rd Int. SAMPE Symp. 33 (1988) 1217

16) H.D. Stenzenberger, P. König, M. Herzog, W. Römer, S. Pierce, M. Canning, K. Fear , 31st Int. SAMPE Symp. 31 (1986) 920.

17) H.D. Stenzenberger, W. Römer, M. Herzog, P. König, 33rd Int. SAMPE Symp. 33 (1988) 1546.

ACKNOWLEDGEMENT

Part of this work was performed under Contract 03M1003E5 from the German Ministry of Research and Technology. This support is gratefully appreciated.

Table 1: Compimide 65 FWR Neat Resin Properties

Property	Unit	Test Temp. °C	Cure		
			Cycle 1	Cycle 2	Cycle 3
Flex. Strength	MPa	23	125	141	115
		200	74	103	62
		250	59	87	-
Flex. Modulus	MPa	23	4242	4194	4078
		200	2218	2707	2104
		250	1756	2452	-
Flex. Elongation	%	23	3.0	3.6	2.9
		200	3.4	4.1	3.0
		250	3.9	4.6	-
Tensile Strength	MPa	23	-	-	79
Tensile Modulus	GPa	23	-	-	4162
Tensile Elongation	%	23	-	-	2.3
Fracture Toughness (G_{Ic})	J/m²	23	388	-	-

Cycle 1: 2h 170°C + 5h 210°C + 10h 210°C
Cycle 2: 2h 170°C + 5h 210°C + 10h 250°C
Cycle 3: 3h 110°C + 2h 160°C + 3h 210°C

Table 2: Properties of Compimide 65 FWR-Laminates

Property	Test Temp. °C	Fiber Type		
		T300/6000/ 50B	TENAX HTA-7	IM6
Fiber Content (% by vol.)		58	62	64
Flexural Strength[1] (MPa)	23	1880	1885	1978
	250	1353	1282	973
Flexural Modulus[1] (GPa)	23	123	123	143
	250	118	132	144
90° Flexural Strength (MPa)	23	105	83	70
	250	58	56	ND
90° Flexural Modulus (GPa)	25	8.27	10.6	9.74
	250	5.46	8.5	ND
Tensile Strength[2] (MPa)	25	1500	ND	1850
	200	1600	ND	2000
0° Short Beam Shear Strength (MPa)	25	110 (99)[3]	108	92
	120	94 (63)[3]	94	ND
	175	82 (46)[3]	77	ND
	200	75	71	46[4]
	250	57	51	33
0 ± 45 Short Beam Shear Strength (MPa)	25	61	57	48
	250	50	47	29
G_{Ic} (DCB-test) (J/m²)	23	111	253	ND

1. Values normalized to 60%v
2. Test method ASTM 2290, NOL rings thickness approx. 1,5 mm.
3. Values in parentheses for wet specimens (500 hours 70°C, 95% rel. humidity)
4. Test temperature 210°C.

ND = Not determined.

Table 3: Properties of RTM - BMI resin candidates

		Formulation		
		I	II	III
Composition		65 p. C796 35 p. TM121	70 p. C796 30 p. TM123 10 p. TAIC	65 p. C796 35 p. TM123
Viscosity at (mPa.S)	90°C	1200	1300	5300
	100°C	400	650	1800
	110°C	150	350	830
	120°C	-	240	420
Viscosity at (mPa.S) (after 2 hours at the test temperature)	90°C	2200	7500	12700
	100°C	750	4500	8300
	110°C	500	2900	7000
	120°C	-	3570	2000
Hot plate gel time at (mins)	140°C		75	60
	150°C	-	40	33
	160°C	-	17	14
	170°C	50	8	8

C796 = COMPIMIDE 796

TM121 = COMPIMIDE TM121

TM123 = COMPIMIDE TM123

Table 4: Properties of RTM - carbon fabric laminates

Property		Modification I(1)	Modification II(2)
Resin content	(vol. %)	60	51
Density	(g/cm^3)	1,49	1,47
SBS-Strength MPa	23°C	39	58
	100°C	30	49
	250°C	22	46
Flexural Strength MPa	23°C	943	760
	100°C	780	720
	250°C	544	690
Flexural Modulus GPa	23°C	58	51
	100°C	57	52
	250°C	56	52

1) Compimide 796/TM121 (65/35)
2) Compimide 796/TM123/TAIC (70/30/10)

Table 5: Mechanical properties of Compimide 796/Compimide TM123 carbon fibre laminates

Fibre type / Fibre size		T300B-50B epoxy	T300-90B no size	T800B-40B epoxy	T800B-BOB polyimide	HTA-7 epoxy	HG XAS-12K no size
0° FS (MPa)	23°C	1 970	1 831	1 833	1 747	1 884	1 713
	250°C	1 300	1 300	1 243	1 268	1 343	1 237
90° FS (MPa)	23°C	111	119	92	99	108	89
	250°C	80	76	69	75	76	63
0° FM (GPa)	23°C	129	125	153	155	126	116
	250°C	133	140	146	182	125	121
90° FM (GPa)	23°C	9·2	8·9	8·6	8·7	9·6	8·6
	250°C	7·5	6·6	9·2	7·3	8·3	6·3
0° SBSS (MPa)	23°C	109 (87)	103 (85)	103 (71)	103 (92)	92 (85)	99 (43)
	120°C	92 (56)	85 (52)	81 (42)	78 (57)	84 (56)	74 (28)
	175°C	76 (43)	72 (38)	70 (27)	68 (40)	75 (38)	65 (22)
	200°C	68	67	60	65	66	60
	250°C	56	49	51	48	53	52
0° ±45 SBSS (MPa)	23°C	80	72	62	81	76	—
	250°C	48	42	52	43	43	—
G_{IC} (DCB-test) (J/m^2)	23°C	140	127	319	319	399	411
$(\pm 25)_2\ 90_s$ EDL							
first failure		257	194	126	202	170	—
ultimate		507	495	522	571	570	—

Abbreviations: FS = flexural strength, FM = flexural modulus, SBSS = short beam shear strength, ( ) = wet, EDL = edge delamination.

Table 6: Mechanical properties of carbon fibre laminates[a]
Resin System: Compimide 796/Compimide TM123/RT
Fibre: Toray Rayon, T800/6000

Property		Control	% Wt, RT		
			13	20	26
Fibre size		Pi	Pi	Pi	Pi
Flexural strength, dry, MPa	23°C,	1747	1769	1631	1687
	250°C	1268	1456	1339	1292
Flexural modulus, dry, GPa	23°C	155	151	150	145
	250°C	182	166	182	152
90° Flexural strength, MPa	23°C	99	85	100	82
	250°C	75	55	52	52
90° Flexural modulus MPa	23°C	8.7	8.96	9.26	8.32
	250°C	7.3	-	-	-
0°, Short beam shear strength, MPa	23°C	103	99	97	100
	177°C	68	68	77	66
	250°C	48	53	54	51
0±45 short beam shear strength, MPa	23°C	81	56	69	69
	250°C	43	45	45	55
G_{Ic}, DCB-test	23°C	319	398	505	471
G_{IIc}, ENF-test	23°C	-	483	689	747
EDL[(±15)$_2$90]s, MPa					
First failure	23°C	202	201	195	204
Ultimate		571	466	514	585
CAI (3.37 J/cm) MPa		134	206	220	248
(6.7 J/cm) MPa		118	160	175	190

(a) RT : Polyhydantoin, Resistherm Bayer AG
control : COMPIMIDE 796 / COMPIMIDE TM 123 = 65/35

Figure 1: Viscosity-Time Correlation for Compimide 65 FWR

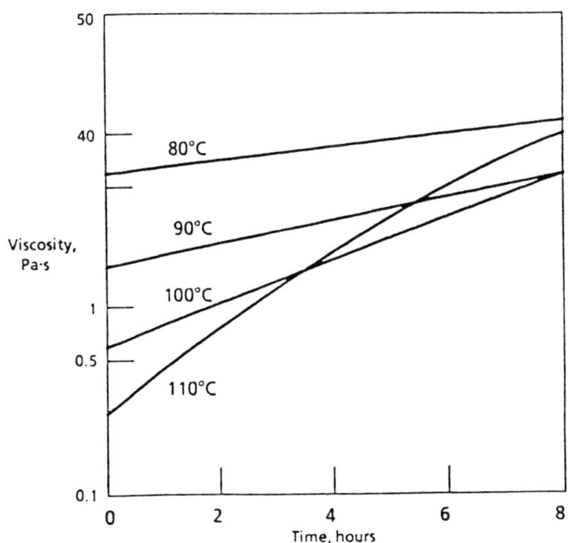

Figure 2: DSC-Scan of Compimide 65 FWR

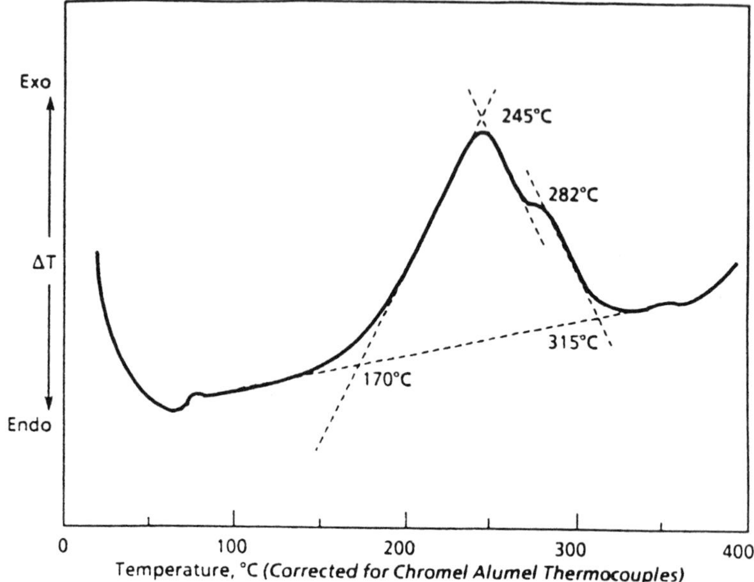

Figure 3: Low pressure autoclave cure cycle for Compimide 796/ Compimide TM123 laminate moulding.

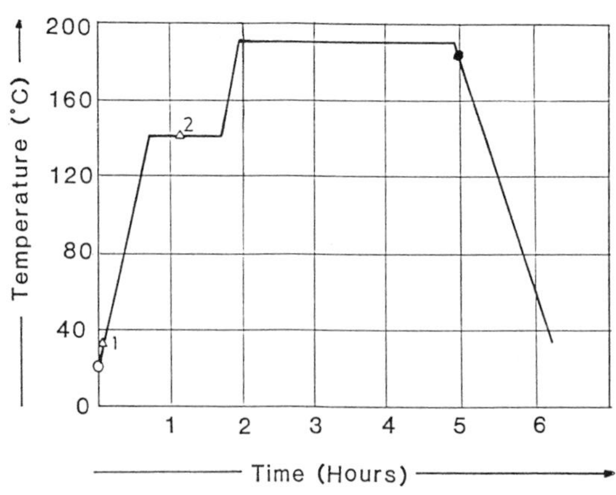

○ Apply vacuum
△1 Apply Pressure of 1bar
△2 Apply Pressure of 4bars
● Vent vacuum and remove pressure

Materials and Processing – Move into the 90's
edited by S. Benson, T. Cook, E. Trewin and R.M. Turner
Elsevier Science Publishers B.V., Amsterdam, 1989

HIGH TEMPERATURE COMPOSITE TOOLING FOR ADVANCED COMPOSITE MANUFACTURE

Dipak Gupta
Composites Division
E. I. Du Pont de Nemours & Company, Inc.

ABSTRACT

The rapid pace of development of high temperature organic resin matrix composite materials in aircraft and aerospace is clearly led by the requirements of the advanced military systems. Composite tooling is the logical choice for building such composite parts. Avimid®N, a high temperature (~400°C) composite tooling technology has been developed in Du Pont to fabricate high quality parts from prepregs based on bismaleimides, polyimides and other thermoplastic composites. Advantages of such composite tooling are light weight for easy handling, low coefficient of thermal expansion that matches the parts, low thermal mass for energy efficiency, and high temperature thermal-oxidative stability to provide long production life.

1. INTRODUCTION

Throughout the manufacturing history, suitable tooling has always been a prerequisite of any fabrication or construction process. The basic requirement has not changed, tooling methodologies have been conceived, developed and modified to accommodate design and materials availability while satisfying product requirements and economic considerations. The advent of new high temperature organic resin matrix composite systems with delamination resistance and high incipient impact energies have necessitated the development of new high temperature tooling technologies.

Traditional tooling materials are unsuitable because of the large thermal excursion needed to consolidate and form these materials. The fabrication of high quality and highly contoured thin section composite parts such as fuselage skins, ribs, spars, etc. requires the development of a new high service temperature (200°-400°C) tooling system with low density, high thermal conductivity and low coefficient of thermal expansion (CTE). Polyimide tooling technology using carbon fiber prepregs has been developed in Du Pont to fabricate high quality parts from prepregs based on bismaleimides, polyimides and high temperature advanced thermoplastics composites.[1,2]

2. NR-150B2 RESIN PROPERTIES

All members of the NR-150 polyimide precursor solution family are based on a proprietary aromatic tetra-carboxylic acid (6FTA) plus stoichiometric quantities of p-phenylenediamine and m-metaphenylene diamine (Figure 1). With the application of heat, the monomers in solution undergo conventional condensation polymerization to form polyimide along with the evolution of solvent and by-product water.[3]

Polyimides from these systems possess an extraordinary degree of thermal oxidative stability. The polymer chain is completely aromatic in the sense that there are no aliphatic hydrogens or unstable linkages. Since stiochiometric concentrations of monomers are employed in the make-up of binder solutions, linear polymerization can occur to form ultrahigh molecular weight polyimides. Since these are linear amophous polymers, they have a significantly greater toughness and resistance to impact damage and fatigue than do conventional, highly crosslinked thermosetting polyimides.

The physical and thermal properties were determined from test specimens taken from compression molded plaques. Table 1 summarizes the properties of the polyimides derived from NR-150B2. High toughness of the resin is retained at high temperatures and after long term isothermal aging. The unique chemical make-up of the molecule also provides excellent resistance to solvents, especially non-polar fluids such as jet fuels, lubricating oils, etc.

FIGURE 1

NR-150 Polyimide Resin

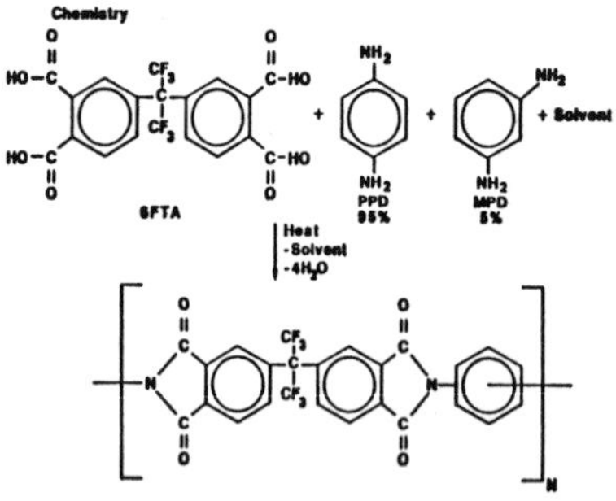

Chemistry

6FTA + PPD 95% + MPD 5% + Solvent → (Heat, -Solvent, $-4H_2O$)

Key features

- Linear amorphous polymer
- Ultra-high molecular weight
- Outstanding thermal-oxidative stability
- Excellent resistance to solvents and moisture

TABLE 1

Unreinforced Resin Properties

Property	Value
Density, g/cc	1.45
Tensile strength, Kpsi (MPa)	16 (110)
Tensile modulus, Kpsi (MPa)	600 (4140)
Elongation at break, %	
- Initial	6.0
- After 100 hours at 700°F (371°C) in N_2	3.8
- After 100 hours at 700°F (371°C) in air	1.5
Fracture toughness, FT-LB f/FT 2 (KJ/m^2)	140 (2.0)
Flexural strength, Kpsi (MPa)	17 (118)
Flexural modulus, Kpsi (MPa)	605 (4175)
Coefficient of thermal expansion, cm/cm/°c x 10^{-6} (24°C-232°C)	90

3. PROPERTIES OF ADVANCED TOOLING LAMINATES

The high temperature stability of all composites, including Avimid®N, is directly affected by the stability of the reinforcing fibers. Composites based on less stable fibers would show a more rapid deterioration resulting from the poorer oxidative resistance of such fibers.

Composite tooling has a very unique combination of mechanical and thermal properties which make it extremely attractive to be used for fabrication of new high temperature organic resin matrix composites. Table 2 is a summary of the typical properties of laminates. The excellent properties of neat NR-150B2 resin are retained in the laminates allowing full realization of the reinforcing fiber's strength and stiffness. The flexural strength and modulus were measured over a wide range of temperatures. High matrix resin glass transition temperature (Tg) resulted in the retention of approximately 90% of the modulus and 65% of the strength at up to 316°C. The reason for the minimal reduction in mechanical properties is indicated by the small weight loss (0.2-0.4%) as measured by the Thermo-Gravimetric Analysis (TGA) at up to 400°C.

4. THERMAL CYCLING IN AIR AT 371°C

The graphite/epoxy prepreg tools have been made over the last 8-10 years with accuracy and reliability. The projected life expectancy of such tool processed in autoclave is between 200-500 cycles. This projection is based on the properties of the resin system and the long term high temperature cycling of cured laminates.

Thermal cycling of laminates of Avimid®N were carried out in a circulating air oven from room temperature to 371°C to simulate fabrication of high temperature composite parts and to demonstrate the material's resistance to matrix cracking and physical properties reduction caused by moisture and thermal transients. The composite panels equilibriated at room temperature were directly inserted into the oven maintained at 371°C. The panels were held at 371°C for 30 minutes and then taken out directly into the room temperature environment and examined under the microscope for surface defects or delamination. Critical examination of the laminates failed to reveal any blisters, matrix cracks or surface erosion.

TABLE 2

AVIMID®N LAMINATE PROPERTIES (TYPICAL) *

PROPERTY	VALUE
Density, g/cc	1.55
Flexural Strength, Kpsi (MPa)	
- 24°C	77 (531)
- 316°C	50 (345)
Flexural Modulus, Mpsi (MPa)	
- 24°C	8 (55,200)
- 316°C	7 (483,000)
Short Beam Shear Strength, Kpsi (MPa)	
- 24°C	10 (69)
- 177°C	7 (48)
TGA Weight Loss (up to 400°C), %	0.3
Glass Transition Temperature (Tg), °C	405
Coefficient of Thermal Expansion	
x-direction, cm/cm/°C x 10^{-6} (24°C-371°C)	4.5

* ALL VALUES NORMALIZED TO 55% FIBER VOLUME

Figures 2 and 3 show the flexural strength and flexural modulus of laminates of Avimid®N at up to 100 thermal cycles. The laminates showed excellent retention of properties when tested at room temperature and at 301°C.

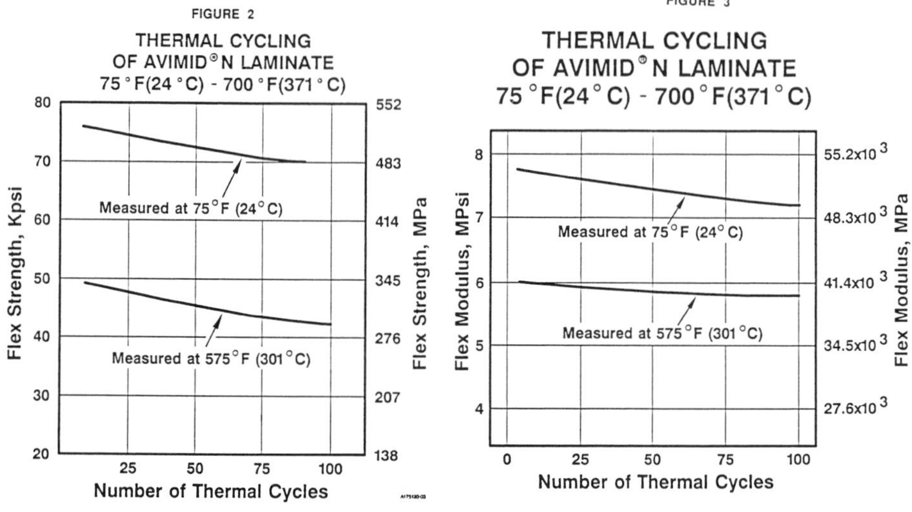

5. SURFACE FINISH OF COMPOSITE TOOLING

The surface finish of the tool is of extreme importance for high surface quality on the part and to allow for easy part removal. An ideal tool surface should have thermal/dimensional stability, smooth surface, uniformly reinforced around curves and tight radii, and durable through many parts production.

We have developed a special woven prepreg (0.1mm cured ply thickness) exclusively for tool surfacing. The advantage of this over a gel coat (i.e. thin resin film) is that the surface layer has the same coefficient of thermal expansion as the tool body it covers. Therefore, there is no thermal stress development between the tool surface and the body when the tool is thermally cycled from room temperature to a very high part curing temperature. Another tool surfacing option is a prepreg based on thin (0.05 mm) short carbon fiber mat with very high resin content (75-80%). This mat is excellent for tool surfaces with complex contours and provides an excellent glossy surface finish.

6. BENEFITS OF COMPOSITE TOOLING

Traditionally, tooling material selection have been determined by what parts to be made, plan to make them and how many parts to be made. While these criteria are still valid, the number and complexity of composite part programs have increased dramatically and as a result additional criteria need to be examined prior to tooling material selection. The notable secondary drivers for the final tool selection should be schedule, production quantity, part size and tolerance, equipment availability, cost, etc. High temperature (~400°C) composite tooling offers various benefits including unequaled thermal stability for precise dimensional control in the production of high temperature advanced composites.

6.1 High Strength/Stiffness With Low Density

Composite tooling is considerably lighter than ceramics, monolithic graphite or metal tools (Figure 4). Figures 5 and 6 compare the flexural strength and modulus of various high temperature tooling materials. Advanced composite tooling is far superior to monolithic graphite and ceramic materials at any temperatures up to 371°C. Lower specific gravity combined with the superior flexural strength and excellent flexural modulus of such tooling make it an ideal material of choice for designing large high temperature tools.

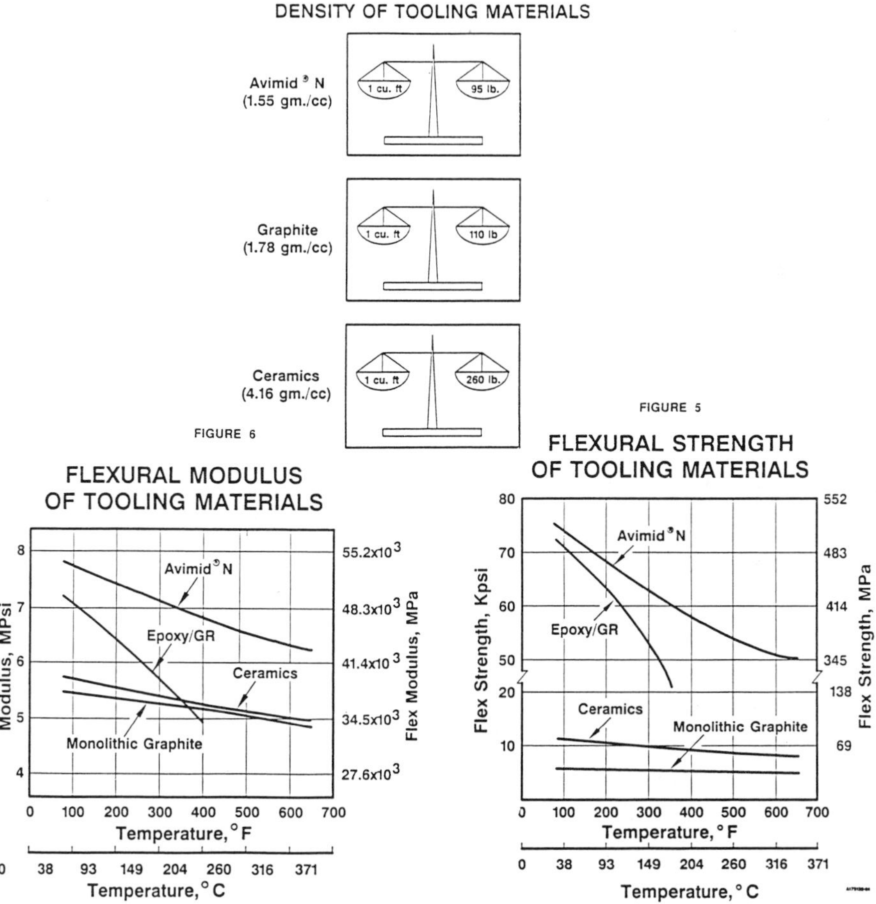

The recommended wall thickness for most tools of Avimid®N is 6mm. A 1.86 m^2 tool face area designed with this material will weigh between 14-18 Kg. A monolithic graphite tool of the same surface area and designed based on equal effective strength (vs. Avimid®N) will weigh 114-136Kg. The weight of a similar ceramic tool will be between 272-318 Kg. The lower weight allows the advanced composite tool to be handled easily, without the use of elaborate or expensive equipment such as pullys, cables or chains.

6.2 Dimensional Stability and Thermal Conductivity

The most critical parameter in the design of a tool and tool performance is the coefficient of thermal expansion. Differential thermal expansion between the tool and the composite part can cause severe damage to the part by fiber disruption, compression buckling and delamination.[4] One approach is to try to calculate the mismatched CTE of the tool in order to account for the tool dimension at the cure temperature. This approach has worked with some success but opens up a host of new problems with regard to discrepant angles and part warpage.

Low thermal coefficient of expansion and high thermal conductivity of advanced composite tooling system (Table 3) make it an attractive choice to fabricate carbon fiber reinforced advanced composites. It is possible to design tooling that is dimensionally correct at room temperature, which can go through a high temperature cure cycle and yield a part just as dimensionally correct. Also, since such composites are more conductive along the carbon fiber direction then through the laminate thickness, hot spots are likely to be less significant and be less of a control concern than in metals or other isotropic tooling materials, such as ceramics and monolithic graphite.

TABLE 3

Thermal Properties of Tooling Materials

Material	Thermal Expansion Coefficient (cm/cm/°C x 10^{-6})	Thermal Conductivity (cal-cm/cm^2-sec-°K)
AVIMID N	4.5	0.17
Monolithic Graphite	4.5	1.65
Ceramics	7.5	0.07
Steel	12.1	1.49
Aluminum	22.5	5.79

6.3 Fast Heat-Up/Cool-Down Rates

As mentioned before, weighting of the various criteria is extremely important in the selection of tooling material system. For example, if the program is schedule driven, then schedule must be considered as one of the prime drivers in tool selection. If the autoclave time/capacity are at a premium, then the tool "heat-up/cool-down rate" becomes a prime driver in order to maximize autoclave loads and to minimize autoclave impact.

The materials suitable for high temperature tooling differ considerably in specific heats and specific gravities which result in large variations in thermal mass. The thermal mass is defined as the product of heat capacity and the apparent density. It is an indication of the heat energy input required for a unit volume of material to achieve a prescribed set of temperature.

Table 4 compares the thermal mass values for various high temperature tooling materials. Metal such as steel and aluminum have, in general, a high specific gravity coupled with moderate or low specific heat. The result is a moderate-to-large thermal mass. Ceramics have the highest specific gravity and the highest heat capacity of all the tooling materials, thus requiring the greatest heat input to achieve a given set of processing conditions.

Avimid®N tooling system has a fairly low specific heat and density. Thus, the energy required to heat such tooling at a given rate is considerable less than all the other tooling materials. The monolithic graphite has fairly low thermal mass, but the large mass of the tool requires high heat input to maintain a desired heating rate or to achieve a prescribed set of temperature.

TABLE 4

Thermal Mass of Tooling Materials

Material	Density (g/cm^3)	Specific Heat (cal/cm^3-°C)	Thermal Mass (cm/cm/°C x 10^{-6})
AVIMID® N	1.55	0.30	0.46
Monolithic Graphite	1.78	0.28	0.50
Ceramics	4.16	0.86	3.58
Steel	7.86	0.11	0.87
Aluminum	2.70	0.23	0.62

6.4 Excellent Vacuum Retention

Composite tooling is vacuum tight, has excellent surface finish and therefore only a release agent needs to be applied before the part lay-up. Also, the composite part can be directly vacuum bagged to the tool because of the absence of undesirable porosity. Monolithic graphite materials is inherently porous and requires sealing for vacuum integrity. The higher the service temperature, the more frequently the tool must be resealed.[5] For process considerations, envelop bagging to a metal subplate is undesirable especially when forming large composite parts.

7. ADVANCED COMPOSITES FABRICATED USING COMPOSITE TOOLING

Table 5 shows the mechanical and thermal properties of Avimid®K, PEEK (APC-2) and bismaleimide (V378A) composite parts fabricated using Avimid®N tool. For each material, the supplier's curing recommendations were followed, with changes made only as necessary to achieve void-free parts with a cured resin content of 32-35% by weight. The C-scans showed uniformly low energy loss and the photomicrographs showed that the voids were small and widely distributed. The densities of the parts were close to the theoretical values. They also had high flexural properties and very little weight loss as determined by Thermo-Gravimetric Analysis (TGA).

TABLE 5

ADVANCED COMPOSITES FABRICATED USING AVIMID®N TOOL

	AVIMID®K-III/AS-4	PEEK APC-2/AS-4	BMIV378A/AS-4 *
Laminate orientation	(+45/0/-45/90)3S	(+45/0/-45/90)3S	(+45/0-45/90)3S
Weight resin, %	33.5	32.0	33.8
Volume fiber, %	58.9	60.6	58.2
Density, gm/cm^3	1.56	1.57	1.55
Flexural strength, MPa	843	872	814
Flexural modulus, MPa	49,000	50,000	48,000
Glass transition (dry), °C	250	173	231
TGA weight loss (up to 400°C), %	0.18	0.83	1.33

* As cured in autoclave at 177°C

8. SUMMARY

Composite tooling system provides engineers with an attractive material for unique design and manufacturing advantages:

- High dimensional accuracy
- High strength/stiffness with low density
- Thermal properties match to parts
- Large and complex tool capability
- Fast heat-up/Cool-down cycles
- High service temperature
- Long tool life

9. ACKNOWLEDGEMENTS

The author wishes to acknowledge the technical assistance of Mr. Robert L. Benedict. A great deal of appreciation goes to Miss Patti Turner in the preparation of the manuscript.

10. REFERENCES

1. Gupta, D., "Avimid®N Composite Tooling For Advanced Aircraft and Aerospace Applications," SME Conference, May 1988, Paper TE88-211.

2. Gupta, D., "Avimid®N: An Innovative Composite Tooling For High Temperature Advanced Composites," 34th International SAMPE Symposium and Exhibition, Reno, Nevada, May 8-11, 1989.

3. Gibbs, H. H. and Breder, C. V., "High Temperature Laminating Resins Based on Melt Fusible Polyimides," Advances in Chemistry Series No. 142, Copolymers, Polyblends and Composites, 1975, p. 442.

4. Keller, B., Castillo, A. and Grasty, J., "Composite Tooling for High Temperature Thermoplastics," SME Conference, May 1988, Paper TE88-212.

5. Rodgers, R. S., "An Evaluation of Chemically bonded Ceramic Autoclave Tooling," SAMPE Seminar (L. A. Chapter), August 1988.

11. BIOGRAPHY

Dr. Dipak Gupta is a Research Associate with the Composites Division of the Du Pont Company where he has been employed since 1970. He received his MS (1968) and PhD (1970) in Chemical Engineering from the University of Pennsylvania. He has been responsible for the development and applications of high performance reinforcing fibers such as Kevlar® Aramid, Liquid Crystal Polyester and ceramics. During the past 3 years, he has been involved in the development of high temperature polyimide composites and composite tooling (Avimid®N). He has published extensively in the areas of polymer viscoelasticity, high performance fibers/composites and holds several United States patents on radial tires and high temperature composites. He is a member of several technical societies including American Institute of Chemical Engineers (AICHE), Society of Manufacturing Engineers (SME), SAMPE and Phi Lamda Epsilon.

Materials and Processing – Move into the 90's
edited by S. Benson, T. Cook, E. Trewin and R.M. Turner
Elsevier Science Publishers B.V., Amsterdam, 1989

ADVANCED METALLICS: MATERIALS AND PROCESSES.

C.A. STUBBINGTON

HI-TEC METALS R & D LTD., CHANDLERS FORD, SOUTHAMPTON SO5 3BZ

1. INTRODUCTION

For many years monolithic metallic materials have dominated the engineering world. Materials used have ranged from low density magnesium and aluminium alloys to high strength steels and super alloys. In airframes high strength aluminium alloys have held a dominant position since the mid-30s. This situation is now changing with the use of organic based composite materials in significant amounts in military aircraft such as the AV-8B and their proposed use in the European Fighter Aircraft. Composite materials comprise some 25% of the structure mass of the AV-8B and current predictions are for CFRP to comprise 60% of the structure mass of combat aircraft designed at the end of the century. This trend from monolithic metallic materials to reinforced organic resins is not confined to military aircraft since there are many indications that new designs of civil aircraft will contain significant amounts of inorganic composites.

Thermoset and thermoplastic matrix reinforced composites are attractive because of their potential for mass saving and because the material in components and structures can be tailored to cope with predicted service stresses. Mass savings achieved are related to applications and although savings in the range 10 to 30% are possible, many fall at the lower end of the range. There is now considerable interest in exploiting the improved damage and environmental tolerance of thermoplastic materials such as polyetherether-ketone (PEEK). PEEK can be formed into complex shapes by heating to around 380°C in an operation similar to that of metal forming, in fact so similar that a technique has been developed to shape PEEK between sheets of super-plastic aluminium alloy[1].

From what I have said so far you could be forgiven for thinking that I am addressing the wrong topic, after all our present session is concerned with advanced metallics. This is not the case but I thought it appropriate to mention briefly important developments which are impacting the choice of materials in new aircraft designs and which give an indication of the challenges which metallic materials must face if they are to remain commercially viable for airframe construction.

2. RECENT AND CURRENT DEVELOPMENTS IN METALLIC MATERIALS

A number of developments, both materials and processes, which have come to maturity in the last decade, will help metallic materials to meet the challenge of non-metallics in the next. For example, aluminium-lithium alloys, superplastic forming and diffusion bonding of titanium alloys and superplastic forming of aluminium alloys. All will give mass savings and, in the case of SPF/DB, the additional benefit of cost savings. There are also a number of materials/processes at the research and development stage. These include, alloys produced by rapid solidification processes, mechanical alloying, metal matrix composites, isothermal and hot die forging, squeeze casting and ARALL.

It is a fact of life that there is a lengthy time-span from the laboratory demonstration of a new material or process to its use in flight hardware, since the development of both is often an iterative process. In the case of a new alloy, after research on laboratory quantities has indicated a promising composition, it is necessary to scale-up the amounts produced through pilot plant quantities to full-scale production. Apart from the process engineering associated with each stage, considerable effort is required to generate the mechanical property data which will provide the justification to move to the next stage. Finally when commercial production quantities have been achieved the alloy must be thoroughly evaluated in wide ranging tests so that the designer has confidence to specify the alloy for his forthcoming projects, or, if circumstances justify it, to retrofit the new alloy into an on-going project. The confidence building final phase of evaluating production material can involve the design, construction and testing of demonstrator components or structures.

3. ALUMINIUM-LITHIUM ALLOYS

The lengthy time-span required for the development and exploitation of a new material is well demonstrated in the case of the low density aluminium-lithium alloys[2-4]. In the UK exploration of alloy compositions to give an acceptable balance of engineering properties and of safe melting and casting techniques to obtain high integrity material was started in the early 1970s. The UK programme to develop aluminium-lithium alloys was aimed at a density reduction of $\sim$10% compared with 2000 and 7000 series alloys, while at the same time matching their engineering properties. A medium strength alloy, 8090, with a density of 2.5 g/ml has been successfully developed in collaborative programmes involving the Royal Aerospace Establishment, British Alcan and potential user industries. After the evaluation of laboratory and pilot plant quantities, full-scale production material is now available in the late 1980s as a result of major investments at each stage by British Alcan. This material is being evaluated and future applications will depend on a successful outcome to the many current programmes in the UK, the US and Europe.

It is accepted that, due to the high cost of lithium, aluminium-lithium alloys will be more costly than 2000 and 7000 series alloys. Although some metal working and forming equipment used to manufacture components in the latter alloys can also be used to fabricate aluminium-lithium alloys, some current fabrication techniques are less suited to these alloys. As a result of the relatively low cost of 2000 and 7000 series alloys some aircraft manufacturers have invested heavily in extensive computer controlled machining facilities capable of machining complex components eg wing planks, from large thick slabs of rolled plate. This technique is less suited to costly aluminium-lithium alloys. We need to avoid the production of large amounts of swarf which, because of its lithium content, will be difficult to recycle into secondary production[5]. Skin-stringer fabrications would use aluminium-lithium alloys more economically and this form of construction, together with net or near-net shape technologies, such as superplastic forming and precision forging, is more appropriate for these alloys. Investment casting is also being explored for net shapes in lithium containing aluminium alloys[6].

4. SUPERPLASTIC FORMING OF ALUMINIUM ALLOYS

The phenomenon of superplasticity occurs in a wide range of alloys with suitable microstructures, including aluminium alloys and superplastic aluminium alloys have been available for many years. The use of these alloys in critical structures has, however, been inhibited by the formation during superplastic forming of cavities which have a deleterious effect on both static and dynamic properties[7]. In recent years British Aerospace has investigated and developed a back pressure forming technique which eliminates cavity formation. This technique will have an important impact on the applications of superplastically formed aluminium alloys. Recent investigations have shown that aluminium-lithium alloys are superplastic and the combination of their low density and high stiffness with this cost-effective fabrication route will be very attractive[8].

Ideally diffusion bonding should be combined with superplastic forming, thus permitting the fabrication of complex multi-layer structures but it is proving difficult to develop a commercially viable diffusion bonding process for high strength aluminium alloys because of an adherent, non-dissolving oxide film[9].

5. TITANIUM ALLOYS

Titanium alloys have high specific strength and good fatigue properties, and are resistant to corrosion in most aircraft operating environments. Unfortunately they are significantly more expensive than aluminium alloys and also tend to be more difficult to machine. These facts have inhibited the application of titanium alloys in airframe components, although they are used as an alternative to steel when mass saving is required and where space is restricted. The high cost of titanium alloy makes them good candidates for net or near-net shape technology such as superplastic forming and diffusion bonding, casting, powder metallurgy and close-to-form forging.

Much effort has been devoted in the last decade to an exploration of the potential of superplastic forming for the production of complex shapes in titanium alloys[10-13]. Because titanium alloys dissolve their oxide films when heated in the temperature range used for superplastic forming, two surfaces in contact under moderate pressure will become bonded together (diffusion bonded). Hence superplastic forming and diffusion bonding can be achieved sequentially or simultaneously (SPF/DB) to produce complex multi-layer components. The advantages offered by SPF/DB are reduced part count, lower costs, reduced mass and increased reproducibility of shape. Reduced part count includes a reduced requirement for fasteners and the associated holes which together can initiate fatigue failures. There is no doubt that titanium alloy structures fabricated by SPB/DB will be increasingly used in military and civil structures as more manufacturers introduce the necessary plant into their production facilities. This will increase the proportion of titanium alloys in airframes from the current level of 5-7% of structure mass. Components fabricated using the SPF/DB techniques were used in the EAP demonstrator aircraft and further applications are planned for the European Fighter Aircraft.

Applications of titanium alloy castings have shown a definite upward trend in recent years for missile, airframe and gas turbine components. This trend is likely to continue since casting in association with hot isostatic pressing (HIP) can give high integrity components of complex shape at reasonable cost.

The ease with which titanium alloys can be diffusion bonded and the increasing availability of large HIPing facilities opens up the possibility of producing components with microstructures tailored to meet different conditions in different parts of a component[14]. For example a creep resistant microstructure could be bonded to one resistant to fatigue, the two having been initially produced by different production or thermo-mechanical processing techniques.

Titanium alloys produced by the ingot route are used extensively in gas turbines where they operate satisfactorily for long periods at temperatures up to 570°C. Over the last two decades IMI (Titanium) Ltd has been particularly successful in alloy development programmes aimed at meeting the demands of the gas turbine industry for improved elevated temperature properties, including creep resistance. Their most recent contributions have been IMI 829 and IMI 834[15]; both will find their place in future designs of gas turbines and will assist in meeting the continuing requirement for higher thrust to weight ratios.

6. RAPID SOLIDIFICATION PROCESSES

Rapid solidification rate processes for the production of novel alloys are being explored in many current programmes. A number of techniques have emerged[16], including gas atomisation, melt spinning, rotating electrode and Osprey spray deposition, to achieve very rapid cooling of liquid metal (rates $>10^5$ C/sec) in the production of powder, ribbon or whiskers. In conventional ingot metallurgy the relatively slow cooling rates mean that the resulting alloys are essentially of equilibrium composition. By increasing the rate of cooling it is possible to refine the microstructure and to increase the amount of alloying addition that can be retained in solution, thus obtaining increases in strength, modulus and elevated temperature stability. For thermal stability elements with low solid state diffusivity are required, ideally combined with high solubility in the liquid and low solubility in the solid states. If, using rapid solidification techniques, low density elements such as lithium or beryllium can be introduced into the aluminium lattice in larger atomic percentages than is possible by ingot metallurgy, then further reductions in density and increases in properties such as modulus will be achieved.

The microstructures of rapidly solidified particulate material have unique properties but there is a real risk of microstructural degradation and hence of loss of properties during compaction processes needed to produce bulk material. Hence compaction conditions, particularly temperature, must be carefully controlled to retain the benefits of rapid solidification into the final product. The compaction of aluminium alloy powders to produce high integrity material is difficult due to the adherent oxide film which forms on rapidly solidified material when it is exposed to small amounts of oxygen. This film prevents the ready bonding of particles during subsequent processing and to achieve satisfactory particle to particle bonding the oxide film must be heavily sheared by, for example, extrusion of pressed compacts[17]. Powder handling, including classification and compacting must be done in superclean, controlled atmosphere facilities to prevent the accidental introduction of damaging inclusions into the final product.

A wide range of rapidly solidified aluminium alloys intended for advanced propulsion systems and for airframe components in supersonic aircraft[18] has been investigated in the UK and the USA.

Novel microstructures in aluminium alloys can also be produced by physical vapour deposition. In this process advantage is taken of the atom by atom quenching from the vapour phase to circumvent the limitations of ingot metallurgy and the difficulty of recombining powder, ribbon or whisker materials produced by rapid solidification processes without degrading their microstructures and thus their engineering properties. A physical vapour deposition technique to produce bulk material directly has been pioneered by workers at the Royal Aerospace Establishment[19,20]. The Al-Cr-Fe alloy system has been extensively investigated and alloys with good room and elevated temperature properties have been produced in laboratory quantities. The elevated temperature stability of these alloys is significantly better than that of 2000 and 7000 series alloys. Alloys produced using PVD will be expensive and major investments in sophisticated plant will be required to produce material in quantities and sizes necessary for aircraft, missile or other applications. There will be competition from metal matrix composites and from materials produced by other rapid solidification techniques for applications requiring high specific stiffness and good elevated temperature properties.

Apart from techniques involving the rapid cooling of liquid metal or evaporation and condensation, mechanical alloying can also be used to produce novel alloys[21-23], including aluminium alloys. This technique has been exploited by INCO in their production of IN-9052 (Al-4.0Mg-0.8O-1.1C) and IN905XL.

7. ENGINEERED MATERIALS

At the end of the 20th century increasing research effort is being devoted to the development and use of 'tailored' or 'engineered' composite materials. As we have seen, this trend has been obvious since the mid-1960s in the case of organic composites and more recently has clearly emerged for metallic materials. Composites are being developed using relatively ductile matrices of aluminium, magnesium and titanium reinforced with continuous or discontinuous high strength, high stiffness carbon, alumina, silicon carbide or boron fibres[24,25]. Metallic matrices are also being combined with high modulus particulate such as silicon carbide.

Particulate reinforced material is attractive in that isotropic properties are obtained and it can be shaped by conventional metal working techniques such as extrusion, forging and rolling. Ingots of SiC particulate reinforced aluminium alloys are being produced by the Dural Aluminium Corporation using a proprietary process which enhances wetting of the particulate by the molten alloy but which inhibits excessive chemical reaction.

Powder metallurgy techniques are being used to fabricate SiC whisker reinforced metal matrix composites and sheet produced by the Advanced Composite Materials Corporation is being evaluated by Lockheed Aeronautical Systems for an advanced tactical fighter[26]. A 2000 series alloy reinforced with 15 v/o SiC has been used with reported fracture toughness values of 63 MPa$\sqrt{m}$. SiC particulate reinforced aluminium matrix composites have also been produced by the powder route and have been qualified for use in precision guidance systems.

Apart from having high specific properties, the coefficient of Thermal Expansion of these composites can be tailored to be similar to that of other metallic materials.

Squeeze casting is an excellent technique for the production of high integrity, monolithic aluminium and magnesium alloy castings with improved microstructures and properties. It is also a convenient and effective way of producing metal matrix composites[27,28]. Random fibre mats, prepositioned in moulds, are infiltrated with metal under pressure, thus achieving total penetration and high quality product free of porosity. Reinforced piston crowns with improved elevated temperature properties have been produced using this technique and many other applications are being explored.

Composites with magnesium or titanium alloy matrices are also being developed[29,30]. Cast magnesium alloy-graphite fibre composites have been produced for joints in space structures. Magnesium wets most potential reinforcements satisfactorily and the Dow Chemical Company has produced composites by mixing molten magnesium with reinforcing particulate. When magnesium matrix composites are fully developed and characterised and available for the manufacture of large components, their inherent corrosion resistance and long-term durability will need to be convincingly demonstrated, otherwise designers will be reluctant to specify these novel materials and aircraft operators will be reluctant to accept them. In benign environments, such as space, the use of reinforced magnesium alloys will be less controversial but, unless they are corrosion resistant, their terrestial fabrication and subsequent storage will need to be carefully controlled and monitored.

Rapid solidification techniques are being used to explore magnesium alloy compositions aimed at improved corrosion resistance. If this can be achieved by the addition of elements not possible using conventional techniques or by the addition of elements in amounts greater than is possible using conventional techniques, then the material produced will be very attractive as a matrix for reinforcement.

Apart from liquid metal infiltration of fibre compacts, the inclusion of reinforcement into molten metal and the powder route, simultaneous spraying of metal and particulate to form metal matrix composite in situ is also possible[31]. As for other routes the material produced can be shaped by conventional metal forming process.

The difficulty of subsequent shaping of MMC billets must not be underestimated, together with the task of achieving an adequate balance of engineering properties. It is almost ironical that metallurgists who struggled for many years to improve the fracture toughness of high strength aluminium alloys and who achieved their goals by using high purity materials to reduce the number of brittle intermetallics in the matrix are now loading matrices with brittle, often angular, particles in a volume percentage far in excess of the intermetallic content of the relatively impure alloys of yesteryear. The research being carried out on reinforcement-matrix interfaces to relate their characteristics as-processed, after fabrication sequences and after prolonged service at elevated temperatures to properties is a vital part of development programmes, since these interfaces will have a major influence on the properties of the composites.

8. ARALL

Composites which involve the lamination of aluminium alloys with Kevlar fibre epoxy prepregs have been researched in Holland[32]. Significant weight savings, in the range 15-40%, are being predicted for airframe components using these laminates. Increased resistance to fatigue crack growth is one of their attractive features.

9. CONCLUSIONS

To summarise I believe that the materials which will be used in the 1990s, particularly in aircraft, missile and space systems are already with us and are those on which extensive data bases have been established on production material.

The 1990s will see continuing efforts to develop alloys by RSR technology and to produce and characterise a wide range of metal matrix composites. Applications for these new materials need to be identified at an early stage and targets set for properties to be achieved since the route to the market place will be via identified components or structures. Considerable investment in plant will be required to bring many of the new materials into commercial production and when this is required potential suppliers and their accountants will need assurance that there is a profitable market for the products.

Considering military airframes: the European Fighter Aircraft is likely to be the last major combat aircraft project in which the UK is involved this century. Choices of materials/fabrication techniques for EFA have been made based on major evaluation and demonstrator programmes. EFA will be in service well into the next century and air defence systems could improve in that time-span to the extent that it might be judged too difficult thereafter for manned aircraft to penetrate them with acceptable losses. Sophisticated aircraft with their very high unit costs might then be abandoned in favour of large numbers of stealthy missiles. The high stiffness, high temperature, high acceleration requirements of such missiles would be worthwhile targets for some materials currently under development.

On the civil side there is likely to be a steady, rather than spectacular, increase in the use of organic composites but conventional ingot route aluminium alloys in the 2000, 7000 and 8000 series (aluminium-lithium alloys are considered conventional) are likely to dominate in large civil (and military transport) airframes well into the next century.

In the context of airframe materials and the search for improved materials/ fabrication processes the case of titanium alloy powder research and development is worth considering. Despite very considerable efforts over the past 20 years to produce satisfactory powder and to compact it to net or near-net shapes the author is unaware of any airframe component fabricated from titanium alloy powder flying as a production item. This, despite the fact that titanium alloy powders can be readily bonded by HIPing at elevated temperatures. During the same time-span titanium castings have made significant progress into flying hardware, as have components fabricated by superplastic forming and diffusion bonding.

The titanium powder position has some relevance when considering the exploitation of RSR routes to novel aluminium alloys. In the case of aluminium alloys the problems are more formidable. Most RSR techniques result in the production of material requiring compaction to bulk product. For aluminium alloys this is difficult due to adherent, non-dissolving oxide films. This oxide must be sheared to obtain satisfactory particle to particle bonding and thus the direct production of net shapes is unlikely. It is true that compacted powder can be extruded to shape but the various stages required, including classification and degassing, will all add to the cost of the final product and clearly if massive airframe components are to be made using the powder route, it will be necessary to handle very large volumes of powder.

In the context of MMC it is suggested that the front runners at the end of the century for consideration for airframe and missle designs in the next are likely to be (i) materials produced by the ingot or spray deposition routes using particulate to obtain modest, near-isotropic property improvements and (ii) reinforced cast-to-shape materials. It will be possible to shape and fabricate the former at reasonable cost and the latter will either require no machining or very little eg of mating faces. Continuous fibre reinforced aluminium alloys will be combined with particulate reinforced sheet in skin-stringer constructions since the point made about reducing the waste of high cost aluminium-lithium alloys by extensive machining operations will apply even more forcibly to metal matrix composites. Aluminium alloys with good elevated temperature properties, produced by rapid solidification techniques, will also find applications if high integrity bulk material can be produced at affordable cost.

REFERENCES

1) Materials Newsletter, Advanced Materials and Processes Vol 130, No.4, p47. October 1986

2) Grimes R., Cornish A.J., Miller W.S. and Reynolds M.A. Metals and Materials Vol 1, No.6, p357. June 1985

3) Peel C.J., Evans B., Grimes R. and Miller W.S. paper 94, North European Rotocraft Forum, Stresa, September 1983

4) Evans B., Proc Conf Advanced Materials Research and Development for Transport p71, Strasbourg, November 1985

5) Metals and Materials Vol 4, No.3, p132. March 1988

6) Webster D., Haynes T.G. and Fleming R.H. Advanced Materials and Processes Vol 133, No.6, p25. June 1988

7) Ridley N. AGARD Lecture Series No.154. Superplasticity. 1987

8) McDarmaid D.S. and Shakesheff A.J. Proc Con Advanced Materials Research and Development for Transport, Strasbourg. November 1985

9) Partridge P.G. AGARD lecture Series No 154. Superplasticity. 1987

10)McDarmaid D.S. Mater Sci Eng 70, 123-129, 1985

11)Ward D.M. Metals and Materials Vol 2, No.9, p560. September 1986

12)Stephen D. AGARD Lecture Series No 154. Superplasticity. 1987

13)Welding and Metal Fabrication March 1986

14)Advanced Materials and Processes Vol 131, No.1, p54. January 1987

15)Metals and Materials Vol 4, No.6, p342. June 1988

16)Savage S.J. and Froes F.H. Journal of Metals p20. April 1984

17)Young-Won K., Griffith W.M. and Froes F.H. Metals/Materials Technology Series 8305-048

18)Aerospace Engineering March/April 1984, 5-10

19)Bickerdike R.L., Clark D., Eastabrook J.N. et al Int J Rapid Solidification 1986, 2, 1-19

20)Gardiner R.W. and McConnell M.C. Metals and Materials Vol 3, No.5 p254 May 1987

21)Erish L. and Donachie S.T. Metals Progress Feb 1982, 22

22)Metallurgica November 1986 p515

23)Bridges P.J., Brooks J.W. and Gilman P.S. Proc Conf Advanced Materials Research and Development for Transport p85, Strasbourg, November 1985

24)Feest A., Metals and Materials Vol 4, No.5, p273. May 1988

25)Lewis C.L. Materials Engineering p.33. May 1986

26)Aviation Week and Space Technology 18 November 1985

27)Chadwick G.A. Materials Science and Technology Vol 4, p181. March 1988

28)Verma S.K. and Dorcic J.L. Advanced Materials and Processes Vol 133, No. 5, p48. May 1988

29)Materials Engineering p60. September 1986

30)Advanced Materials and Processes Vol 133, No.6, p18. June 1988

31)Willis T.C. Metals and Materials Vol 4, No.8, p485. August 1988

32)Vogelsang L.B. and Gunnink J.W. Delft Univ of Technology, Department of Aerospace Engineering Report LR-400, August 1983

Materials and Processing – Move into the 90's
edited by S. Benson, T. Cook, E. Trewin and R.M. Turner
Elsevier Science Publishers B.V., Amsterdam, 1989

FAILURE MECHANISMS IN TITANIUM ALUMINIDE/SiC COMPOSITES

B.N. Cox, M.R. James, D.B. Marshall,
W.L. Morris, C.G. Rhodes and M. Shaw

Rockwell International Science Center, 1049 Camino Dos Rios,
Thousand Oaks, CA 91360

Thin sheets of titanium aluminide alloy reinforced by continuous SiC fibers have been tested under monotonic and cyclic loading at room temperature. The principal failure mechanisms have been determined. The effects of residual stresses, the state of the fiber/matrix interface, the bridging of matrix cracks by fibers, and the environment have been investigated. Implications for composite fabrication and lifetime prediction are discussed.

1. INTRODUCTION

Titanium aluminide composites are being developed for various high temperature applications in aerospace, for which they offer excellent specific strength and stiffness. However, our understanding of the mechanisms by which they fail in hostile environments under mechanical and thermal loading is very limited. One feels intuitively that they must lie somewhere between conventional, ductile metal matrix composites and brittle ceramic-ceramic composites, since the intermetallic matrix possesses only limited ductility and yet is not entirely brittle. Thus, critical questions for processing improvements remain to be answered: for example, is it desirable to have a weak interface, as is sought in brittle composites, or a strong interface, the usual ideal for ductile metal matrix composites; do residual stresses work for or against survival; and what are the principal effects of a hot oxidizing environment?

Various experiments to answer these and similar questions have been carried out on specimens cut from a thin sheet of titanium aluminide alloy reinforced by continuous SiC fibers. These experiments have revealed that the possible modes of failure are indeed very complex. Nevertheless, some generalizations are already permissible, and they are presented below.

2. SUBJECT MATERIAL AND INITIAL RESIDUAL STRESSES

The panel was manufactured by Textron Specialty Metals‡ by consolidating an alternating stack of four Ti-25Al-10Nb-3V-1Mo (super-α_2) alloy foils and three SiC (SCS6) fiber mats. Its final thickness was ~ 650 μm. The three unidirectional plies of

‡ Textron Specialty Metals, Lowell, Massachusetts

145 μm diameter fibers constituted 36% of its volume. The matrix consisted of α_2 grains separated by continuous β phase.

The composite sheet was very irregular, apparently because of significant movements of the fibers during consolidation. In some areas in sections cut normal to the fibers, only two layers of fibers were found over distances of up to 6 fiber diameters. The thickness of the outermost layer of alloy covering the first row of fibers was on average ~ 80 μm, but occasionally as little as 35 μm. This was found to be very important in fatigue reliability (see below). The in-plane spacing between fibers was also irregular. Where fibers were unusually close together, radial cracks were found between them, often linked to circumferential interface cracks. Similar observations have also been made in Ti-6Al-4V-2Fe/SCS6 composites.[1] Radial cracks were also occasionally found within the fibers, but these played no apparent role in any failure mechanism.

Young's modulus for the matrix was determined by measurements of the Rayleigh wave velocity and by analyzing the longitudinal composite modulus by the rule of mixtures.[2] The best value for the modulus was found to be 80 ± 10 GPa, considerably lower than values found for the monolithic alloy.

Initial residual stresses in the super-α_2 composite were measured by x-ray diffraction.[2] They were found to be significant in both the longitudinal direction (parallel to the fibers), where they were +230 ± 30 MPa; and the in-plane transverse direction, where they were -130 ± 10 MPa. These stresses are about half those that would exist if cooldown from consolidation at 1850°F was accompanied by elastic strain only. Further measurements made as the outer layer of alloy was electro-polished away led to the conclusion that the stresses in these outer layers are approximately uniform.[2]

3. MONOTONIC BENDING AND UNIAXIAL TENSION

Some specimens cut from the panel were tested in monotonic bending. The sample surface was polished away after unloading to reveal the fibers. It was found that the fibers had suffered periodic cracking at strains greater than ~ 0.6%. Since the tensile longitudinal residual stresses measured in the matrix imply compressive residual stresses in the fibers, the net stress in the fibers at failure was only ~ 2 GPa, about 30% lower than the strengths quoted by the manufacturers for virgin SCS6 fibers.

The period of the fiber cracking (~ 400 μm) implies a critical shear stress for frictional sliding at the interface of 180 MPa. At the occasional points where good interfacial bonding persisted, the fiber cracks grew directly into the matrix (Figure 1(a)). On the other hand, in the more common instance of interfacial debonding and sliding, the stress intensification of a broken fiber was dissipated in multiple, small matrix cracks arrayed along the interface (Figure 1(b)). Complete failure of the specimen occurred at a strain of ~ 1.0%.

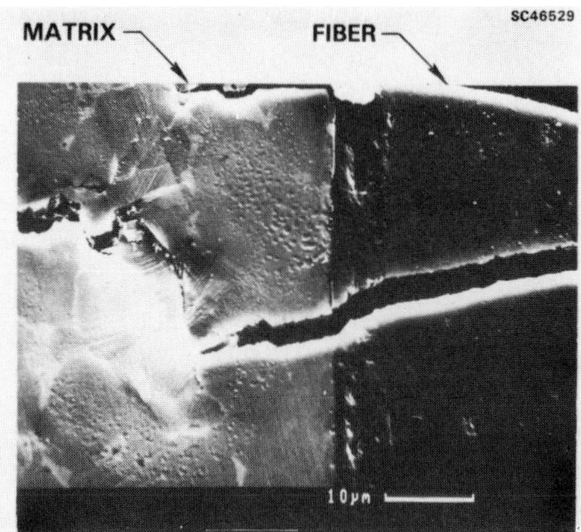

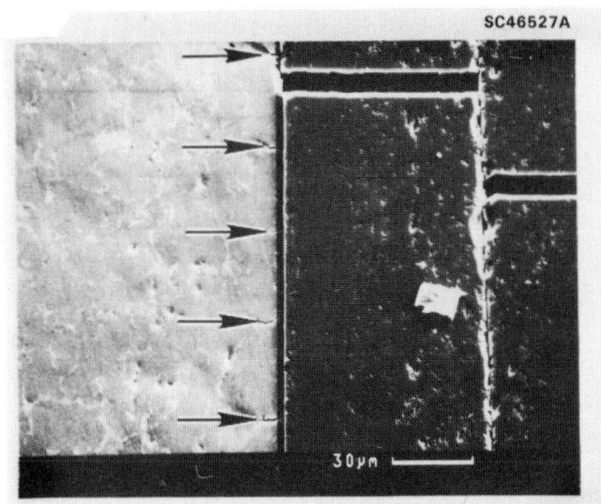

FIGURE 1 (a) A crack in super-α_2/SCS6 that has extended directly from the fiber to the matrix. (b) Periodic matrix cracking (arrows) seen near cracks that have been deflected by interfacial debonding and sliding.

More details of these measurements can be found in Ref. 3.

4. ROOM TEMPERATURE FATIGUE FAILURE

Fatigue experiments on the super-α_2/SCS6 composite were carried out on miniature tapered cantilever beam specimens. All experiments were conducted at room temperature in air under fully reversed loading (R = -1), with the stress axis parallel to the fibers. Some experiments were conducted on pristine specimens and others on specimens that had first been thermally cycled in air.

In the pristine specimens, the appearance of the first cracks on the surface was preceded by substantial localized surface strain fields, which were easily measured by stereoscopy or automated displacement field mapping.[4] This observation and destructive sectioning showed that the cracks had initiated subsurface at the fiber/matrix interfaces (Figure 2). Initiation occurred most readily in places where the fibers lay unusually close to the specimen surface. After initiation, the cracks grew out to the specimen surface, propagating first at 45° to the applied stress axis and then in Mode I normal to the applied stress axis. Cracks occasionally coalesced, forming either a longer Mode I crack or, in the case of one fiber that lay very near the surface, a zigzag crack running

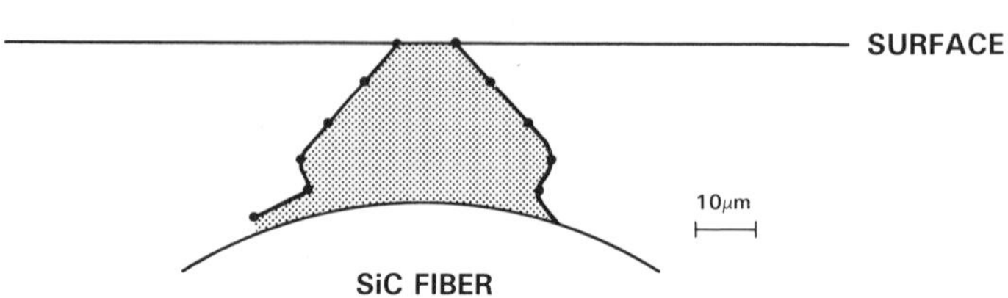

FIGURE 2 Profile of a fatigue crack in super-α_2/SCS6 obtained by removing alloy by progressive polishing.

along the fiber. Destructive examination revealed that cracks sometimes grew subsurface around fibers and sometimes along interfaces. However, post test etching revealed no fibers broken transversely by fatigue cracks. Typical crack paths through the matrix microstructure followed the α_2/β interfaces, with α_2 grains remaining mainly uncracked.

Growth rates for small cracks are shown in Figure 3 for two applied strain amplitudes in the outer layer of the matrix: 0.3% and 0.5%. Each datum in Figure 3 shows the average of several growth rate measurements, the original data possessing an order of magnitude more scatter. The applied stress amplitude was calculated for Figure 3 using the modulus of the matrix measured in the composite, 80 GPa.[2] The local stress amplitude was taken to be the sum of this and the measured longitudinal residual stress (230 MPa). The net stress intensity factor range, ΔK, was calculated under the assumption that the cracks were semicircular. The initial decrease in crack growth rate with increasing crack length may be an artifact of the cracks initiating subsurface. When the cracks are first seen on the surface, they are really much bigger than they appear (Figure 2) and ΔK is therefore underestimated. As 2c increases, the semicircular shape assumed in calculating ΔK is achieved more closely, and the growth rate, when expressed as a function of ΔK, appears to decline. The growth rates at the higher ΔK end of Figure 3 are comparable to those found in monolithic titanium aluminides. This might suggest that growth rates can be predicted from monolithic alloy data provided the local stress in the matrix is known.

However, the situation is really more complicated than this: crack mouth opening displacement (c.m.o.d.) measurements show significant hysteresis, which is probably caused by bridging of the crack by fibers interacting with the matrix via friction. A measurement of c.m.o.d. for a 55 μm crack is shown in Figure 4. For such a small crack,

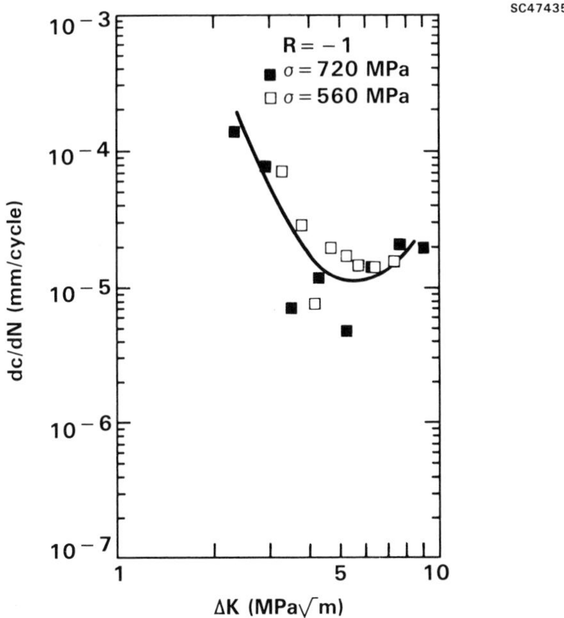

FIGURE 3 Small fatigue crack growth rates in super-α_2/SCS6 for two strain amplitudes, at which the net stress amplitude in the outer layer of the matrix had the values shown.

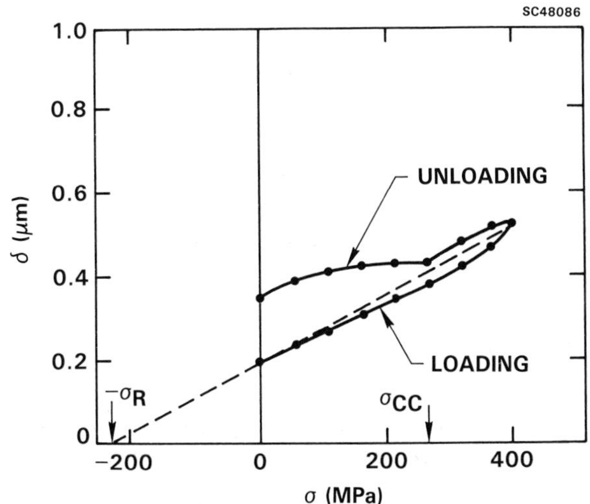

FIGURE 4 Crack mouth opening displacements measured for a 55 μm fatigue crack in super-α_2/SCS6 as a function of the applied matrix stress.

the hysteresis is a large fraction of the total c.m.o.d. For larger cracks (> 200 μm), the absolute hysteresis remains similar in magnitude, but it is consequently a smaller fraction of the total c.m.o.d. It has been shown elsewhere that the net stress intensity factor for a bridged crack passes through a minimum as the crack grows at constant applied load.[5] Thus closure from crack bridging presents an alternative explanation of the minimum in Figure 3.

The longitudinal residual stress can be deduced from Figure 4 by extrapolating the linear loading curve back to zero opening displacement. The value obtained, σ_R = 220 ± 50 MPa, is consistent with that measured by x-ray diffraction. The magnitude of the bridging pressures and hence their effect on the net stress intensity factor acting on the crack tip are deducible from Figure 4 and measurements of strain in the bridging fiber.[6] Knowledge of the bridging pressure is a prerequisite to predicting growth rates of both long and small trans-fiber cracks.

Room temperature mechanical fatigue following thermal cycling was tested by first exposing specimens to isothermal heating at 800°C in air. These specimens showed very different initiation and growth habits. Cracks initiated on the specimen surface, then grew very straight in Mode I. Measurements of c.m.o.d. now showed no measurable hysteresis. Microstructural analysis showed appreciable oxidation of the matrix phase at the fiber/matrix interfaces throughout the specimen. This oxide had presumably locked up the interfaces, preventing frictional sliding.

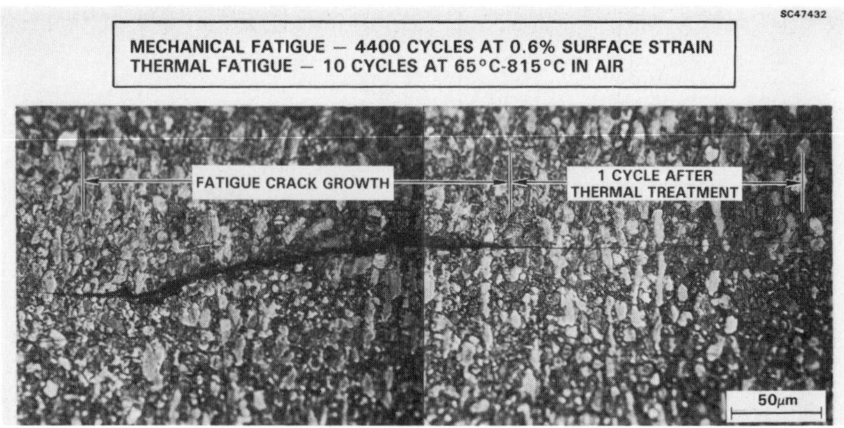

FIGURE 5 The effect of thermal cycling on room temperature fatigue crack growth in super-α_2/SCS6.

In another test, a fatigue crack was initiated and grown to 2c ≃ 240 μm (over 4400 cycles at 0.6% applied outer layer strain) in a pristine specimen. The specimen was then subjected to 10 thermal cycles between 65 and 815°C in air in the absence of load. Upon subsequent reloading in the fatigue machine, the fatigue crack doubled in length on the first cycle (Figure 5). The mode of growth was that of a very brittle material: the crack extension was very straight, cutting through the microstructure. Once again, c.m.o.d. measurements displayed no hysteresis.

5. SUMMARY

1. Continuous-fiber intermetallic composites contain significant residual stresses, which play a prominent role in the mechanics of failure. Residual stresses act to increase strength under monotonic loading, but probably increase vulnerability to fatigue.

2. Both ultimate strength under monotonic loading and fatigue crack growth resistance in super-α_2/SCS6 composites are enhanced by the weakness of fiber/matrix interfaces. Frictional sliding at interfaces is possibly an important source of toughness in fatigue crack growth. Weak interfaces may enhance fatigue lifetime by allowing matrix cracks to be bridged by the fibers or deflected out of Mode I growth. They minimize the damage of fiber breakage under monotonic loading (or spike overloads during fatigue) by preventing fiber cracks from propagating directly into the matrix.

3. Neither the super-α_2 alloy matrix nor the fiber/matrix interfaces in super-α_2/SCS6 composites can tolerate a hot, oxidizing environment.

4. The regularity of the spatial distribution of reinforcing fibers is critical to fatigue lifetime. The first and most damaging fatigue cracks initiate at locations where the fibers lie unusually close to the specimen surface.

ACKNOWLEDGEMENT

This work was supported by Rockwell International's Rocketdyne Division under Air Force Contract No. F33657-87-C-2214.

REFERENCES

1. M.R. James, W.L. Morris, and B.N. Cox, "Local Stress Relaxation in a Ti-6Al-4V-2Fe/SCS6 Composite During Thermal Cycling," in Proc. Seventh Int. Conf. Fracture, Houston, TX, March, 1989, ed. H.L. Marcus.

2. B.N. Cox, M.R. James and R.C. Addison, "Determination of Residual Stresses in Thin Sheet Titanium-Aluminide Composites," submitted to Acta Metall.

3. D.B. Marshall and M. Shaw, "Failure Mechanisms in Titanium Aluminide/SiC Composites under Monotonic Loading," submitted to Acta Metall.

4. M.R. James, W.L. Morris and B.N. Cox, "A High Accuracy Automated Strain Field Mapper," submitted to Exptl. Mechanics.

5. D.B. Marshall and B.N. Cox, Acta Metall. 35, 2607-19 (1987).

6. B.N. Cox, D.B. Marshall and K.A. Marsh, "The Determination of Crack Bridging Forces," submitted to Int. J. Fracture.

Materials and Processing – Move into the 90's
edited by S. Benson, T. Cook, E. Trewin and R.M. Turner
Elsevier Science Publishers B.V., Amsterdam, 1989

THE MECHANICAL AND PHYSICAL PROPERTIES OF SILICON CARBIDE REINFORCED AA2124 AND AA8090 MANUFACTURED USING A POWDER METALLURGY ROUTE

C W BROWN AND W S MILLER

BP Research, Chertsey Road, Sunbury-on-Thames, Middlesex. TW16 7LN

1. ABSTRACT

Metal matrix composites based on AA2124 and AA8090 are currently produced using a powder metallurgy route by BP. Some properties of these composites, such as modulus, thermal expansivity and conductivity vary linearly with SiC content. Other properties, notably strength and ductility, also vary with SiC content but in a more complex manner. In these cases there is a strong influence of production route and semi-fabrication procedure.

This paper compares the mechanical and physical properties of a range of metal matrix composites made via different production routes. It is shown that powder metallurgy can lead to optimised property combinations which are superior to those of any other manufacturing technique.

2. BACKGROUND

The incorporation of silicon carbide particles in aluminium alloys leads to significant increases in elastic modulus even at small additions. For example, a SiC volume fraction of only 20% will lead to a 50% increase in stiffness. However there are many other effects caused by these ceramic particles; principally,

- increased density
- reduced toughness
- reduced thermal expansivity,
- increased internal interface for precipitation,
- increased quenched-in dislocation density.

It is proposed here that all of these effects can be grouped into either of two separate categories. These categories arise from the direct and indirect influences of the ceramic second phase on the composite behaviour.

A composite of two constituents mixed in a particular volume fraction will have an average density, thermal conductivity and elastic modulus dependent directly on the volume fraction of each phase. Changes in particle size and distribution are unlikely to have any effect on this average although particle shape may lead to some anisotropy in an aligned composite. These properties are therefore virtually independent of microstructure and hence processing route.

Indirectly, the presence of a ceramic will also lead to modifications in the matrix microstructure and hence composite strength, ductility and formability. This effect arises mainly from the control of recrystallisation and grain growth shown by the fine reinforcing particles. However there is also a further factor resulting from the differing thermal expansion coefficients of the reinforcement and the matrix. This strain mismatch causes prolific dislocation generation on thermal cycling which as well as controlling the retained work level also affects precipitation in the matrix. By these mechanisms, the same parameters as above of particle size, shape and distribution can therefore indirectly control many key mechanical properties.

The extent of both the direct and indirect control of properties is restricted by the ability of the production process to incorporate the desired reinforcement in the matrix. There are three main processes generally capable of achieving some of these goals, viz

a) melt stirring

b) spray deposition

c) powder blending

An increasing complexity of product is possible in the order a) to c). Indeed, the overlap of these areas in terms of produceable composite is small and direct comparisons of their potential are difficult. This report avoids such a problem by producing a range of composites via one method (powder blending). These results are then used to interpret a comparison of slightly dissimilar composites from different manufacturers made via the different production routes.

3. EXPERIMENTAL RESULTS

Silicon carbide grits of a range of average sizes (1μm, 2μm, 3μm, 7μm, 23μm, 100μm) were blended with a single aluminium alloy powder (8090 Composition: 2.51 Li, 1.47 Cu, 0.75 Mg, 0.12Zr) to produce six composite microstructures. The powders were then compacted by canning and hot isostatically pressing (HIPping) to full density. These billets were flattened by forging and subsequently hot rolled to 2mm sheet for property evaluation. The room temperature tensile testing was performed on flat "dog-bone" test pieces with extensometer monitoring of a 20mm gauge length. The cross-head speed was 2mm min^{-1}.

The tensile strengths and elastic moduli of these composites are given in Table 1.

TABLE 1. The Effect of Particle Size on The Mechanical Properties of 16v/o SiC in AA8090*

Av:SiC Size	Proof Strength	Ultimate Tensile Strength	Elastic Modulus	Averages of
1μm	528 MPa	571 MPa	107 GPa	10 tests
2μm	495 MPa	549 MPa	104 GPa	8 tests
3μm	493 MPa	535 MPa	105 GPa	12 tests
7μm	448 MPa	522 MPa	103 GPa	4 tests
23μm	435 MPa	481 MPa	100 GPa	8 tests
100μm	410 MPa	421 MPa	100 GPa	9 tests

* All properties are measured in a heat treated condition of solution treated and artificially aged.

It is clear that within the accuracy of the testing technique there is an intrinsic control over modulus from the constant volume fraction of reinforcement. However it is also clear that the particle size can have a signficant influence on tensile strength following heat treatment. Humphreys[1] has shown this strength control to come from two major indirect effects of the ceramic particles. These are the dislocation density produced dependent on the solution treatment temperature and the matrix grain size control resulting from particle induced recrystallisation. The main feature here however is the demonstrated degree of independence between the directly and the indirectly controlled properties which gives the potential for structure optimisation in these composites.

4. COMPARISON OF DIFFERENT PRODUCTION ROUTES

There are currently three major production routes used for the manufacutre of particulate metal matrix composites.

The potentially lowest cost option is to stir the particulate reinforcement into a bath of molten aluminium alloy which is then allowed to solidify. Proprietary techniques have been developed[2] to achieve a reasonably even distribution in the ingot particularly for lower volume fractions of coarser reinforcement. The ingots are then usually extruded prior to use to further improve the distribution and hence the indirectly controlled mechanical properties.

A second option is to incorporate the particulate into the atomising spray during the spray deposition process[3]. The ceramic particles are then entrained in the depositing matrix and a composite product is formed directly. Again this process works best for low volume fractions of coarser reinforcement where ceramic/ceramic contacts can be avoided. These products are also usually extruded before evaluation, in this case to remove porosity as well as improve distribution.

The third option is to mix together in powder form the range of constituents required in the composite. Canning and extruding or canning and HIPping then produces a fully dense material without loss of reinforcements or matrix alloying elements. This control allows a wide range of reinforcement sizes and contents to be used.

A comparison of mechanical properties in similar materials from these separate routes is given in Table 2.

TABLE 2. The Influence of Manufacturing Route on Composite Properties

Production Route	Composition	Elastic Modulus	Proof Strength	Ultimate Tensile Strength	Elongation to Failure
Stir Cast[2]	AA2014 +20v/o of 23μm SiC	106 GPa	471 MPa	490 MPa	2.5%
Co-spray[3]	AA2014 +10v/o of 13μm SiC	81.2 GPa	457 MPa	508 MPa	1.8%
Powder[4]	AA2124 +20v/o of 10μm SiC	103 GPa	399 MPa	550 MPa	7.0%
Powder[5]	AA2124 +17v/o of 3μm SiC	100 GPa	417 MPa	590 MPa	6.0%

All properties measured longitudinally on extrusions except 5 which is from cross rolled sheet.

Conclusions from this data are hampered by the chemistry variation and a lack of heat treatment information in the manufacturers claims. It is clear from our work here however that the varying particle sizes are not sufficient to explain the differences observed. There needs to be an underlying effect of distribution control or matrix homogeneity that leads to improved properties, particularly ductility, for the powder based routes.

It is also interesting to note that all results are quoted for heavily worked material. Properties for unworked product are rarely available, presumably because of distribution and porosity problems. A powder production route which can offer high quality ingot or billet which does not require a great deal of extra work will therefore be attractive for extending the range of applications for these composites.

5. CONCLUSIONS

1. SiC particles control composite properties separately by their physical presence and by their influence on the matrix microstructure.

2. Particle size has a significant influence on composite strength but has no effect on stiffness.

3. Powder based processing routes can lead to the most attractive combinations of composite strengths and ductilities through their control on microstructure.

ACKNOWLEDGEMENT

The authors would like to acknowledge the invaluable contribution of Ms L Lenssen and Mr P Daniell in performing the experimental work. They would also like to thank BP for the provision of laboratory facilities and permission to publish this paper.

REFERENCES

1) F.J. Humphreys, Proc of the 9th Int. Symp. on Metallurgy and Material Science, RISO, September 1988.

2) Dural Aluminium Composites Corporation, San Diego, publicity brochure.

3) T.C. Willis, J. White, R.M. Jordan, I.R. Hughes, Proc of the Int. Conf. on PM Aerospace Materials, Luzern, November 1987, paper no 29.

4) DWA Inc, Chatsworth, publicity brochure.

5) BP private data. 1988.

Materials and Processing – Move into the 90's
edited by S. Benson, T. Cook, E. Trewin and R.M. Turner
Elsevier Science Publishers B.V., Amsterdam, 1989

AEROSPACE MATERIALS: TRENDS AND POTENTIAL

W. Bunk*, P. Esslinger**, H. Kellerer***

*DFVLR, Institut für Werkstoff-Forschung, D-5000 Köln 90
**MTU München GmbH, D-8000 München 50
***MBB GmbH, D-8000 München 80

1. MATERIALS FOR STRUCTURES

1.1. General philosophy

The development of aerospace materials is driven by a fascinating mix of scientific, technical, economical and also military motivations. In this paper we will try to give a short review of the current situation in materials for airframes and engines, our present goals and limitations and of the philosophies and strategies employed in the pursuit of these goals. What are those goals, what are the requirements for our materials? First of all: safety, of course. Materials and materials science are crucial to the safety of aircraft. Second: performance, naturally (figure 1). But these two main requirements are often contrasted with cost considerations: cost of material and cost of manufacturing processes. The materials scientist and the materials engineer are challenged to find the right balance within this magic rectangle. It may be safe to say that the emphasis is shifting away today from maximum performance to safety and cost and this is true not only for commercial but also for military aircraft. In a time of - temporarily - decreasing fuel prices, reduction in the direct operating cost of aircraft is obtained not only by weight reductions but by decreasing purchasing and maintenance. This has an immediate impact on materials and process development. Science and technology are influenced today by a host of conflicting interests (figure 2).

Also there seems to exist a very different approach between western and far eastern materials R and D teams. More individualistically thinking western material scientists often neglect to invest time in advanced ideas of processing but are satisfied by a publication or patent. In contrast to this observation their far eastern colleagues are evenly motivated to devote themselves to the science of processing. They do not know of a gap between faculties of science and engineering, but cooperate with other disciplines to serve the national goal of winning the battle between the old fashioned classical route of manufacturing of a material or a component and the advanced, but still somewhat exotic alternative approach. One may hope that restrictions of knowledge exchange by politicians will be overcome in learning that such an attitude will demotivate scientists and diminish chances and opportunities.

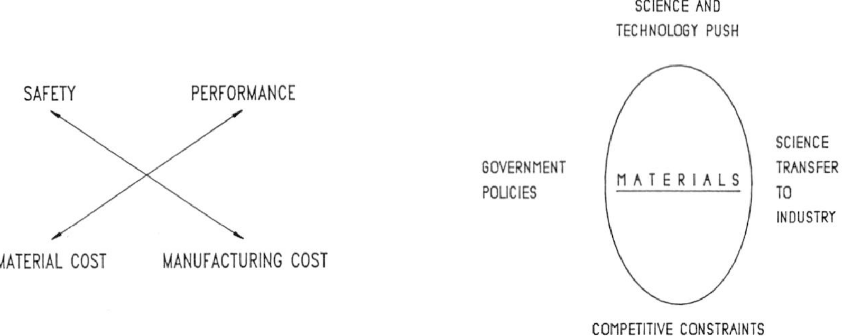

FIGURE 1
Technology drivers in materials development

FIGURE 2
Social factors influencing materials development

1.2. Competition between metals and polymer composites

When talking about metallic materials we also have to spend a few words on competition. Metals will hardly be able to beat CFRP's on terms of specific strength or specific stiffness or fatigue behaviour. However, there are several other areas where they may have definite advantages:

First of all: CFRP's are generally expensive in both material cost and in manufacturing. The hopes for cheap carbon fibres so far have not been realized and even high performance resins command staggering prices.

Second: The ductility of todays resins reduces CFRP design strains to somewhere around 0.5 %, roughly one tenth of the values commonly accepted for metals. This is a severe limitation for many applications. In addition, impact behaviour is relatively poor, and due to the nature of the impact damage, it may not be visible by simple inspection from the outside.

Third: A definite advantage of metals, at least compared to todays resins, is the temperature limitation of CFRP's. Most epoxy resins should not exceed 150 ^{o}C. Today 200 ^{o}C can be considered the maximum temperature for structural applications. These temperature restrictions are aggravated by the uptake of moisture under normal service conditions which results in softening of the matrix. Moisture, however, is a problem also for metals and their corrosion may be causing more difficulties than lowering of the glass transition temperature in polymers.

A yet not fully evaluated advantage of metals, at least most metals, is a fact contrary to composites: they can easily be recycled. Considering the ever increasing cost of materials, this is especially important for the scrap generated during manufacturing.

An important point to be considered is the development potential of metals. We know that todays structural alloys result in mechanical properties which amount to only 30 % of the values to be expected from atomistic considerations. This value should be compared to the 70 % realized in todays carbon fibres. Therefore, we are convinced that metals do have a definite potential for further developments.

1.3. Economical considerations

Looking at the drawbacks of todays CFRP's: metals will offer technical advantages in ductility and impact behaviour and temperature resistance above 200 ^{o}C. Metals are still considerably cheaper than polymer composites and MMC's will probably offer more specific stiffness for the money than reinforced plastics. In addition we have proven manufacturing processes for complicated thick section components which are difficult to match in CFRP.

However, aluminium-lithium is reaching the cost level of composites; fibres for continuous reinforcement of light metals are in an inacceptable price range, powder metallurgy is expensive. In order to compensate for increasing material cost we have two strategies:

- *Increase the fly to buy ratio*. This is extremely important in Al-Li. The classical way of carving components out of thick plate just does not work, when the base metal is three times as expensive as aluminium and scrap cannot be recycled. We have to increase our efforts to develop technologies for producing net shapes without chips. Superplastic forming may be a solution, or precision casting, new forging routes, in some areas even powder metallurgy.
- *Decrease the cost of joining and assembling*. The CFRP's drastically show how reduction in the number of detail parts can reduce overall manufacturing cost. In the airbus fin, switch to plastics reduces the number of fasteners by 90 %. Superplastic forming again may help, especially SPF + DB, roll bonding, weld bonding; but all these processes are restricted to relatively few applications.

A less obvious but important point to reduce cost of metallic structures is surface protection. Airlines look at *life cycle cost*: given the drastically increased life time of our aircraft, maintenance and corrosion repair are rapidly increasing in importance.

Our paint schemes and especially our anodising processes still offer a high theoretical potential for considerable improvement. Can this potential be realized? Can we build cathodic protection into our oxides? Will inhibitors be a solution?

In this area we should look closely at what our friends in the electronic industries are doing: it might appear that some of their PVD and CVD processes might be upscaled for our purposes.

As in polymers, where it is difficult to predict hot/wet behaviour over 20 years, we do not yet have meaningful corrosion tests, which reflect the real life of an aircraft. What is the meaning of a salt-spray test? How can we quantitatively measure the quality of a coating? There is still a lot to do!

1.4. Three main thrusts in metals for structures

Our current light metal alloys give mechanical properties which are around 30 % of theory. In the long run there may be chances for considerable improvements. However, based on todays possibilities, metals will not be able to match the specific, weight related, mechanical properties of polymer composites.

Nevertheless, we can predict increases in strength, modulus, corrosion resistance and fatigue. The main problem is that, usually, improving one property decreases the other and it is the overall balance which is difficult to optimize. In the competition with CFRP's we see three main developments:

- *Aluminium-Lithium*. Why has lithium been tried so late as an alloying element to decrease density and increase Young's modulus of high-strength aluminium materials? Processing obstacles have been the major reason for this. For designers of superlight metals lithium is a very important alloying element[1] (figure 3). R+D activities are in progress to generate alloys with higher contents of lithium by RST for further reduction of density.
- *Powder metallurgy*. There is still some controversy on the future of PM for light alloys. We feel that the potential is high, to realize alloy compositions which are not feasible with ingot metallurgy. Considerable improvements may be expected in corrosion behaviour and in elevated temperature strength.

Using rapid solidification technology (RST) material scientists learned to increase the solubility fields in phase diagrams under non-equilibrium conditions, however. Transition elements like Fe and V can be kept in metastable solution and after controlled precipipation used as dispersion hardened material with a higher temperature capability[2] (figure 4).

- *Metal matrix composites (MMC)*. It is proposed that based on the length/diameter ratio of the reinforcing fibres three classes of MMC's should be distinguished as is shown in figure 5: on the low end side, there is the reinforcement with short particles, for instance silicon carbide. Dispersion hardening effects and hindering of grain growth lead to an increase in stiffness, to improved elevated temperature properties and to better wear behaviour. Casting or plasma spraying could be attractive manufacturing procedures resulting in reasonably priced components. Continuous fibre reinforcement yields extremely interesting mechanical properties. Boron fibres, SiC, Al_2O_3, and carbon have been extensively investigated as reinforcements. Tensile properties of 1400 N/mm^2 at room temperature and 800 N/mm^2 at 400 ^{o}C have been realized with aluminium based alloys, already several years ago.

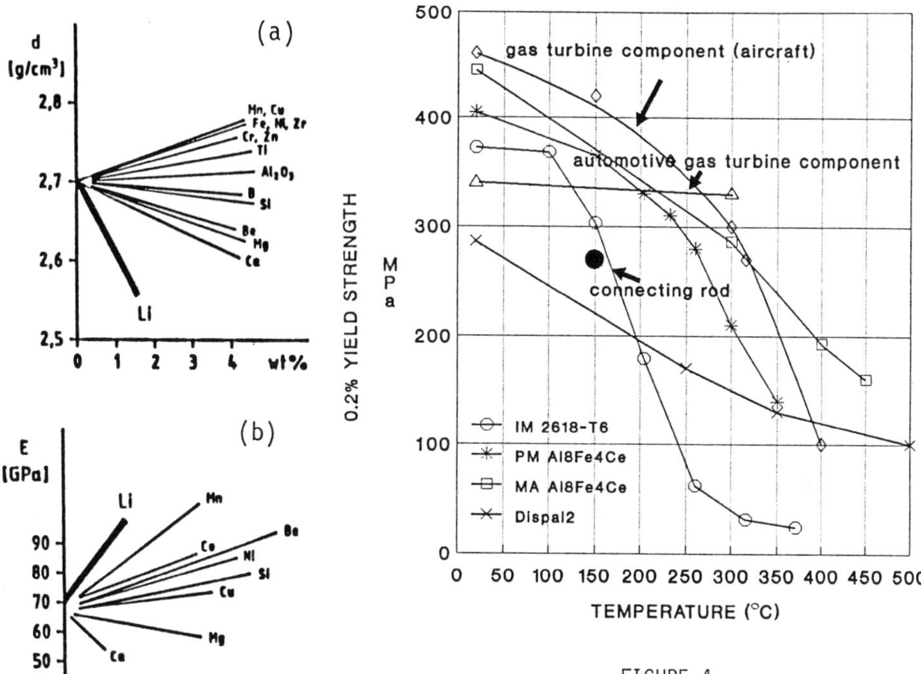

FIGURE 4
Advanced aluminium alloys for high temperature applications

FIGURE 3
Effect of alloying elements to aluminium on (a) density and (b) elastic modulus

However, the fibres available today are still far too expensive for widespread applications. In our opinion, most of the fibres involving CVD processes will not meet the desirable cost limits: boron and SiC deposited on tungsten cores or carbon coated with diffusion barriers belong to this category. Transformation of organo-metallic precursors or melt spinning of ceramics may offer interesting approaches for economically alternative fibres. Unfortunately, most initiatives for the development of new fibres come from Japan and the US; european efforts in this field appear to be restricted to very few companies. Typical examples for the tensile properties which can be realized with particle- or continuous fibre reinforcement can be seen in figure 5.

For a higher temperature resistance SiC-fibre reinforced Ti-alloys are developed and lead to about 100 % higher strength provided specially coated SiC-fibres are implemented to control the reaction between fibres and the Ti-matrix

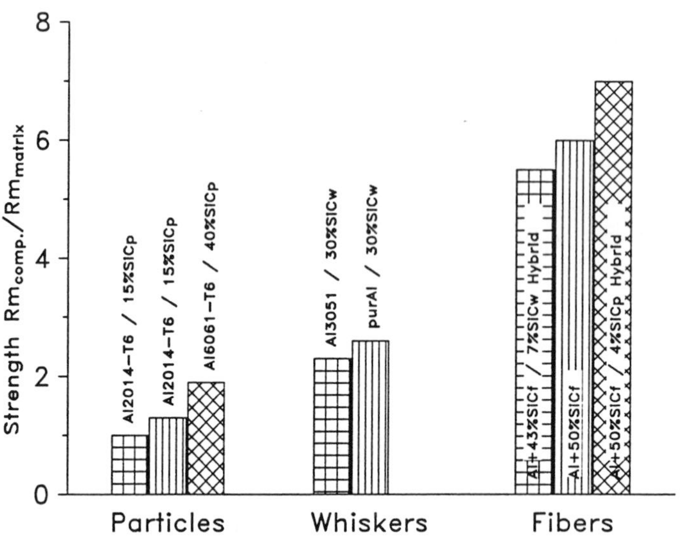

FIGURE 5
Reinforcement of metal matrices

in the interface[3]. MMC processed by diffusion bonding or powder metallurgical techniques are produced to achieve satisfactory properties at elevated temperatures up to about 800 ^{0}C (figure 6).

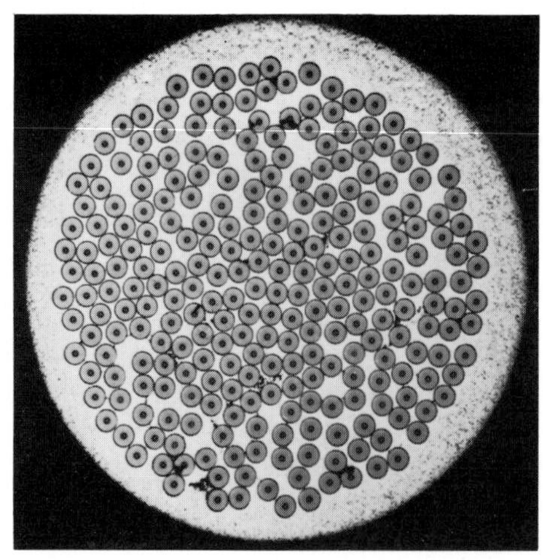

FIGURE 6
SiC-fibre (40 vol.%) reinforced Ti6Al4V

It has been shown that properties approaching those of continuous fibres can also be realized with discontinuous shorter fibres if they are perfectly aligned in one direction and if the l/d ratio exceeds a certain critical value. This could be a very attractive alternative because some fibre manufacturing processes lend themselves better to the fabrication of shorter length. Unfortunately, contrary to polymers, component manufacture still presents problems: fibres tend to break during deformation processes and thus fall below the critical l/d ratio.

2. MATERIALS FOR JET ENGINES

Static and dynamic loads caused by high and varying centrifugal forces under thermal conditions in chemically aggressive combustion gases characterize the high demands for advanced materials for jet engines. The greatest challenge at the present time is the SAENGER engine project, illustrated in figure 7.

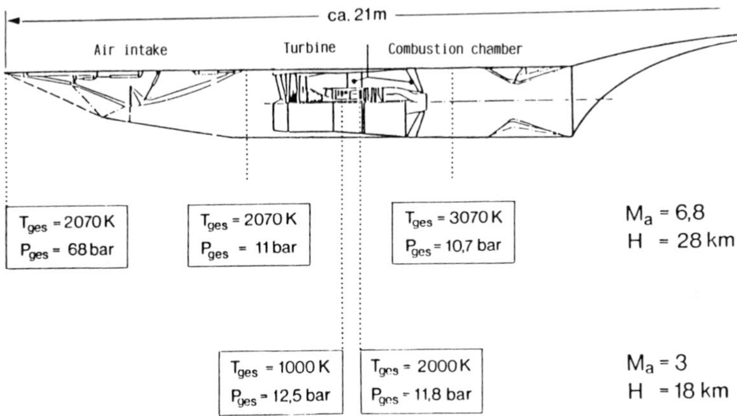

FIGURE 7
Rough sketch of an engine for the "SAENGER" project

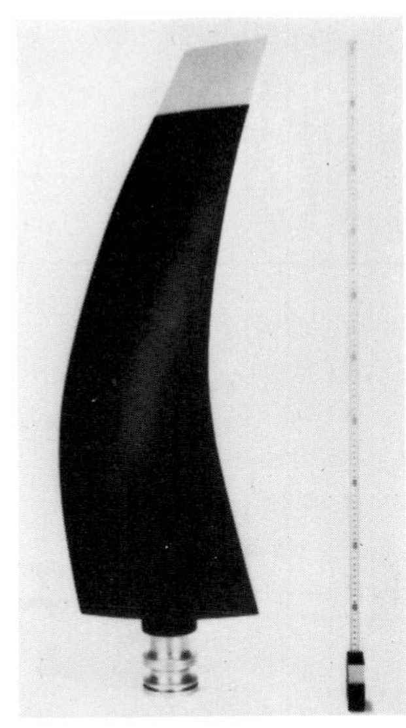

FIGURE 8
FRP-propfan blade (experimental)

However, less exotic developments are also an interesting challenge for the engineering of materials for jet engines. Figure 8 gives an example of the attempt to verify the fabrication of large fan blades with fibre reinforced composites.

Spectacular new designs, however, do not change the fundamental principles of selection and design of appropriate materials for jet engine construction. Diagrams as shown in figure 9 still serve as a first assessment for the choice of materials. For the design, data as in table 1 coordinate material properties with specific engineering parts. The table also illustrates that selection of materials and materials modelling finally can only be a compromise between contradicting demands. In reality, an ideal material never exists. This is also a strong argument for the hybrid structure of certain components (figure 10).

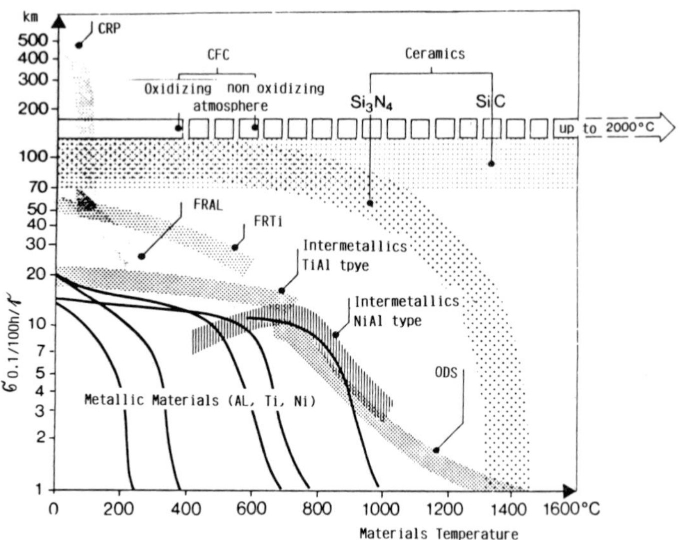

FIGURE 9
Specific creep strength of engine material candidates

	Specific Strength $\sigma_{0.2}/\gamma$	Specific Creep Strength	LCF Strength	HCF Strength	Thermal Fatigue Strength	Oxid./Corr. Resistance	Wear Resistance	Specific Weight	Specific Modules E/γ	Thermal Expansion	Thermal Conductivity	Impact Resistance	Castability	Machinability	Formability	Welding/Soldering Properties	REMARKS
Disks																	
LPC	•		•			0	X	0					•[1]	X	X	X	
HPC	•	0	•			0	X	0						X	X	X	• = very important
Turbine	•	0	•		X	0	X	0						X	X	X	o = important x = has to be considered
Blades																	
LPC			X	•		X	X	0				0		X	0	0[2]	
HPC		0	X	0		0	X	0		X		X		X	X		
Turbine		•	X	X	0[3]	•	X	0		X	X	X	0	X			1 = blisks in small engines 2 = for hollow fan blades 3 = for vanes
Casings																	
Compressor	X		X	X		0		0	0	0	X	•		X	0	0	
Comb. Chamber	X	0	X	X	0	•		X	X	X	X				0	0	
Turbine	X	X	0		0	0		0	0	0	X	•		X	0	0	

TABLE 1
Important material properties for engine parts

FIGURE 10
Hybrid blisk (cast bladed ring, PM bore) prepared for inertia welding

2.1. Compressor materials

Titanium is still the alloy base preferred for compressor components. Strong efforts in engine design are aimed at a compressor completely equipped with titanium alloy, which has not been achieved so far because of high temperatures in higher compressor stages (peak values in the range of 600 to 630 ^{0}C). The English-German development of the alloy IMI 834 sets new standard regarding the limit of service temperatures which until recently has been assumed to be 500 to 550 ^{0}C (figure 11). With the new alloy, a service temperature as high as 600 ^{0}C may be considered. Even a higher value can be expected, in case the development of appropriate protective coatings against oxidation is successful. There is already evidence that development of PM-alloys with even higher strength at elevated temperatures (figure 11) hardenend by oxide dispersions (e.g. erbium oxide) is possible.

IMP-materials based on Ti_3Al and TiAl are expected to contribute an even bigger step ahead. Problems related to specific high temperature strength and oxidation resistance do not exist here or appear to be manageable at least. The question is, how to deal with the problems given by the brittle nature of the material, i.e. sufficient damage and defect tolerance, production in reliable quality and safety in respect to spontaneous failure. Although efforts to increase ductility have been successful so far, progress has only been made on account of a sensitive loss in high temperature strength. For this reason, the author's opinion is that development and application of IMP-materials should not be oriented by the example of metals, but by that of structural ceramics.

Ni-based ordered alloys or intermetallic compounds like Ni_3Al and NiAl have been analyzed by C.T. Liu of ORNL and O. Izumi, Tohoku University, Japan, since

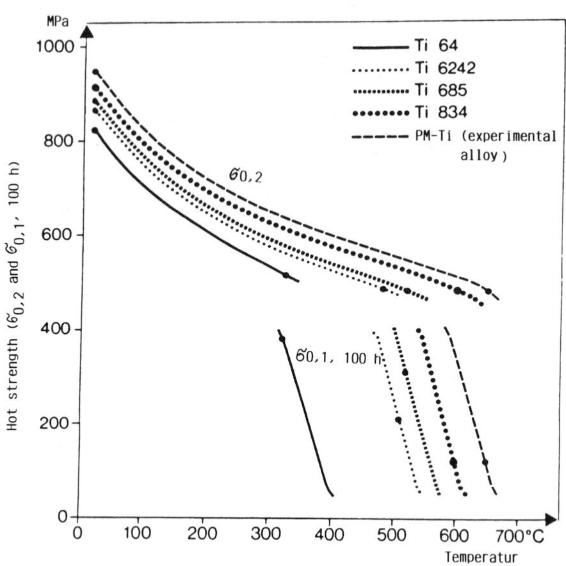

FIGURE 11
Hot strength of some popular and new developed titanium alloys

the 70s. Strategies to overcome their inherent low ductility seem to be well understood but no substantial interest has been documented by industry so far.

Work on Ti-based intermetallics was pioneered by H.A. Lipsitt, AFWAL in Dayton, OH, some ten years ago. Some results on Ti_3Al-based alloys - meanwhile commercially available as "alpha 2 aluminide" or "super alpha 2 aluminide" - have been published, while publications on TiAl seem to be much more restricted. It is assumed that Ti-aluminides will most probably be used for new aerospace vehicles because of their potential in low density and satisfying creep properties at temperatures up to 850 ^{0}C including remarkable oxidation resistance. R+D projects for fibre reinforcement of these materials are on their way.

2.2. Materials for combustion chambers and turbines

There is no approved material that could withstand the prevailing temperatures (up to 1800 ^{0}C in the combustion chamber) at a useful stress level. Therefore, cooling is inevitable in the hottest regions of the turbine. The development of an effective cooling system often is a demanding task in construction and manufacturing technology by itself. However, the designers demand for materials with still higher strength at elevated temperatures still persists, since any cooling consumes undesired amounts of power that decrease the efficiency.

The present limit with respect to service temperatures of about 1050 ^{0}C is reached by single crystalline nickel based casting alloys. Not considering ODS-materials, which also pose inherent problems and deliver only a marginal gain in service temperature, only four material groups offer possibilities for a considerable step ahead:

- monolithic structural ceramics
- fibre reinforced ceramics (carbon-carbon inclusive)
- IMP-materials
- refractory metals

Both material groups based on ceramics incorporate the inherent problems of brittle materials as have been mentioned already above with titanium aluminides. Great efforts are being undertaken to solve these problems. One of the solutions could be obtained by the application of hybrid components in order to admit only compressive stresses (figure 12 e.g.). There are two "fatal" disadvantages which still prevent the application of carbon-carbon in jet engines: extremely low oxidation resistance and extraordinarily long manufacturing times producing thicknesses even in the range of millimeters.

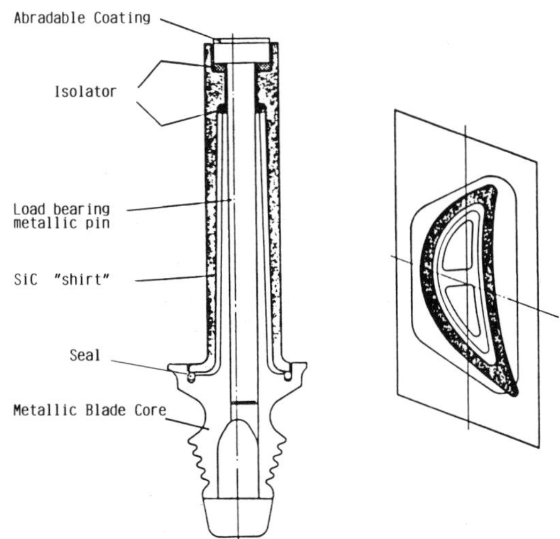

FIGURE 12
Metal/ceramic hybrid blade (experimental)

Since monolithic ceramic materials lack sufficient intrinsic toughness, several reinforcements using fibres or whiskers to avoid catastrophic cracks by crack branching and depletion have been tried thus increasing fracture energy. The structure and composition of the interface is of great importance to allow weak bonding and "pull out" of the toughening component under loading conditions. Si_3N_4 as a specially sintered matrix was successfully applied in very fast turbocharger rotors and other components in autoengines[4]. It is hoped to develop this material for aeroengine applications by introducing fibres or whiskers to achieve a more damage tolerant structure (figure 13).

SiC/SiC composites have been developed by SEP and AEROSPATIAL, France, by braiding SiC fibres to 3-D-preforms and subsequently infiltrate SiC by chemical vapour deposition to form a matrix of a high temperature resistant composite up to 1100 ^{0}C or even higher (figure 14).

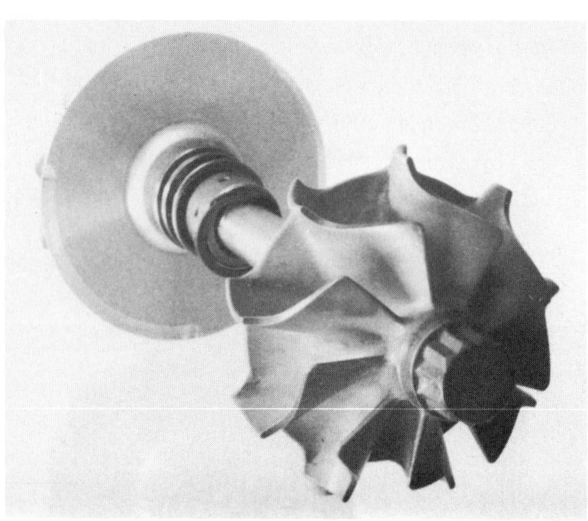

FIGURE 13
Si_3N_4 ceramic turbocharger rotor (Nissan)

FIGURE 14
SiC/SiC rotor (SEP)

The application of refractory metals still is opposed by one or a combination of the following disadvantages:

- extremely low oxidation resistance
- brittleness
- high specific weight

Niobium only has the first of the disadvantages mentioned and offers the best possibilities compared with all possible candidate materials. However, one prerequisite for its application is the combination of an effective coating with a certain oxidation inertia obtained by alloy formation of the base material.

A large shift in the share of the individual groups of materials may be expected in the next 10 to 15 years in case the efforts of the materials research scientists are successful. An example for a prognosis in this respect is given in figure 15.

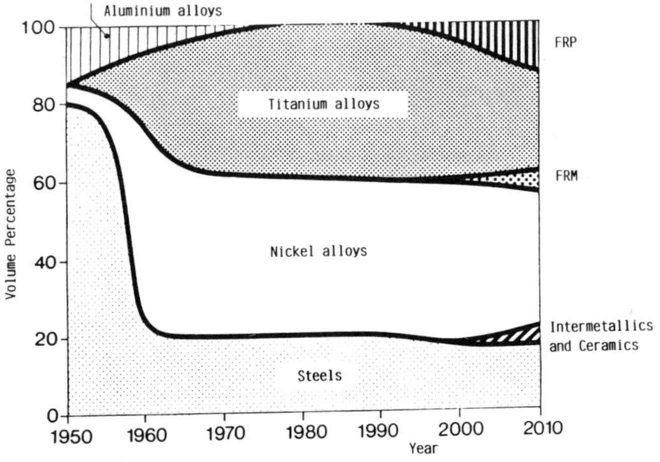

FIGURE 15
Material volume percentages in aero engines

Hot structures: So far we have been talking about structures for subsonic or supersonic aircraft. Projects like HERMES, NASP, HOTOL and SAENGER with their extreme temperature requirements during hypersonic flight and reentry will face engineers in the US and also in Europe with new challenges: large structural parts for service temperatures up to 1800 ^{o}C. A moving ramp in a scramjet-engine could be seen as a typical example.

What are the materials which are available? Intermetallics? Monolithic ceramics? Fibre reinforced ceramics: e.g. SiC/SiC (temperature limit 1400 ^{o}C)?

Let us look at fibre reinforced graphite as a potential candidate material already used for rocket engines, exhaust cones and nose tips for reentry bodies. Now transfer of this know-how to airframe application is being considered with an extremely interesting tensile strength at 1000 ^{o}C of 1000 N/mm^2. Limitations like very low transverse strength can be overcome by 3D reinforcement hopefully. Long processing times for thick structures and oxidation resistance for long times up to 1500 ^{o}C are areas of intensive R+D (figure 16)[5]. Main thrust of development in the future will consist in multilayer arrangements to resist outward carbon diffusion, inward oxygen diffusion, erosion and differences in thermal expansion. Load introduction into these components will be a special problem.

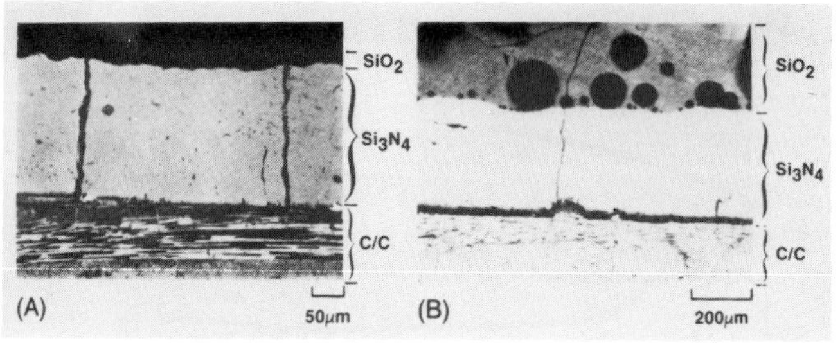

FIGURE 16
Improvement of oxidation resistance of carbon-carbon by CVD coatings of Si_3N_4: SiO_2 scale stability on CVD Si_3N_4 - flowing air environment; (A) 2 h at 1760 ^{o}C, (B) 3 h at 1843 ^{o}C

3. CONCLUSIONS

We live in an era where polymers, metals and ceramics appear to compete with each other. However, it may well be that the real competition is not between these three, but between the classical isotropic and the emerging anisotropic, i.e. fibre reinforced materials, whose properties are tailored to the designers' needs.

MMC's, CFRP's and carbon-carbon's are still often seen as different classes of materials: we have to understand them as one group, the only difference being the matrix.

Future research will be focussed on different topics in the different classes of materials. Coming back to our initial picture, figure 1, one might see the following trends (figure 17).

ACKNOWLEDGEMENT

The authors acknowledge discussions with colleagues of their labs.

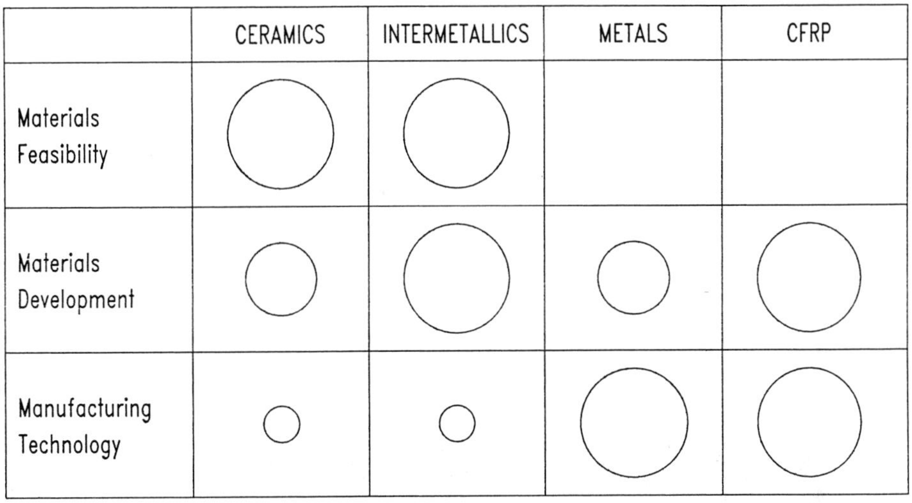

	CERAMICS	INTERMETALLICS	METALS	CFRP
Materials Feasibility				
Materials Development				
Manufacturing Technology				

FIGURE 17
Estimated level of effort

REFERENCES

1) M. Peters and K. Welpmann, Metall 39 (1985) 1141-1144.

2) M. Thumann, Proc. "Symp. Materialforschung 1988" (1988) 530-556.

3) R. Leucht, H.J. Dudek and G. Ziegler, Z. Werkstofftechn. (1987) 27-32.

4) M. Taguchi, Adv. Cer. Mat. (1987) 754-762.

5) J.R. Strife and J.E. Sheehan, Cer. Bull. (1988) 369-374.

Materials and Processing – Move into the 90's
edited by S. Benson, T. Cook, E. Trewin and R.M. Turner
Elsevier Science Publishers B.V., Amsterdam, 1989

DEVELOPMENT OF A HIGH PERFORMANCE BISMALEIMIDE PREPREG

Zef BREUKERS, Rogier BROUWER, François ESSERS, Detlef SCHUDY
DSM Advanced Composites Research
P.O. Box 18, 6160 MD Geleen, The Netherlands

1.SUMMARY

DESBIMID glass fabric prepregs, based on bismaleimide resin chemistry, are created as a new category of materials suitable for aircraft interior panels. They combine excellent heat release and fireworthy properties with high mechanical performance. This paper discusses the development of/glass fabric prepregs and presents various preliminary material data from both laminates and sandwich panels.

2.INTRODUCTION

Commercial aviation is among the safest means of public transport. In case of an uncontrolled aircraft fire - one of the greatest hazards regarding overall survivability - the fireworthiness of interior materials is crucial with respect to passenger safety. These interiors are lined and furnished with a large quantity of synthetic and natural organic materials that will burn when subjected to fire and that will develop smoke and toxic gases.

In the seventies Airbus Industries therefore introduced the ATS 1000.001 specification for flammability, smoke and toxic gas emission as an instrument to select adequate materials.

In the same period the Federal Aviation Administration (FAA) started programs to develop improved fireworthy cabin interiors and defined new regulations and tests based on radiant energy. After a series of evaluations, the Ohio State University (OSU) heat release test, ASTM E 906, was selected as the best candidate for materials evaluation. This test is conducted for five minutes. The peak heat release over the five minute period and the integrated total heat release for the first two minutes are recorded for each sample. The FAA has set the heat release performance criteria for aircraft interior panels at 65/65. Panel systems that can meet these requirements as well as the (vertical) bunsen burner test as defined in FAR 25.853 are then design options for new fireworthy interiors. Every component of the composite interior panel can have a positive or negative effect on OSU heat release results but especially resins provide an important negative contribution.

DSM developed a DESBIMID bismaleimide resin system - suitable for hot-melt prepregging - that meets the FAA mandated flammability tests and the Airbus Industries ATS 1000.001. This prepreg is unique in combining these flammability characteristics with superior mechanical performance and without any trade-off in resin processability or resin toxicity. This paper will discuss the resin chemistry, processing, optimization of a 150 ^{o}C cure cycle and mechanical and flammability data on both laminates and sandwich panels.

3.RESIN CHEMISTRY

DESBIMID is a thermosetting resin system, based on bismaleimide chemistry, that has been developed by DSM Research.

Addition of maleic anhydride to 4,4'-methylenedianiline in acetone results in an instantaneous precipitation of the bismaleamic acid. Cyclohydration is carried out by means of acetic anhydride and a small amount of catalyst. The ultimate product is solid at room temperature. The composition of this solid is rather complex: bis-maleimides, -isomaleimides and -acetamides as well as combinations can be formed[1].

This basic resin system is soluble and compatible with a whole range of reactive comonomers and can be used for resin transfer moulding and filament winding applications as well as for hot-melt prepregging applications.

The hot-melt prepregging resin system contains a reactive acrylic comonomer and an elastomeric toughening agent. The free radical polymerization reaction can be initiated by using peroxydes. By changing peroxyde type and by adjusting the concentration, the cure cycle can be tailored to customer's wishes. The results presented in this paper are based on a typical 150 ^{o}C cure cycle that is in Figure 1.

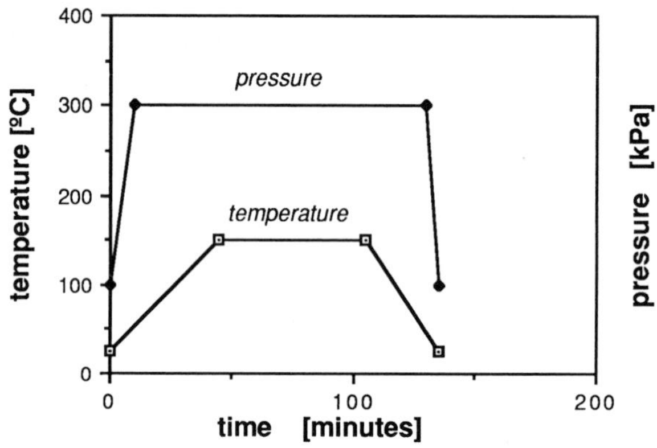

Figure 1. Typical cure cycle. During the whole cycle the product vacuum is kept at a constant level of 80 kPa.

Very recent studies however provided prelimenary data that point to comparable results for 120 ^{o}C curing using a modified initiation mechanism.

The gelation time of the prepreg resin decreases from more than two hours at 50 ^{o}C to approximately 50 minutes at 80 ^{o}C. Figure 2 shows the gel times as a function of temperature. These values were obtained by isothermal dynamic mechanical measurements on a DuPont 983 DMA assuming the gelation time at the moment at which G' equals G".

temperature	approximate gel time
60 °C	100 min.
70 °C	80 min.
80 °C	50 min.

figure 2. Approximate gel times, (G' = G" criterion), measured on a DuPont 983 DMA

4.RESIN PROCESSING

DESBIMID hot-melt prepreg formulations can be composed using the basic bismaleimid resin system, reactive acrylic comonomers and an elastomeric toughening agent. At room temperature these materials are semi-solids.

Resin processing starts with release paper coating at 50 to 60 °C followed by resin impregnation by sandwiching the reinforcing glass fabric and two resin coated release papers at 60 to 80 °C while applying pressure. In an on-line knife-over-plate coating process the prepreg production speed varies from 2 to 5 meters per minute. This rate increases to 5 to 15 meters per minute when using off-line reverse-roll coating depending on both the resin viscosity and the impregnation temperature. A typical resin viscosity profile is shown in Figure 3.

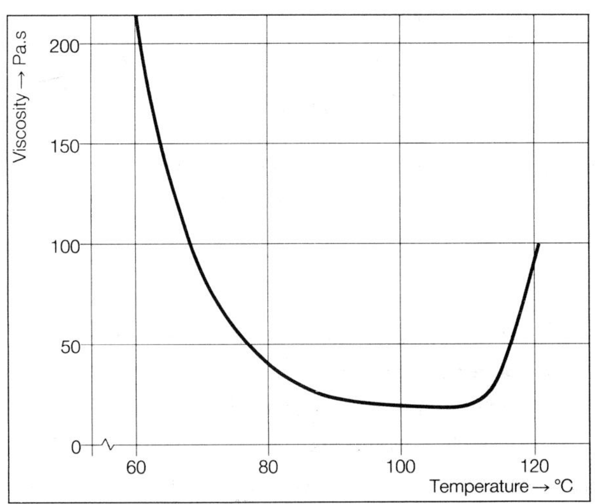

figure 3. A typical DESBIMID bismaleimide resin cure cycle

5.PREPREG CURING

The prepreg curing temperature that is used in this study is 150 °C during 1 hour. In order to determine the optimum curte cycle with respect to auto-adhesion properties of prepregs onto core materials, -flatwise tensile strength- , experiments were performed according to the Box-Behnken statistical approach [2]. The three independent variables in this model -all determining the chemo-rheological characteristic of the resin- are the prepreg resin content, the autoclave heating rate and the autoclave pressure. The dependent variable -to be optimized- is the sandwich flatwise tensile strength according to DAN 406 and MIL-STD 401B. The product vacuum is kept at a constant level of 80 kPa in and during all experiments. A 96 kg/m^3 9.6 mm thick Nomex® core is used in all samples. In the experimental set-up the prepreg resin content varies from 40 to 50 weight percent; the autoclave heating rate from 1 to 5 °C/min and the autoclave pressure from 200 to 600 kPa. After the statistical analysis three dimensional plots are generated by keeping one independent variable at a constant value.

Figure 4 shows a surface response curve and Figure 5 a contour response curve of the sandwich flatwise tensile strength as a function of the autoclave heating rate and pressure at a constant resin content of 46 weight percent. This 46 weight percent resin content is a compromise between optimum auto-adhesion and optimum OSU/flammability properties. From these figures it can be concluded that good auto-adhesion can be obtained at low heating rates and high pressures.

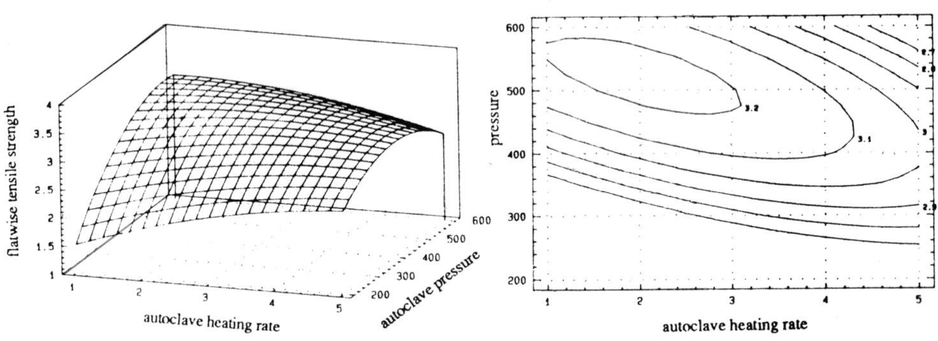

figure 4.
Response surface of the flatwise tensile strength as a function of the autoclave pressure and the autoclave heating rate at a fixed resin content of 46 weight percent.

figure 5.
Response contour plot of the flatwise tensile strength as a function of the autoclave pressure and the autoclave heating rate at a fixed resin content of 46 weight percent.

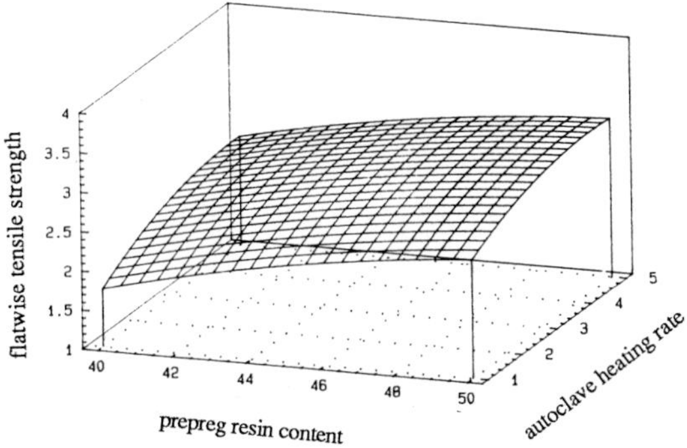

figure 6.
Response surface of the flatwise tensile strength as a function of the prepreg resin content and the autoclave heating rate at a fixed autoclave pressure of 300 kPa

At a fixed autoclave pressure of 300 kPa and at a constant resin content a heating rate increase always provides a positive effect as can be seen in Figure 5 and Figure 6 while at a pressure of 500 kPa an flatwise tensile strength optimum at approximately 2 to 3 ^{o}C/min can be seen.
In order to join aircraft industry standards, high heating rates (4 to 5 ^{o}C/min) in combination with relatively low autoclave pressures (300-400 kPa) give the best results for DESBIMID prepregs.

6.MECHANICAL PERFORMANCE

Figure 7 shows the physical and mechanical properties of DESBIMID laminates and sandwich panels. A US style 7781 glass fabric typewas used in the experiments. The laminate resin content is approximately 40 weight percent.The presented data are prelimenary and can be improved significantly by further fine-tuning.

7.OSU HEAT RELEASE RESULTS

As discussed in the introduction to this paper, the Federal Aviation Administration's Ohio State University Heat Release test [3] is the most important selection criterion for new aircraft interior panels. The peak heat release over a five minute period must be less than 65 kW/m^2 ; the integrated total heat release during the first two minutes must be maximum 65 kW.min/m^2.
DESBIMID glass fabric prepregs are very suitable with respect to the O.S.U. heat release test. 3 mm Laminates and 5 and 14 mm sandwich panels faced with 2 DESBIMID prepregs on each side, all easily passed the test, as can be seen in Figure 8. These data were obtained by an independent qualified institute.

Physical properties			
Neat resin gel time [2])	min	**approx 5**	
Prepreg weight	g/m^2	**550**	
Resin content	weight %	**46 ± 1**	ASTM D 3529
Volatile content	weight %	**< 1**	ASTM D 3530
Laminate mechanical properties			
Interlaminar shear strength (ILSS)	MPa	**48**	ASTM D 2344
Tensile test:			ASTM D 3039
ultimate tensile strength	MPa	**460**	
tensile modulus	GPa	**24**	
ultimate tensile strain	%	**1.9**	
Toughness: [3])			
fracture energy (G_1c)	J/m^2	**370**	Double Cantilever Beam Method
Flexural test:			ASTM D 790
ultimate flexural strength	MPa	**540**	
flexural modulus	GPa	**28**	
ultimate flexural strain	%	**2.0**	
Compression strength	MPa	**350**	ASTM D 3410

Sandwich mechanical properties [4])				
Flexural strength	N	**305**		DAN 406
Peel strength	N	**38**		DAN 406
Flatwise tensile strength	MPa	**3.2**		DAN 406

1. The data have been obtained from laminates made according to the recommended cure cycle. The values are based on specimens containing a 40 mass percent resin content. A US style 7781 fabric was used.
2. Maesured at 150 degrees Celsius
3. Laminate thickness of 2 mm
4. Nomex® honeycomb core (thickness 12.7 mm ; cell size 3.2 mm ; specific weight of 48 kg/cubic metre

figure 7. Physical and mechanical properties of DESBIMID laminates and sandwiches

Flammability properties				
OSU heat release rate:				FAR 25.853, Appendix F Part IV
laminate: [5])				
total in 2 min	kW.min/m^2	**2**		
peak in 5 min	kW/m^2	**58**		
sandwich:				
total in 2 min	kW.min/m^2	**18**		
peak in 5 min	kW/m^2	**28**		
Flammability: [5])				ATS 1000.001
burn length	mm	**34**		
afterflame time	s	**3**		
afterglow time	s	**0**		
drip flame time	s	**0**		
Smoke emission: [5])		**non-**		ATS 1000.001
maximum specific optical density:		**flaming**	**flaming**	
in 1.5 min	–	**0**	**0**	
in 4 min	–	**0**	**51**	
Toxic gas emission: [5])				ATS 1000.001
measured in 4 min:				
HCN	ppm	**<2**	**<2**	
CO	ppm	**<10**	**<10**	
$NO + NO_2$	ppm	**5**	**5**	
$SO_2 + H_2S$	ppm	**<20**	**<20**	
HCL	ppm	**<15**	**23**	
HF	ppm	**<15**	**<15**	

figure 8. O.S.U. and Flammability, Smoke and Toxic Gas Emission properties both from laminates and sandwiches.

8.FLAMMABILITY RESULTS

Laminate and sandwich panel flammability was tested according to the vertical bunsen burner test from FAR 25.853 and ATS 1000.001. The results of this test are illustrated in Figure 8. Both 3 mm laminates as well as 14 mm sandwich panels passed the requirements.

9.SMOKE AND TOXIC GAS EMMISION RESULTS

Smoke and toxic gas emission according to the ATS 1000.001 specification were measured in a NBS smoke box on 3 mm laminates and 14 mm sandwich panels. For both configurations the test was passed easily as can be seen in Figure 8.

10.REFERENCES

1) L.M. Dané and H.R. Brouwer, "A resin transfer moulded graphite bismaleimide composite Engine Cowling beam", 33rd International SAMPE Symposium, March 7-10,1988, Anaheim CA, page 1217.

2) George E.P.Box, William G. Hunter and J. Stuart Hunter, "Statistics for Experimenters. An introduction to design, data analysis and model building", John Wiley & Sons, 1978, ISBN 0-471-09315-7.

3) James M. Peterson, "Development of the Total Design", Aeroplas Conference "Plastics in Civil Aircraft Interiors", October 26 and 27, 1988, Brussels, Belgium.

11.CONCLUSION

DESBIMID glass fabric prepregs are very suitable materials for future aircraft interiors. They easily meet the FAA's O.S.U. most stringent heat release requirements as well as the ATS 1000.001 flammability, smoke and toxic gas emission specifications.

Materials and Processing – Move into the 90's
edited by S. Benson, T. Cook, E. Trewin and R.M. Turner
Elsevier Science Publishers B.V., Amsterdam, 1989

THERMAL PROTECTION SYSTEMS BASED ON SILICONE SOLID GUM RUBBER WITH VARIOUS FIBERS, FABRICS, AND ADDITIVES

Doug J. BUTLER

ICI-Fiberite, Product Manager, Insul|rite, 23271 Verdugo A-1, Laguna Hills, California 92653, U.S.A.

Silicone rubber has rather unique characteristics over a wide temperature range as compared to the organics. This, along with its ablative characteristics, makes it an excellent thermal protection material in the environments anticipated for aircraft and propulsion systems in the near future. A family of polydimethylsiloxane (PDMS)-based compounds, called Insul|rite, is in production by ICI/Fiberite intended to meet these systems' requirements. Fibers, fabrics, filaments, reinforcements, and additives have been specifically selected for these compounds to assure optimum systems consideration: (1) calendered—PDMS/discontinuous fiber for blanket- or tape-wrap; (2) granular--PDMS/discontinuous fiber for compression molding; (3) broadgoods—PDMS/fabric-impregnated for blanket- or tape-wrap; and (4) impregnated filament—PDMS/continuous tow, impregnated for filament winding. Specific applications include: internal/external insulation, blast tubes, nozzles, and exit cones for propulsions and launch systems (solid, liquid, ramjet, hybrids); aerodynamic and survivability protection for aircraft; and fire protection. This paper will present rationale related to systems improvements that can be achieved with these materials.

1. INTRODUCTION

1.1. Background

Thermal protection materials for most of our current aircraft and propulsion systems have evolved from organic rubber and resins available in the '40s and '50s. The basic materials were taken from commercial industries and modified for aerospace applications. Silicone rubbers were unique at this time, and very expensive. They also suffered from bad publicity in aerospace related to contamination, and were considered as release material - not to bond to substrates or to be bonded to. Even as silicone rubbers came into wide use in industry, aerospace was not able to look seriously because resources were devoted to continued evolution, qualifications, and production using the organic-based thermal materials. As problems developed with organic systems, aerospace was able only to modify and tighten controls - not change. On the commercial side, silicone rubber sales have experienced tremendous growth not only in the United States, but especially in Japan and Western Europe.

1.2. Design rationale

Aircraft and propulsion designers have long recognized the technical advantages of silicone rubbers for higher and lower temperature applications. Described simply: the silicone-oxygen linkage in the silicone polymer chain is the same Si-0-Si bond in quartz, sand, and glass contributing to outstanding high temperature properties of the silicones and their resistance to oxidation by ozone, corona, and weathering. The organic rubbers and resins with double carbon bonds are more easily cleaved by ozone, ultra-violet light, heat, and other environmental conditions. Generalized comparisons between silicone and organic rubbers are shown in Figure 1.

```
CH3    CH3    CH3    CH3
 |      |      |      |
-Si - O - Si - O - Si - O- Si - O
 |      |      |      |
CH3    CH3    CH3    CH3

 H      H  H  H      H
 |      |  |  |      |
-C - C = C - C - C - C = C - C
 |  |         |  |  |
 H CH3        H  H CH3
```

Liquid, two-part, room temperature vulcanizing (RTV's) silicone rubbers have been used fairly extensively as thermal materials, and other applications. There are some important controls related to mixing, casting (pouring, injection), bonding, curing, and other materials contact to be observed. Many of these stringent controls may be alleviated by using high temperature vulcanizing (HTV's) gumstock silicone rubber, replacing the RTV's.

1.3. Silicone thermal materials systems philosophy

Recent work has demonstrated that the thermal systems designer can be supported to achieve the overall design requirements essential for advanced systems, including long life and reliability. The gumstock polydimethylsiloxane (PDMS)- based

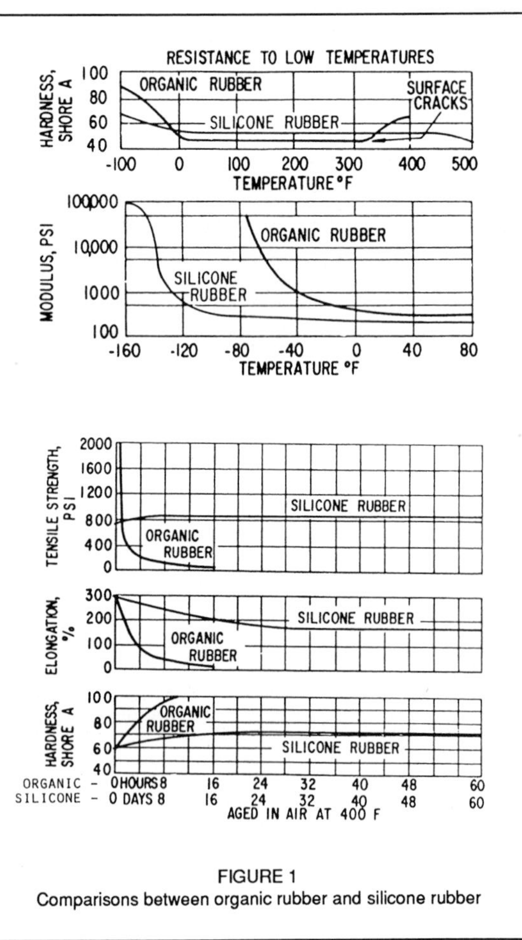

FIGURE 1
Comparisons between organic rubber and silicone rubber

compounds are available with support summarized as follows:

- Design Function - Customer

- Materials Compatibility
- Dissimilar Materials Bonding
- Compound Producibility
- Fabrication Methodology
- Storage and Handling Environment

2. COMPOUNDS DESCRIPTIONS

2.1. General

The PDMS compounds based on high temperature gumstock (not liquid RTV's) have been formulated with various loadings of powders, fibers, fabrics, and filaments to complement design function and fabrication methods. Reduced temperatures are not required for shipment and storage. When stored in closed containers at or below 80°F (25°C), the compounds have a shelf life of at least six months. Acceptable test panel moldings can extend the useful life. Pressure is not required for the vulcanization reaction; however, low pressure is necessary to flow, densify, and shape the compound to final configuration. High pressure can be used if desired. Vulcanization (cure) temperature is 320°-380°F (160°-190°C). Cure time is dependent on precise temperature and part configuration: a 1/16-inch (1.6mm) thick test panel will cure in 15 minutes at 320°F (160°C). Post-cure at 350°-400°F (175°-200°C) will clear out reaction by-products. There are no unusual or excessive precautions required for handling and processing these compounds. The specific families of compounds are described, following.

2.2. Sheet (tape); discontinuous oriented carbon fibers

Two versions (differing by specific gravity) of this family with oriented carbon fibers are in the form of calendered sheet, tape, or extruded ribbon. The designer may specify thickness, width, cured, or uncured. Specific applications might be as an internal insulator to protect a substrate from low velocity, high temperature combustion gases; or exterior thermal protection in aerodynamic heating environments. The ability to position the fibers for maximum design function provides the designer with greater flexibility.

2.3. Molding compound; discontinuous, random carbon fibers

Two ranges of specific gravity are available in this family also. The granular forms can be readily cold-formed and compression-molded into complex configurations. By controlling the mold loading method, very good thermal performance in medium velocity combustion gas flow regimes is obtained. Decomposition of the compound yields a very tenacious char layer, thus improving the insulative characteristics.

2.4. Sheet (tape); impregnated, coated fabrics

These PDMS compounds are impregnated, coated fabrics of carbon, silica, quartz, and aramid. The same processes for fabrication of organic resin-impregnated composites apply; however, the cure cycles and pressure (autoclave) requirements may be simplified and reduced. The stiffer carbon fabrics may be used for “near” structural applications; or in hybrid structures. Polymer selection and loading would help support this. Thermal performance, because of the strong char development, is excellent. The silica and quartz compounds perform best in an oxidizing atmosphere such as that of liquid engines. The aramid system has application for ballistic hardening in aerospace and civil use (police protection).

2.5. Filament; impregnated, continuous tow or yarn

These compounds are PDMS-impregnated, coated tows or yarns of carbon, aramid, and glass. Established filament winding processes for resin-impregnated structures apply. These compounds can be fabricated into pressure vessels; or combined with resin systems to produce hybrid thermal/structural components. Civil uses as pressure vessels might include safer, lighter life support tanks for firemen and divers.

3. FABRICATION METHODOLOGY

3.1. Compression molding

Low pressure is needed to form the compounds. Components can be formed in matched metal molds; or the compound can be cured and bonded simultaneously to substrates with proper surface preparation and primer selection. The material can be cold-formed with a press then press-cured; or clamped and placed in an oven for cure. If part is not “net molded”, machining is very easy.

3.2. Tape or blanket wrap

Standard tape wrapping equipment used for organic resins is applicable. The part may be cured in an autoclave without the temperature ramping and cooldown cycles used with resins. Or, it may be instantaneously cured during the wrap cycle with proper selection of pressure/temperature rollers. Oven postcure will complete the cure operations. The compounds can also be cured and bonded in-situ to substrate; or secondarily bonded with adhesives.

3.3. Rotary cure

The discontinuous fiber and fabric sheet (tape) compounds may be continuously cured at specified thickness for the calendered material or specified number of laminates for the fabric materials. This would permit secondary bonding with room temperature adhesives if there are high temperature restrictions associated with system integration; or other manufacturing restraints.

3.4. Extrusion

The discontinuous fiber compounds may be extruded into desired cross-sections; and supplied in cured or uncured form. A typical fabrication approach for installing the compound to the wall of a structural cylinder would be to wrap a continuous ribbon with tapered cross-section into the structure then install a pressure bladder and cure/bond the compound.

4. SPECIALTY COMPOUNDS

Specialty compounds based on the silicone gumstock polymer can be prepared in concert with the systems designer or his customer. This can be done by the proper selection and evaluation of polymers, additives, reinforcements, fibers, fabrics, and curing agents.

4.1. Silicone gumstock polymers

The basic lowest cost polymer, polydimethylsiloxane, can be modified with groups such as vinyl, phenyl, and/or fluoride. Depending on the design need, these modified polymers can be used to improve oil resistance, crystallization temperature, mechanical strength, and compression set. Mobile molecules such as plasticizers used in organic rubbers are not necessary. This results in improved systems compatibility and long term reliability.

4.2. Additives, reinforcements

To assure compatibility with the high performance characteristics of the silicone polymer and with other materials in the system, only inorganic solids and reinforcements are selected. Some that have seen application are: glass, silica, quartz, carbon, silicon carbide, and iron oxide powders. Many others are available to support specific design criteria.

4.3. Fabrics

The baseline compounds described have realized applications with only a miniscule sampling of available fabric materials, styles, and types. The world of possibilities here would best be addressed in detailed discussions with the designer who wishes to pursue a concept.

4.4. Curing agents

The silicone gumstock compounds are normally heat-cured using one of the following organic peroxides: BIS(2,4-dichlorobenzoyl) peroxide; di-benzoyl peroxide; di-cumyl peroxide; and di-tertiary butyl peroxide. The particular peroxide must be selected to match the polymer and fabrication methodology.

5. RESULTS

Sub-scale ablation testing using a standard test vehicle has been performed with these compounds to establish their performance relative to each other; and to other ablative/insulative materials. A schematic of the test vehicle, along with its performance characteristics are presented in Figure 2.

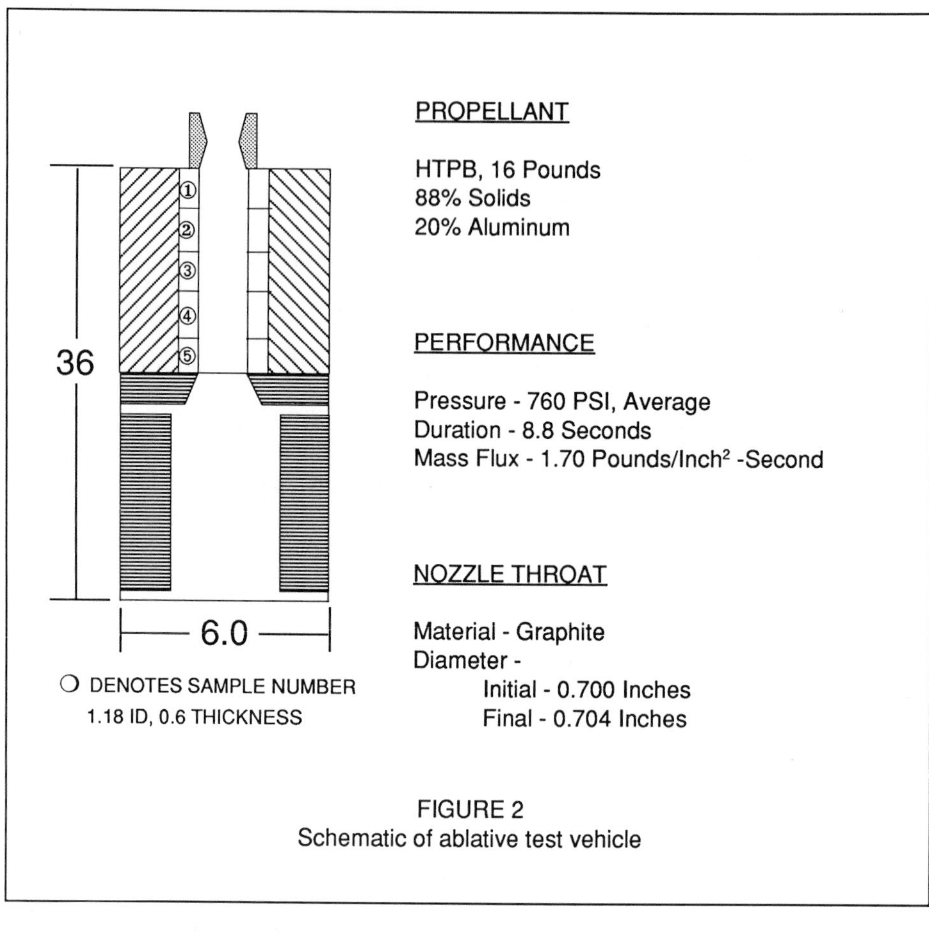

FIGURE 2
Schematic of ablative test vehicle

Results from these comparative tests are presented in Table I.

6. CONCLUSIONS

Future systems in conceptual stages; or in development for high temperatures, long duration reliability, can take advantage of the versatile nature of the silicone gumstock polymer. Compounds using this polymer can complement systems development in such areas as: bonding and compatibility; unicomposite and hybrid composite structures and pressure vessels: lower cost processes; unique nozzles (rubber); thrust vector control; low smoke; insensitive munitions; electrostatic discharge; ballistic hardening; laser hardening; sub-sonic, sonic, and hypersonic thermal protection.

Table I. Ablation data from standard test vehicle

Compound	Specific Gravity	Ablation Rate
Discontinuous Fibers: Calendered Sheet		
NBR/Asbestos	1.28	
EPDM/Aramid	1.14	
PDMS/Carbon	1.16	
Discontinuous Fibers: Granular		
EPDM/Asbestos	1.28	
PDMS/Carbon	1.24	
Fabrics		
Phenolic/Carbon	1.48	
PDMS/Carbon	1.23	
PDMS/Carbon Knit	1.33	
PDMS/Carbon Knit	1.26	
Phenolic/Silica	1.80	
PDMS/Silica	1.56	
PDMS/Silica	1.45	
PDMS/Quartz	1.49	

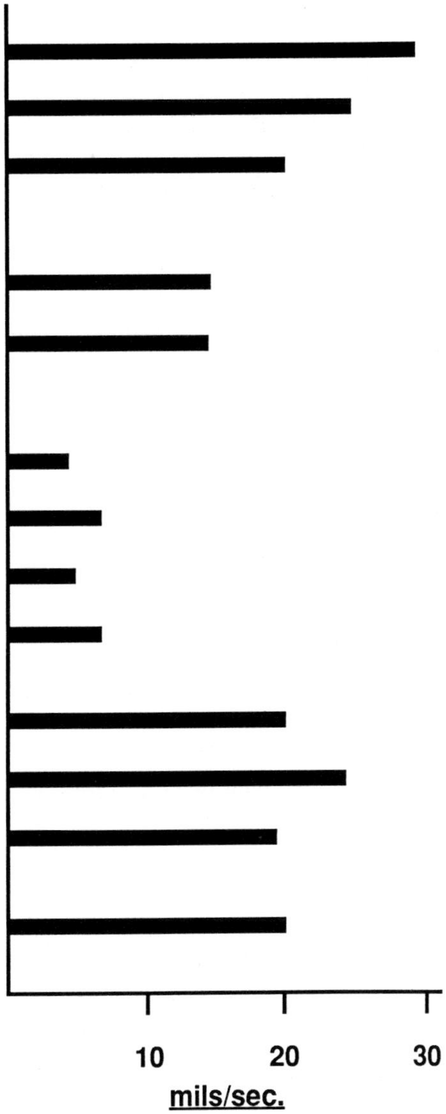

Materials and Processing – Move into the 90's
edited by S. Benson, T. Cook, E. Trewin and R.M. Turner
Elsevier Science Publishers B.V., Amsterdam, 1989

A NEW FAMILY OF HIGH PERFORMANCE YARNS FOR COMPOSITE APPLICATIONS

R.J.Coldicott & T.Longdon

Courtaulds Research, P.O.Box 111, 72 Lockhurst Lane, Coventry, UK

Conventional fibre processing techniques have been adapted to develop a method for processing spun staple yarns based on high performance fibres. The new process offers an economic route to very fine carbon yarns. The process has been further developed to produce intimate co-spun blends of reinforcing fibres and thermoplastic matrix fibres, eg. carbon/PEEK, which show enhanced fibre dispersion compared with alternative mixing methods such as co-weaving, co-serving and co-mingling. The yarns possess a range of attractive features; they permit lightweight, drapeable fabrics to be produced for composite end use and in carbon/carbon composites these spun yarns lead to significantly increased interlaminar shear performance over continuous filament carbon. The blended yarns, due to greater fibre intimacy, produce improved thermoplastic composites with minimal voids and resin rich areas on consolidation. Both families of yarns show a high translation of tensile properties from the parent continuous tow to yarn, and both process readily on conventional textile machinery. Performance data is presented, and applications are discussed. The carbon/PEEK blend has exciting potential as a bonding material for enhancing the induction welding of thermoplastic component parts.

1. INTRODUCTION

1.1 Carbon Yarns

The manufacture of continuous carbon fibre tows is now well established, with a number of international suppliers offering a wide variety of PAN and pitch based fibres. Typically these tows contain from 12000 (12k) down to 1000 (1k) filaments, and are presented to the composite manufacturer in the form of UD tapes or biaxial woven fabrics. In the latter case textile weavers have had to develop specialised techniques for handling and processing these very stiff fibres. Moreover, the production costs for continuous carbon tows rise steeply as the number of filaments in the tow is reduced : 1k tows are extremely expensive, and very few materials are available below 1k.

To overcome this problem Courtaulds have applied their textile expertise to develop a method for the production of very fine spun staple carbon yarns for high performance end uses. Carbon yarns down to 100 filaments (0.1k) have been spun, permitting the production of very lightweight, thin UD and woven materials for composite manufacture. The yarns process well on textile machinery. The Heltra process produces 100% carbon yarns, Grafitex, for use in thermoset composite production.

1.2 Hybrid Yarns

The technique is also used to spin Filmix yarns, intimately blended carbon (and other reinforcement fibres) with thermoplastic fibres (PEEK, PEI, PPS), for the production of advanced thermoplastic composites.

The last few years have witnessed a rising tide of interest in the potential for high performance thermoplastic resin systems in advanced composite applications. These matrix materials are perceived as having a number of advantages over the conventional epoxy prepregs which presently dominate commercial manufacturing :

* Shorter cycle times in processing
* Infinite prepreg shelf life at room temperature
* Ease of component repair
* Improved performance under hot/wet conditions
* Good chemical resistance
* Increased damage tolerance

However, until recently materials on offer in the market place have suffered a variety of problems. The melt or solution impregnated products produce stiff, boardy prepregs, which, whilst giving good quality flat or near-flat composite structures, are unable to conform to mould shape in more complex geometries incorporating curvature and/or deep-draw. On the other hand attempts to mix reinforcement and matrix in drapeable fibre form by co-weaving, co-serving, and co-mingling suffer to varying degrees from limited quality of fibre dispersion and tend to yield composites with resin rich areas and voids.

In addition, as these materials mix thermoplastic fibres with 3K or heavier carbon tows, they do not permit the production of very lightweight, thin prepreg fabrics.

A novel process has been developed to offer a solution to these problems in the form of spun blended yarns. A range of fine yarns is available which can be converted to lightweight highly drapeable fabrics well suited to thermoplastic composite structure fabrication. Additionally the materials possess exciting potential for jointing thermoplastic parts.

2. TECHNOLOGY

2.1. The Heltra Process

By innovative modification of fibre handling and processing techniques the process converts continuous filament tow to lightweight spun staple yarns. These yarns consist of discontinuous fibres bound together into a well oriented coherent bundle by the insertion of a degree of twist. Since the average fibre length is high, 75-100mm (3-4"), and the yarn structure is carefully controlled, very good tensile property translation can be achieved from the feedstock tow through to spun yarn, (see Table 1). The yarn technology also leads to highly consistent yarns with minimal fibre damage. Both single component and blended multi-component yarns are available.

TABLE 1

Property Translation - Resin Impregnated Tow Test

Material	Tensile Strength	Tensile Modulus	Fibre Density
Continuous Filament 12K tow	4.2 GPa (610 KSI)	235 GPa (34.1 MSI)	1.80 g/cm^3 (0.065 lb/in^3)
Grafitex 80 Tex 2-ply Yarn	3.5 GPa (512 KSI)	228 GPa (33.1 MSI)	1.80 g/cm^3 (0.065 lb/in^3)
% Retained	84	97	100

Since this process converts comparatively inexpensive heavy continuous filament tow into fine staple yarns the method is a very economical route to fine yarns. Indeed, single yarns as fine as 50 filaments have been produced, giving a total of 100 filaments for a 2-ply yarn (0.1K/58 denier/6.5tex). The yarns are usually offered in plied form to balance the unresolved twistliveliness in singles yarns.

A unique feature of a staple spun yarn when compared to a continuous filament tow is the presence of fibre that protrudes from the main body of the yarn in the form of little hairs. The degree of hairiness is dependent on the twist of the yarn, a yarn with low twist possessing a greater degree of hairiness than one with high twist. This feature offers an attractive increase in interlaminar shear strength in some composite forms.

2.2. Carbon Yarns

A versatile family of 100% carbon yarns has been developed, which can be tailored to specific application requirements. Typical variables that can be employed in yarn production include the following:

* source fibres, from a range of high modulus, intermediate modulus and high strain carbon
* mass per unit length
* number of plies
* twist of both the single and plied yarn

* coating, which can include epoxy sizes, PTFE, graphite and metal
* hairiness

Grafitex yarns can be constructed to ensure good processability through to fabric on conventional textile machinery.

2.3. Hybrid Yarns

Hybrid yarns have been produced in a wide variety of blended fibres :

Reinforcing Fibre	Matrix
Carbon	PEEK
Glass	PEI
Aramid	PPS
Quartz	Nylon
Ceramic	Polypropylene
	Liquid Crystal Polymers

Most yarns spun to date have been blends of a single reinforcing fibre with a single matrix fibre, but the new process will permit the production of multicomponent hybrid yarns if required. Carbon yarns blended with thermoplastic PEEK fibres are proving particularly attractive to high performance composite end users, and some initial indications of their mechanical properties are given below (section 4).

The process results in very intimate blending of reinforcing fibre and matrix fibre. Fig.1 compares Filmix and co-mingled carbon/PEEK yarn cross sections : the superior dispersion in the Filmix yarn is very evident. This property leads to even consolidation to composite with an absence of resin rich areas and minimal void formation (see Fig 2). Moreover, the insertion of twist overcomes one of the other problems associated with co-mingled yarns, their tendency to separation of reinforcing and matrix fibres during textile processing.

The spun yarn structure is very versatile. Fibre type and origin, reinforcing fibre volume fraction and yarn construction can all be varied to suit the application or customer. The availability of very fine yarns leads to the realisation of very lightweight, drapeable, post thermoformable prepreg fabrics, suitable for fabricating complex shapes with compound curvature.

Current work on carbon and hybrid materials is centred on compiling property data bases and identifying optimal conditions for converting prepreg materials through to composite.

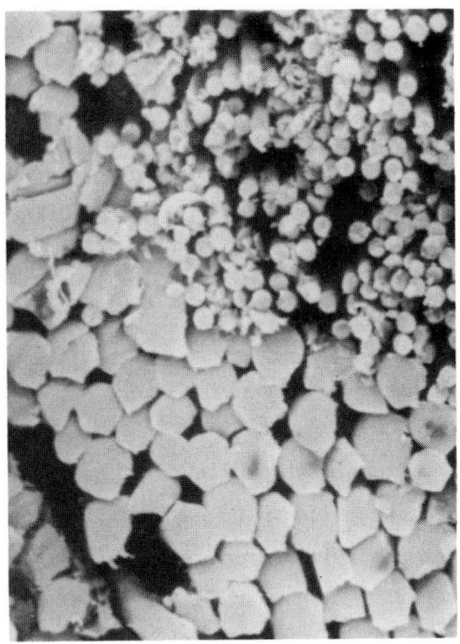

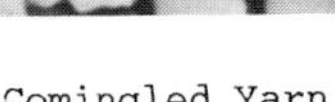

Comingled Yarn

Filmix Yarn

Figure 1
Comparison of Intimacy
of Fibre Blending

3. CONVERSION TO COMPOSITE

There are two stages in producing composite parts from Heltra yarns. First they must be converted into a fibre assembly or prepreg, typically by a textile process such as weaving. Thereafter the prepreg material is plied up in the required orientations before consolidating under heat and pressure.

3.1. Textile Processing

The importance of good processability on textile machinery to ultimate composite properties has been emphasised in a recent paper[1]. Fibre damage must be avoided whilst retaining orientation and minimising separation of reinforcing and matrix fibres. Textile sizes, applied to improve processability can seriously affect the quality of the fibre-matrix interfacial bond in the final composite.

Grafitex and Filmix carbon/PEEK yarns have been processed by an extensive range of textile routes. Woven tied-unidirectional and bidirectional fabrics have been produced in weights ranging from 400 g/m^2 down to 70 g/m^2. Unlike continuous tow material, Heltra yarn permits fabric weaving from beam as well as from warp creel. This offers the weaver attractive savings in terms of reduced floorspace requirements, rewinding costs, and loom downtime. The yarns have been braided in a range of constructions for use in aerospace applications and can even be knitted and stitched satisfactorily without yarn damage. In all cases good processability has been achieved without the need to apply a textile finish to the yarn.

3.2. Conversion To Composite

The emphasis to date has been on the production of composite parts from the Filmix carbon/PEEK blend material. PEEK has a high melting point (342 C) and yields very viscous melts; it therefore presents significant processing challenges to the composite manufacturer.

Again the intimacy of blending of carbon and PEEK deriving from the Heltra process offers a potential benefit in terms of reducing the thermal and mechanical work required to flow the thermoplastic component uniformly through the reinforcing fibres. Good quality composites have been produced from carbon/PEEK Filmix material using compression moulding, hot stamping and high temperature autoclaving techniques. Fig. 2 shows a cross-section of such a composite produced from bi-directional material. Note the uniformity, with an absence of resin rich areas and voids.

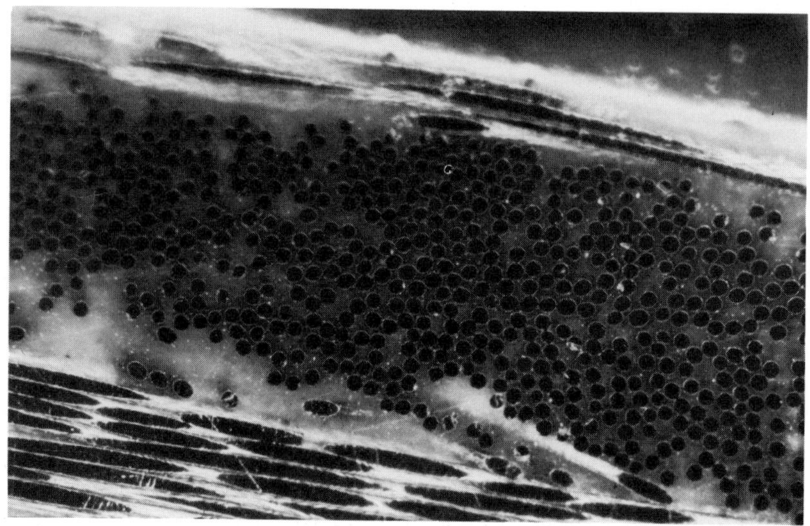

Figure 2
Section through a Filmix
Carbon/PEEK Composite

4. MECHANICAL DATA

At present databases are being compiled on property data from a range of different fibre types, yarn counts and twists, and for different fabric constructions for both the carbon and hybrid yarns. Table 2 shows mechanical data for a carbon/PEEK unidirectional fabric of $70g/m^2$ used in an aerospace application. In particular, note the very low measured void content.

TABLE 2

Results for Consolidated U.D.
Filmix Composite - Carbon Volume Fraction = 52%

U.D. Filmix Apollo 53-750/PEEK 150			
Property	(V_f = 52%)	Value	
Tensile Strength	0°	1.2 GPa	174 KSI
	90°	82 MPa	11.9 KSI
Tensile Modulus	0°	162 GPa	23.5 MSI
	90°	8.5 GPa	1.23 MSI
Poisons Ratio		0.84	0.84
Flexural Strength		1.09 GPa	158 KSI
Flexural Modulus		130 GPa	18.9 MSI
Compressive Strength	0°	675 MPa	97.9 KSI
Compressive Modulus	0°	127 GPa	18.4 MSI
In plane shear strength		223 MPa	32.3 KSI
In plane shear modulus		11.8 GPa	1.71 MSI
Interlaminar shear strength		99 MPa	14.4 KSI
Measured void content		<0.5%	<0.5%

5. APPLICATIONS

5.1. Aerospace

With the combination of lightweight and mechanical performance Heltra materials offer a level of performance which should naturally lead to their use in aerospace, particularly where very thin, lightweight structures with complex curvature are required.

Hybrid yarns have been woven into ultrafine unidirectional fabrics with weights as low as 70 g/m^2 which give a consolidated ply thickness of less than 0.05mm (2 thou). A fabric of this weight has been used to construct a concept aerospace panel which possesses excellent stiffness whilst remaining very thin (Fig. 3). The panel demonstrated firstly the deep draw capability of the yarn, in that it possessed eight stiffening dimples of one inch depth, and secondly the advantage of ultrafine fabrics, in that the number of plies for a prescribed thickness could be increased. This allowed a precise tailoring of both stiffness and thermal expansion to the customers specification.

Figure 3
Consolidated Carbon/PEEK Space Frame

Knitted structures from Filmix and Grafitex yarn allow the construction of complex 3-D shapes and joints. These components can then be consolidated under the action of heat and pressure to form the final composite.

5.2. Joining of Thermoplastic Composites

One advantage gained with thermoplastic composites is that joints may be achieved using a welding process. This overcomes problems of poor bonding between thermoplastic matrices and adhesives, and also eliminates the stress concentrations often associated with mechanical fastenings. Several techniques are available, including vibration, ultrasonic, electrical resistance and induction heating and the success of each method has depended on the substrate material and joint geometry. Of these, the most promising method for automated weld production is induction heating, but recently published work[2] has highlighted problems in that the method fails to generate sufficient energy transfer to achieve polymer fusion in unidirectional APC-2, a commercially significant carbon/PEEK material.

Our recent work has demonstrated that by adding a thin layer of Heltra Filmix woven fabric at the joint interface of APC-2 composites, induction heated joints can be produced with significantly enhanced weld strengths. Table 3 presents the results of tests on a series of different weld types used to joint various APC-2 laminates. Clearly the presence of a Filmix layer at joint interfaces dramatically improves the weld lap shear strength for induction welded samples, and leads to substantially stronger bonds than alternative joining techniques. Moreover the use of a Filmix interlayer leads to faster heating rates and reductions in the time taken to achieve bonding.

TABLE 3

Lap Shear Bond Strengths for APC-2 Flat Laminates Joined Using Various Bonding Techniques

	Adhesive Bonding	Graphite Resistance Welding	Vibration Welding	Induction Welding Without Filmix		Induction Welding With Filmix	
Laminate	UD	UD	UD	UD	QI	UD	QI
Mean lap shear Strength (MPa)	15.7+	23.9+	39.3+	No Weld	9.0	51.1	35.9

Sample size 100 x 25 x 1 mm, with 25 mm joint overlap
Test method ASTM D-2919 (D1002)
UD means uni-directional
QI means quasi-isotropic
+ Figures from ref. 2

Since the fabric layer is responsible for generating the necessary heat at the joint interface possibilities exist for extending the technique of induction heating to the bonding of non-conducting fibre reinforced thermoplastic composites such as glass and aramid. The rapid, automated welding of APC-2 and other thermoplastic materials now appears to be a real possibility allowing quick and simple assembly procedures to be adopted in production.

5.3. Carbon/Carbon Composites

Carbon/carbon composites are materials that consist of carbon fibre embedded in a carbon matrix system. The aim is to combine the advantages of fibre reinforced composites such as high strength, stiffness and toughness with the refractory properties of structural ceramics. These materials are chemically inert and biocompatible, possess good thermal stability, have a high resistance to thermal shock, and, most importantly, exhibit high strength and stiffness at high temperatures. Unfortunately the carbon matrix is often very brittle. This results in poor interlaminar properties and may result in the 'dusting out' of the matrix under fatigue loading.

Continuous filament carbon fibre tow gives excellent warp tensile performance, but has this design limitation of weak interlaminar properties. Previously known spun staple materials such as spun oxidised PAN fibre in processing to carbon/carbon yield increased interlaminar shear strength, but at the cost of a 60% loss in tensile performance. Moreover they have the associated disadvantage of exhibiting high shrinkage in carbonisation (typically 35%).

Recent independent research has demonstrated that the new carbon yarns offer increases of over 70% in interlaminar shear strength while retaining nearly 70% of tensile strength, and exhibiting low shrinkage in carbonisation (typically 5%). This is a result of the hairy structure the yarn possesses, effectively providing additional 'through the thickness' reinforcement in the matrix. Continuing development is also producing yarns for ablative and friction applications as well as investigating carbon braided tubes as a novel building block for carbon/carbon structures.

5.4. Sports and Leisure

An epoxy sized carbon plain weave fabric of $140g/m^2$ has been used to successfully construct a fairing, seat and petrol tank for a Formula 1 motorcycle. The primary requirements were those of stiffness and light weight, since the components were fabricated from a single ply of fabric. There is additionally a need for damage tolerance in such a structure, and whilst the single carbon skin performed without problem it is expected that a blend of carbon/kevlar will be used in subsequent developments to enhance this aspect. The new process can easily produce a blended carbon and kevlar yarn to maintain the single ply arrangement which is favoured, due to the weight constraints.

Filmix co-spun blends should naturally find applications in the high-tech end of sports and leisure. The carbon/PEEK fabrics with their potential for light, very thin and stiff structures with good damage tolerance, are also attractive to end users in motorsports and powerboating.

Carbon, glass and Kevlar blended with nylon and polypropylene will offer the right combination of performance and cost to satisfy leisure applications like skis, racquets, golf clubs and fishing rods, where the advantages of thermoplastics for volume production can be utilised.

5.5. Composite Tooling

With the advent of high performance epoxy thermoset matrix systems composite component manufacturers have been given the opportunity to construct their tooling from materials which are directly comparable with those that are used in the component, offering a better match in thermal expansion, and the ability to construct tools without incurring costly and time consuming machining. These advantages of prepreg tooling can be increased by using Grafitex yarns, where the ultrafine fabrics can improve the surface finish, and the increased interlaminar shear properties of the yarns reduce the likelihood of delamination.

5.6. Medical

Designers have recognised the advantages of carbon fibre reinforced composites in medical applications for a while, where high strength and stiffness combined with light weight and good fatigue resistance are supplemented by the low absorption of X-rays. This has allowed radiographers to reduce the patients exposure to radiation whilst obtaining better images. Carbon fibre X-ray patient support couches have been manufactured for the past 15 years, being stiff enough to support people in cantilever with very little deflection, and hence allowing access to the patient from almost any angle. With ultrafine fabrics now being available through the utilisation of Grafitex yarns, other applications in the radiological field are being found. X-ray film cassettes can be fabricated from a knitted carbon structure, :, and mammography plates can also be constructed from these fine carbon yarns.

5.7. Packing Yarns

The packing yarn industry has witnessed a substantial increase in the useage of both carbon and oxidised PAN fibres with the demise of asbestos. Consequently Heltra has developed a range of packing yarns from both fibre types that possess significant advantages over continuous filament tows in that the yarns are easier to braid, they have better bulk packing characteristics due to their circular cross-section, and they have a greater ability to absorb lubricants and coatings.

5.8. Conductive Yarns

The conductive properties of carbon can be increased by coating with metallic films, and nickel coated carbon fibres have been processed to form yarns that can be woven into conductive fabrics for use in EMI shielding applications. For increased electrical conductivity, carbon yarns can be plied with copper wire.

6. CONCLUSIONS

The new process provides an economic route to lightweight single and multicomponent spun yarns. Yarns can be produced from a wide range of fibre types and yarn construction can be tailored to meet the end users precise requirements. Lightweight, drapeable prepreg fabrics can be constructed and are ideally suited for fabricating very thin laminates and complex components with compound curvature.

Thermoplastic composites produced from Filmix yarns exhibit very uniform dispersion after consolidation, owing to the intimacy of fibre blend.

Grafitex yarns have been shown to produce enhanced interlaminar shear performance in carbon/carbon composites.

The carbon/PEEK material has significant potential for the joining of thermoplastic composites by induction welding.

REFERENCES

1) E. Lynch, SAMPE J.,25(1),17 (1989)
2) D.M. Maguire, SAMPE J.,25(1),11 (1989)

MATERIALS SCIENCE MONOGRAPHS (Advisory Editor: C. LAIRD)

Vol. 1 Dynamic Mechanical Analysis of Polymeric Material (Murayama)
Vol. 2 Laboratory Notes on Electrical and Galvanomagnetic Measurements (Wieder)
Vol. 3 Electrodeposition of Metal Powders (Călușaru)
Vol. 4 Sintering — New Developments (Ristić)
Vol. 5 Defects and Diffusion in Solids. An Introduction (Mrowec)
Vol. 6 Energy and Ceramics (Vincenzini)
Vol. 7 Fatigue of Metallic Materials (Klesnil and Lukáš)
Vol. 8 Synthetic Materials for Electronics (Jakowlew, Szymański and Włosiński)
Vol. 9 Mechanics of Aerospace Materials (Nica)
Vol. 10 Reactivity of Solids (Dyrek, Haber and Nowotny)
Vol. 11 Stone Decay and Conservation (Amoroso and Fassina)
Vol. 12 Metallurgical Aspects of Environmental Failures (Briant)
Vol. 13 The Use of High-Intensity Ultrasonics (Puškár)
Vol. 14 Sintering — Theory and Practice (Kolar, Pejovnik and Ristić)
Vol. 15 Transport in Non-Stoichiometric Compounds (Nowotny)
Vol. 16 Ceramic Powders (Vincenzini)
Vol. 17 Ceramics in Surgery (Vincenzini)
Vol. 18 Intergranular Corrosion of Steels and Alloys (Čihal)
Vol. 19 Physics of Solid Dielectrics (Bunget and Popescu)
Vol. 20 The Structure and Properties of Crystal Defects (Paidar and Lejček)
Vol. 21 Interrelations between Processing, Structure and Properties of Polymeric Materials (Seferis and Theocaris)
Vol. 22 Atmospheric Deterioration of Technological Materials: A Technoclimatic Atlas (Rychtera)
Part A: Africa
Part B: Asia (excluding Soviet Asia), Australia and Oceania
Vol. 23 Plasma Metallurgy (Dembovský)
Vol. 24 Fatigue in Materials: Cumulative Damage Processes (Puškár and Golovin)
Vol. 25 Sintered Metal–Ceramic Composites (Upadhyaya)
Vol. 26 Frontiers in Materials Technologies (Meyers and Inal)
Vol. 27 Joints with Fillet Welds (Faltus)
Vol. 28 Reactivity of Solids (Barret and Dufour)
Vol. 29 Progress in Advanced Materials and Processes: Durability, Reliability and Quality Control (Bartelds and Schliekelmann)
Vol. 30 Non-Ferrous Metals and Alloys (Sedláček)
Vol. 31 Defect Recognition and Image Processing in III–V Compounds (Fillard)
Vol. 32 The Si–SiO_2 System (Balk)
Vol. 33 Perspectives on Biomaterials (Lin and Chao)
Vol. 34 Silicon Nitride in Electronics (Belyi et al.)
Vol. 35 High Tech — The Way into the Nineties (Brunsch, Gölden and Herkert)
Vol. 36 Composite Systems from Natural and Synthetic Polymers (Salmén, de Ruvo, Seferis and Stark)
Vol. 37 Copper Indium Diselenide for Photovoltaic Applications (Coutts, Kazmerski and Wagner)
Vol. 38 High Tech Ceramics (Vincenzini)
Vol. 39 Ceramics in Clinical Applications (Vincenzini)
Vol. 40 Electron Microscopy in Solid State Physics (Bethge and Heydenreich)
Vol. 41 Looking Ahead for Materials and Processes (De Bossu, Briens and Lissac)
Vol. 42 Materials Data for Cyclic Loading. Parts A–E (Boller and Seeger)
Vol. 43 Technical Mineralogy and Petrography. Parts A and B (Szymański)
Vol. 44 Defect Recognition and Image Processing in III–V Compounds II (Weber)
Vol. 45 Solid State Electrochemistry and its Applications to Sensors and Electronic Devices (Goto)

Vol. 46 Basic Mechanisms in Fatigue of Metals (Lukas and Polak)
Vol. 47 Surface and Near-Surface Chemistry of Oxide Materials (Nowotny and Dufour)
Vol. 48 Creep in Metallic Materials (Čadek)
Vol. 49 Hardness Estimation of Minerals, Rocks and Ceramic Materials (A. Szymański and J.M. Szymański)
Vol. 50 Stereology of Objects with Internal Structure (Saxl)
Vol. 51 Experimental Methods in Mechanics of Solids (Szczepiński)
Vol. 52 Inorganic Phosphate Materials (Kanazawa)
Vol. 53 Advances in Epitaxy and Endotaxy (Schneider, Ruth and Kormány)
Vol. 54 Crystal Engineering (Desiraju)